The Engineering of Sport 6

Eckehard Fozzy Moritz and Steve Haake (Eds.)

The Engineering of Sport 6
Volume 3: Developments for Innovation

 Springer

Eckehard Fozzy Moritz
SportKreativWerkstatt GmbH
Herzogstraße 48
D-80803 München
Germany
efm@SportKreativWerkstatt.de
www.SportKreativWerkstatt.de

Steve Haake
Centre for Sport and Exercise Science
Collegiate Hall
Sheffield Hallam University
Sheffield S10 2BP
UK
s.j.haake@shu.ac.uk

Library of Congress Control Number: 2006927112

ISBN-10: 0-387-34680-5
ISBN-13: 978-0387-34680-9

Printed on acid-free paper.

Printed in the United States of America.   (EB)

9 8 7 6 5 4 3 2 1

springer.com

Preface

What you are holding in your hands is probably the best overview of activities in sports engineering available at the time of printing; i.e. the state of the art in summer 2006. It is the result of so many people's work to whom we are indebted that it is difficult to name them: there are the authors, the scientific advisory board, the scientific committee, the theme patrons, the publisher and printer, the advisors of whatever kind – and, here we have to make an exception, there is Ingo and Amanda. Nobody who has been part of the production of this book could have done without them, at the very least us: they handled issues you wouldn't even believe could turn up with efficiency and charm. Thanks, Ingo Valtingoier; thanks, Amanda Staley.

In the accumulation of the contributions and the preparation of the proceedings we encountered one development that we were very happy about: the sports engineering community keeps growing – in the number or researchers and experts involved, but also in the breadth of disciplines and institutions contributing. This should definitely be interpreted as a positive development – even though in the evaluation of contributions this lead to a number of intricate discussions. Is sports engineering primarily science? Is it engineering? Is it science and engineering helping sports? Some reviewers had differing views on that: if it is science, you need method, data, and discussion; if it is engineering, you need method and an outcome with some demonstrable usefulness, if it is an aide to sports then whatever has been done needs demonstrable relevance. As a consequence, some contributions very well done from an engineering perspective have been turned down by hardcore scientists, and vice versa; in some cases we tried to intermediate, in others it may have been bad luck for the contributors. We think sports engineering will have to live with this variety of perspectives and interests; it is rather the appeal of this field in the process of finding itself. Openness combined with consistent reasoning will be needed to progress from here; somewhere in-between academic traditions and Feyerabend's famous "Anything goes".

As a quick glimpse behind the scene, besides the disciplinary quarrels sketched above some "cultural" clashes could also not be avoided. One German reviewer put his comments in a very direct way that was hard to bear for the British author; some East Asian authors had a hard time in focusing their writing on the most interesting results and were thus bluntly thrown out; some well-known members in one community have seen their abstract turned down by experts from another area who did not know about the writer's fame… these anecdotes point to just a couple of more issues the sports engineering community will have get to grips with in the not too distant future.

As the result of various influences in these proceedings you will find a number of new topic areas indirectly related to but important to sports engineering. One area of concern we like to especially highlight here is the topic of sustainability, which may serve as an important yardstick for the future development of sports engineering and hopefully other industrial activities. Furthermore, you will find contributions on trends, cultural influences, human factors and on neural network modeling. Finally, according to the special emphasis of this conference we were successful in seeking a large number of papers in the area of innovation and design, including economic perspectives and proposals for novel design approaches. To our regret, even though we had tried hard we could get no contributions on industrial design – this area with so much relevance to sports equipment apparently is still a step-child in our community.

In the assembly of these proceedings we have endeavored to realize some novel approaches. First of all, we used "theme patrons" for different topic areas who not only helped acquire contributions but were also asked to write a synopsis of the contributions in "their" fields. This will hopefully increase the use value for readers, who by just reading the synopses can have a basic idea about developments in certain fields, and can then scan contributions on a much better knowledge basis. This is a first step towards converting the proceedings into a sort of handbook which hopefully will be taken up by future editors.

Then, as we tried to increase the relevance of sports engineering to sports, we have asked authors to take special care to illustrate the respective relevance, and to put their contribution into a sports-related category rather than a discipline-oriented category. Therefore, one volume of these proceedings has been named "developments for sports"; it is the biggest and could have even been bigger. The second volume is termed "developments in disciplines", which consists mainly of contributions focusing on modeling and measurements. A third volume has been named "developments for innovation", a tribute to this special focus of this conference (being organized by a center for innovation in sports), and to the fact that we could accumulate an amazing number of contributions in this field.

Finally, we hope that the reader will appreciate the outcome, and we'll be very happy to receive comments of whatever kind, be it criticism, proposals for improvement or grappa casks and flower arrangements.

Eckehard Fozzy Moritz
Stephen Haake

Editors
July 2006

Contents

3 Materials

6 Human Factors

Contributors

Simon C. Adelmann
University of Birmingham, UK

Michiyoshi Ae
University of Tsukuba, Japan

Uzoma Ajoku
Loughborough University, UK

Shinichiro Akiyama
Toyota Motor Corporation, Japan

Firoz Alam
Royal Melbourne Institute of Technology, Australia

Pär-Anders Albinsson
Swedish Defence Research Agency, Sweden

Enrique Alcántara
Universitat Politécnica de Valéncia, Spain

Brady C. Anderson
University of Calgary, Canada

Lauren Anderson
Loughborough University, UK

Ross Anderson
University of Limerick, UK

Dennis Andersson
Swedish Defence Research Agency, Sweden

Yiannis Andreopoulos
The City College of New York, USA

Ali Ansarifar
Loughborough University, UK

Ayako Aoyama
Tokyo Institute of Technology, Japan

Takeshi Asai
Yamagata University, Japan

Andrew Ashcroft
University of Cambridge, UK

Alan Ashley
United States Ski Association, USA

Mirco Auer
Swiss Federal Institute for Snow and Avalanche Research Davos, Switzerland

Andreas Avgerinos
Democritus University of Thrace, Greece

Arnold Baca
University of Vienna, Austria

Sarah Barber
University of Sheffield, UK

Franck Barbier
Université de Valenciennes, France

Matthew R. Barker
Auckland University of Technology, New Zealand

Joseph Beck
United States Air Force Academy, USA

Nicolas Belluye
Decathlon, France

Alexey, Belyaev
Perm State Technical University, Russia

Göran Berglund
Sandvik Material Technology, Sweden

Nils Betzler
Otto von Guericke University Magdeburg, Germany

Marc Bissuel
INSA Lyon, France

Kim Blackburn
Cranfield University, UK

Jane R. Blackford
University of Edinburgh, UK

Kim B. Blair
Massachusetts Institute of Technology, USA

Stephan Boerboom
Technische Universität München, Germany

Harald Böhm
Technische Universität München, Germany

Róbert Bordás
Otto von Guericke University Magdeburg, Germany

Pierre-Etienne Bourban
Ecole Polytechnique Fédérale de Lausanne (EPFL), Switzerland

Jean-Daniel Brabant
INSA Lyon, France

Alan N. Bramley
University of Bath, UK

Ken Bray
University of Bath

Desmond Brown
University of Ulster, UK

Steve Brown
University of Wales Swansea, UK

Mark-Paul Buckingham
University of Edinburgh, UK

Jeremy Burn
Bristol University, UK

Mike P. Caine
Loughborough University, UK

Matt J. Carré
University of Sheffield, UK

David J. Carswell
University of Wales Swansea, UK

Catherine J. Caton
University of Birmingham, UK

Chaochao Chen
Kochi University of Technology, Japan

Lance Chong
University of Illinois, USA

Simon Choppin
University of Sheffield, UK

Jeffrey J. Chu
Simbex, USA

Steffen Clement
AUDI Sport, Germany

Etienne Combaz
Ecole Polytechnique Fédérale de Lausanne (EPFL), Switzerland

Mario Comín
Universitat Politécnica de Valéncia, Spain

Alex Cork
Loughborough University, UK

James Cornish
University of Birmingham, UK

Robert Cottey
HEAD Sport AG, Austria

Aimee C. Cubitt
University of Bath, UK

Kieran F. Culligan
Massachusetts Institute of Technology, USA

David Curtis
Sheffield Hallam University, UK

Dave Custer
Massachusetts Institute of Technology, USA

Tim Deans
Bristol University, UK

Jeroen Dethmers
Universiteit Maastricht, Netherlands

Neil Dixon
Loughborough University, UK

Sharon J. Dixon
University of Exeter, UK

Jamie Douglas
International Tennis Federation, UK

Patrick J. Drane
University of Massachusetts Lowell, USA

Melanie Dumm
Technische Universität München, Germany

Juan Vicente Durá
Universitat Politécnica de Valéncia, Spain

Colin Eames
United States Air Force Academy, USA

Markus Eckelt
University of Applied Sciences Technikum Wien, Austria

Jürgen Edelmann-Nusser
Otto von Guericke University Magdeburg, Germany

Frank Einwag
Klinik für Orthopädische Chirurgie und Unfallchirurgie Bamberg, Germany

Carl F. Ettlinger
Vermont Safety Research, USA

Paul Ewart
University of Waikato, New Zealand

Emanuela Faggiano
University of Padova, Italy

Mathieu Fauve
Swiss Federal Institute for Snow and Avalanche Research Davos, Switzerland

Owen R. Fauvel
University of Calgary, Canada

Peter Federolf
University of Salzburg, Austria

Monika Fikus
University of Bremen, Germany

Christian Fischer
Ecole Polytechnique Fédérale de Lausanne (EPFL), Switzerland,

Peter R. Fischer
University of Augsburg, Germany

Keith Fitzpatrick
University of Limerick, UK

Paul Fleming
Loughborough University, UK

Ingmar Fliege
Technical University Kaiserslautern

Matthieu Foissac
Decathlon, France

Kathryn Franklin
University of Glamorgan, UK

Philippe Freychat
Decathlon, France

Piergiuseppe Fumei
University of Padova, Italy

Franz Konstantin Fuss
Nanyang Technological University, Singapore

Javier Gámez
Universitat Politécnica de Valéncia, Spain

Nico Ganter
Otto von Guericke University Magdeburg, Germanymagdeburg.de

Paul Gebhard
Technische Universität München, Germany

Alexander Geraldy
Technical University Kaiserslautern

Anton Gerrits
TNO, Netherlands

Alexandros Giannakis
CSEM - Swiss Center for Electronics and Microtechnology, Switzerland

Maria Giannousi
Democritus University of Thrace, Greece

Paul J. Gibbs
Loughborough University, UK

Christophe Gillet
Université de Valenciennes, France

Juan Carlos Gonzáles
Universitat Politécnica de Valéncia, Spain

Simon Goodwill
University of Sheffield, UK

Philippe Gorce
Toulon University, France

Rae, Gordon
University of Glamorgan, UK

Reinhard Gotzhein
Technical University Kaiserslautern

Richard M. Greenwald
Simbex, USA

Thomas Grund
Technische Universität München, Germany

Guglielmo Guerrini
Italian Kayak Federation, Italy

José María Gutierrez
Universitat Politécnica de Valéncia, Spain

Stephen J. Haake
Sheffield Hallam University, UK

Christian Hainzlmaier
Technische Universität München, Germany

Nick Hamilton
Sheffield Hallam University, UK

Dong Chul Han
Seoul National University, Korea

R. Keith Hanna
Fluent Europe Ltd., UK

Andy R. Harland
Loughborough University, UK

John Hart
Sheffield Hallam University, UK

Thomas Härtel
Chemnitz University of Technology, Germany

Ulrich Hartmann
Technische Universität München, Germany

Andreas Hasenknopf
MLD, Germany

Dieter Heinrich
University Innsbruck, Austria

Ben Heller
Sheffield Hallam University, UK

Mario Heller
University of Vienna, Austria

Christian Henneke
SportKreativWerkstatt GmbH, Germany

Martin Herbert
Bristol University, UK

Falk Hildebrand
Institute for Applied Training Science (IAT) Leipzig, Germany

Norbert Himmel
Institut für Verbundwerkstoffe GmbH, Germany

Frédérique Hintzy
Laboratoire de Modélisation des Activités Sportives, France

Nobuyuki Hirai
University of Tsukuba, Japan

Yuusuke Hiramatsu
Meijo University, Japan

Philip Hodgkins
Loughborough University, UK

Martin Hofmann
Otto von Guericke University Magdeburg, Germany

Frank Hoisl
Technische Universität München, Germany

Christopher E. Holmes
Loughborough University, UK

Yoshihisa Honda
Kinki University, Japan

Joe Hopkins
Western Michigan University, USA

Neil Hopkinson
Loughborough University, UK

Nicolas Horvais
Laboratoire de Modélisation des Activités Sportives, France

Yohei Hoshino
Hokkaido University, Japan

Kenji Hosokawa
Chubu University, Japan

Mont Hubbard
University of California, Davis, USA

Andrew Hytjan
University of Colorado at Boulder, USA

Yesim Igci
Princeton University, USA

Hiroshi Iida
Polytechnic University Kagawa, Japan

Yoshio Inoue
Kochi University of Technology, Japan

Carl Johan Irander
Sandvik Material Technology, Sweden

Jon Iriberri Berrostegieta
Performance Enhancement Centre, Basque Government, Spain

Gareth Irwin
University of Wales Cardiff, UK

Aaron Ison
Cascade Engineering, USA

Andrea Isotti
University of Padova, Italy

Koji Ito
Japan Institute of Sport Sciences, Japan

Takuzo Iwatsubo
Kansai University, Japan

Thomas Jaitner
Technical University Kaiserslautern

Daniel A. James
Griffith University, Australia

David M. James
University of Sheffield, UK

Iain James
Cranfield University, UK

Mike J. Jenkins
University of Birmingham, UK

Marke Jennings-Temple
Cranfield University, UK

Alexander W. Jessiman
Simbex, USA

Tomohiko Jin
Toyota Motor Corporation, Japan

Robert J. Johnson
University of Vermont, USA

Clifton R. Johnston
University of Calgary, Canada

Roy Jones
Loughborough University, UK

André Jordan
Otto von Guericke University Magdeburg, Germany

Laura Justham
Loughborough University, UK

Hank Kaczmarski
University of Illinois, USA

Hiroyuki Kagawa
Kanazawa University, Japan

Michael Kaiser
Institut für Verbundwerkstoffe GmbH, Germany

Nico Kamperman
TNO, Netherlands

Peter Kaps
University Innsbruck, Austria

Shozo Kawamura
Toyohashi University of Technology, Japan

Ian C. Kenny
University of Ulster, UK

David G. Kerwin
University of Wales Cardiff, UK

Andreas Kiefmann
Technische Universität München, Germany

Cheol Kim
Kyungpook National University, Korea

Moo Sun Kim
Seoul National University, Korea

Sun Jin Kim
Seoul National University, Korea

Wendy Kimmel
University of California, Davis, USA

Efthimis Kioumourtzoglou
Democritus University of Thrace, Greece

Bob Kirk
University of Sheffield, UK

Sebastian Klee

Isabella Klöpfer
Technische Universität München, Germany

Karin Knoll
Institute for Applied Training Science (IAT) Leipzig, Germany

Klaus Knoll
Institute for Applied Training Science (IAT) Leipzig, Germany

Ted Knox
Wright Patterson Air Force Base, USA

Cheolwoong Ko
University of Iowa, USA

Osamu Kobayashi
Tokai University, Japan

Yukinori Kobayashi
Hokkaido University, Japan

Jan Koch
Technical University Kaiserslautern

Hannes Kogler
Fischer GmbH, Austria

Sekiya Koike
University of Tsukuba, Japan

Philipp Kornfeind
University of Vienna, Austria

Giorgos Kotrotsios
CSEM - Swiss Center for Electronics and Microtechnology, Switzerland

Johan Kotze
HEAD Sport AG, Austria

Christian Krämer
Technische Universität München, Germany

Maximilian Krinninger
Technische Universität München, Germany

Michael Krohn
Hochschule für Gestaltung und Kunst Zürich, Switzerland

Andreas Krüger
Otto von Guericke University Magdeburg, Germany

Thomas Kuhn
Technical University Kaiserslautern

Herfried Lammer
HEAD Sport AG, Austria

Nicholas Lavery
University of Wales Swansea, UK

Paul Leaney
Loughborough University, UK

Manryung Lee
Kyungin Women's College, Korea

Woo Il Lee
Seoul National University, Korea

Peter Leeds-Harrison
Cranfield University, UK

Sébastien Leteneur
Université de Valenciennes, France

Chris Lewis-Jones
Delcam plc, UK

Udo Lindemann
Technische Universität München, Germany

Daniel Low
University of Exeter, UK

Peter Lugner
Vienna University of Technology, Austria

Richard Lukes
University of Sheffield, UK

Anton Lüthi
Swiss Federal Institute for Snow and Avalanche Research Davos, Switzerland

Reiner Lützeler
RWTH Aachen University, Germany

Jani Macari Pallis
Cislunar Aerospace Inc., USA

Lionel Manin
INSA Lyon, Francefr

Graeme Manson
University of Sheffield, UK

Jan-Anders E. Månson
Ecole Polytechnique Fédérale de Lausanne (EPFL), Switzerland

Giuseppe Marcolin
University of Padova, Italy

Brett A. Marmo
University of Edinburgh, UK

Antonio Martínez
Universitat Politécnica de Valéncia, Spain

Natividad Martínez
Universitat Politécnica de Valéncia, Spain

Tom Mase
Michigan State University, USA

Steve Mather
University of Nottingham, UKk

Sean Maw
University of Calgary, Canada

Alex J. McCloy
University of Ulster, UK

Mark McHutchon
University of Sheffield, UK

Andrew McLeod
Cranfield University, UK

Hossain Md.Zahid
Toyohashi University of Technology, Japan

Kenneth Meijer
Universiteit Maastricht, Netherlands

Daniel Memmert
University of Heidelberg, Germany

Roberto Meneghello
University of Padova, Italy

Imke K. Meyer
University of Bremen, Germany

Michael Michailov
National Sports Academy, Bulgaria

Véronique Michaud
Ecole Polytechnique Fédérale de Lausanne (EPFL), Switzerland

Thomas Milani
Chemnitz University of Technology, Germany

Paul Miller
University of Colorado at Boulder, USA

Stuart Miller
International Tennis Federation, UK

Guillaume Millet
Université Jean Monnet Saint-Etienne, France

Hirofumi Minamoto
Toyohashi University of Technology, Japan

Sean R. Mitchell
Loughborough University, UK

Chikara Miyaji
Japan Institute of Sport Sciences, Japan

Yusuke Miyazaki
Tokyo Institute of Technology, Japan

Taketo Mizota
Fukuoka Institute of Technology, Japan

Stuart Monk
University of Birmingham, UK

Ana Montaner
Universitat Politécnica de Valéncia, Spain

John Morgan
Bristol University, UK

Eckehard Fozzy Moritz
SportKreativWerkstatt GmbH, Germany

Rhys Morris
University of Wales Cardiff, UK

Martin Mössner
University Innsbruck, Austria

Maximilian Müller
Technische Universität München, Germany

Masahide Murakami
University of Tsukuba, Japan

Werner Nachbauer
University Innsbruck, Austria

Daiki Nakajima
Kansai University, Japan

Motomu Nakashima
Tokyo Institute of Technology, Japan

Takeshi Naruo
Mizuno Corporation, Japan

Alan M. Nathan
University of Illinois, USA

Dirk Niebhur
Technical University Kaiserslautern

Günther Niegl
University of Vienna, Austria

Christian Nolte
University of Augsburg, Germany

Claudius Nowoisky
Otto von Guericke University Magdeburg, Germany

Wubbo Ockels
Delft University of Technology, Netherlands

Stephan Odenwald
Chemnitz University of Technology

Yuji Ohgi
Keio University, Japan

Shigemichi Ohshima
Meijo University, Japan

Atsumi Ohtsuki
Meijo University, Japan

Hiroki Okubo
National Defense Academy, Japan

Steve R. Otto
R&A Rules Limited, UK

Riccardo M. Pagliarella
Royal Melbourne Institute of Technology, Australia

Jürgen Perl
University of Mainz, Germany

Stéphane Perrey
Université de Montpellier, France

Christiane Peters
Technische Universität München, Germany

Nicola Petrone
University of Padova, Italy

Neil Pettican
Cranfield University, UK

Jon Petzing
Loughborough University, UK

Andrew Phillips
University of Bath, UK

John Plaga
Wright Patterson Air Force Base, USA

Christopher J.G. Plummer
Ecole Polytechnique Fédérale de Lausanne (EPFL), Switzerland

Alexander Romanovich Podgaets
Delft University of Technology, Netherlands

Jaime Prat
Universitat Politécnica de Valéncia, Spain

Céline Puyaubreau
Decathlon, France

Franck Quaine
Université Joseph Fourier Grenoble, France

José Ramiro
Universitat Politécnica de Valéncia, Spain

Robin Redfield
United States Air Force Academy, USA

Martin Reichel
University of Applied Sciences Technikum Wien, Austria

Hansueli Rhyner
Swiss Federal Institute for Snow and Avalanche Research Davos, Switzerland

Matthieu Richard
PETZL, France

Claudio Robazza
University of Padova, Italy

Bryan C. Roberts
Loughborough University, UK

Jonathan Roberts
Loughborough University, UK

Markus A. Rohde
University of Siegen, Germany

Jouni A. Ronkainen
Loughborough University, UK

David Rosa
Universitat Politécnica de Valéncia, Spain

Steve Rothberg
Loughborough University, UK

Maxime Roux
Decathlon, France

Daniel Russell
Kettering University, USA

AntonSabo
University of Applied Sciences Technikum Wien

Takahiro Sajima
SRI Sports Limited, Japan

Reiko Sakashita
Kumamoto University, Japan

Toshiyuki Sakata
Chubu University, Japan

Pierre Samozino
Laboratoire de Modélisation des Activités Sportives, France

Yu Sato
Chubu University, Japan

Nicholas Savage
Royal Melbourne Institute of Technology, Australia

Hans Savelberg
Universiteit Maastricht, Netherlands

Michael Schiestl
University Innsbruck, Austria

David Schill
United States Air Force Academy, USA

Kurt Schindelwig
University Innsbruck, Austria

Erin Schmidt
Loughborough University, UK

Heinz-Bodo Schmiedmayer
Vienna University of Technology, Austria

Alexander Schneider
Turn Till Burn GmbH, Switzerland

Isabelle Schöffl
University of Erlangen-Nuremberg, Germany

Volker R. Schöffl
Klinik für Orthopädische Chirurgie und Unfallchirurgie Bamberg, Germany

Stefan Schönberger
Technische Universität München, Germany

Herwig Schretter
HTM Tyrolia, Austria

Andreas Schweizer
Kantonsspital Aarau, Switzerland

Carsten Schwiewagner
Technische Universität München, Germany

Nathan Scott
The University of Western Australia, Australia

Brian P. Self
United States Air Force Academy, USA

Terry Senior
Sheffield Hallam University, UK

Veit Senner
Technische Universität München, Germany

Kazuya Seo
Yamagata University, Japan

Sonali Shah
University of Illinois at Urbana-Champaign, USA

Rebecca H. Shaw
University of Massachusetts Lowell, USA

Jasper Shealy
Rochester Institute of Technology, USA

Alison L. Sheets
University of California, Davis, USA

James A. Sherwood
University of Massachusetts Lowell

Kyoko Shibata
Kochi University of Technology, Japan

Jun Shimizu
Japan Institute of Sport Sciences, Japan

Peter Shipton
Cranfield University, UK

Hitoshi Shiraki
University of Tsukuba, Japan

Anton Shumihin
Perm State Technical University, Russia

Gerard Sierksma
University of Groningen, Netherlands

Lloyd Smith
Washington State University, USA

Peter Spitzenpfeil
Technische Universität München, Germany

Carolyn Steele
Loughborough University, UK

Darren J. Stefanyshyn
University of Calgary, Canada

Gunnar Stevens
University of Siegen, Germany

Victoria H. Stiles
University of Exeter, UK

Valeriy Stolbov
Perm State Technical University, Russia

Martin Strangwood
University of Birmingham, UK

Wolf Strecker
Klinik für Orthopädische Chirurgie und Unfallchirurgie Bamberg, Germany

Martin Strehler
SportKreativWerkstatt GmbH, Germany

Claude Stricker
AISTS – International Academy of Sports Science and Technology, Switzerland

William J. Stronge
University of Cambridge, UK

Aleksandar Subic
Royal Melbourne Institute of Technology, Australia

Maria José Such
Universitat Politécnica de Valéncia, Spain

Cory Sutela
SRAM Corporation, USA

Soichiro Suzuki
Kitami Institute of Technology, Japan

Masaya Takahashi
Sumitomo Light Metal, Japan

Hironuri Takihara
Toyohashi University of Technology, Japan

Ming Adin Tan
Nanyang Technological University, Singapore

Angelo Tempia
Royal Melbourne Institute of Technology, Australia

Eva Tenan
University of Padova, Italy

Dominique Thévenin
Otto von Guericke University Magdeburg, Germany

Mark Timms
Hot Stix Technologies, USA

Daniel Toon
Loughborough University, UK

Marcus Trapp
Technical University Kaiserslautern

Masaya Tsunoda
SRI Sports Limited, Japan

Sadayuki Ujihashi
Tokyo Institute of Technology, Japan

Sándor Vajna
Otto von Guericke University Magdeburg, Germany

Rafael Valero
AIJU, Technological Institute of Toys, Spain

Sergey Vasilenko
JSC Aviadvigatel – Perm Engine Company, Russia

Pedro Vera
Universitat Politécnica de Valéncia, Spain

Johan Verbeek
University of Waikato, New Zealand

Nicholas Vernadakis
Democritus University of Thrace, Greece

Alex Vickers
Cranfield University, UK

Laurant Vigouroux
Université Joseph Fourier Grenoble, France

Jeff Vogwell
University of Bath, UK

Jörg F. Wagner
University Stuttgart, Germany

Klaus Wagner
Institute for Applied Training Science (IAT) Leipzig, Germany

David Walfisch
Massachusetts Institute of Technology, USA

Eric S. Wallace
University of Ulster, UK

Tom Waller
Loughborough University, UK

Andy Walshe
United States Ski Association, USA

Simon Watkins
Royal Melbourne Institute of Technology, Australia

Pek Chee We
Royal Melbourne Institute of Technology, Australia

Christian Webel
Technical University Kaiserslautern

Matthew Weber
University of Colorado at Boulder, USA

Sheldon Weinbaum
The City College of New York, USA

Andrew West
Loughborough University, UK

Cory West
Hot Stix Technologies, USA

Miles Wheeler
University of Colorado at Boulder, USA

Josef Wiemeyer
Technische Universität Darmstadt Germany

Bart Wijers
Terra Sports Technology, Netherlands

Paul Willems
Universiteit Maastricht, Netherlands

Simon Williams
University of Glamorgan, UK

Markus A. Wimmer
Rush University Medical Center Chicago, USA

Erich Wintermantel
Technische Universität München, Germany

Clive Wishart
Bristol University, UK

Kerstin Witte
Otto von Guericke University Magdeburg, Germany

Gavin Wood
Cranfield University, UK

Ian C. Wright
TaylorMade-adidas Golf Company, USA

Qianhong Wu
Villanova University, USA

Volker Wulf
University of Siegen, Germany

Bernd Wunderlich
Otto von Guericke University Magdeburg, Germany

Masanori Yabu
SRI Sports Limited, Japan

Tetsuo Yamaguchi
SRI Sports Limited, Japan

Connie Yang
Loughborough University, UK

Keiko Yoneyama
Tokyo Institute of Technology, Japan

Takeshi Yoneyama
Kanazawa University, Japan

Colin Young
Loughborough University, UK

Allen Yuen
University of Calgary, Canada

Jack Zable
University of Colorado at Boulder, USA

Michael F. Zäh
Technische Universität München, Germany

Eleni Zetou
Democritus University of Thrace, Greece

Andreas Zimmermann
University of Siegen, Germany

Werner Zirngiebl
Praxisklinik für Orthopädie und Sportmedizin, München, Germany

1 Innovation

Synopsis of Current Developments: Innovation

Eckehard Fozzy Moritz
SportKreativWerkstatt GmbH, Germany, efm@sportkreativwerkstatt.de

The importance of innovation and innovation-related research to my great pleasure has strongly increased in the sports engineering community during the last decade. Whereas just a couple of conferences ago I was feeling like a lone voice crying (the innovation doodle) in the wilderness, delightedly teaming up with those very few reporting on systematically designed pieces of sports equipment, by this year we had to split the topic area: You will not only find a number of contributions on various perspectives on innovation in the following, but also a good share of innovative design oriented contributions summarized in another synopsis by Caine, and further trans-disciplinary papers opening up new perspectives on sports engineering in yet other topic areas.

Contributions and topics: An overview

The most stunning, and at the same time delightful, development to my mind is the "infiltration" of innovation researchers from the fields of economics into the sports engineering community. On the following pages you will find two contributions from this perspective: Giannakis and Stricker giving some quite practical insights into strategic considerations for market introduction and marketing of innovations in consumer sports products, Shah highlighting the role of lead users in the origination of innovations, and hence the huge potential for industry in cooperating with leading figures of sports communities in the development of new products.

Three contributions from various disciplinary backgrounds explicitly propose the further development and application of new systematic approaches in the generation of innovations: Krüger et al are discussing a new model aiming at an improved con-textualization of technology development. Even though still quite mechanistic, this approach is an important opening up of traditional models en route to a more holistic view on innovation. Gerrits et al highlight the need and current practices for cus-tomization in sports products. Still with a strong focus on manufacturing aspects, this yet adolescent project will surely lead to important methodological conceptions on how to realize customization in the years to follow. Last not least, a very daring approach is being introduced by Podgaets et al: Through the application of novel mathematical approaches in the solution of sports engineering related problems more parameters and hence more and more diverse perspectives can be considered in a quantitative fashion in design projects and innovation ventures in sports engineering.

A third category of contributions is dedicated towards issues in sustainability. Müller et al propose the development of a new traffic system fostering the use of non motorized vehicles, in an interesting mix of technical system and industrial design. Subic and Paterson are tackling the problem of the ever-growing environmental impact of sports equipment design and production, and are introducing Life Cycle Assessment as a useful tool to improve on the current all but sustainable situation. Finally, Hanna and myself are discussing sustainability issues on a far more general level, and are proposing a more balanced, that is, less technology- and profit-oriented view on sports engineering developments and holistically sustainable innovation as one solution to bridge economic, social, and environmental concerns.

Towards future developments: potentials, predictions, prayers

Even though the contributions summarized above have originated in many different disciplines, more or less they all carry the same message: There is a need to take a more holistic view on sports engineering developments; there are more and more approaches evolving how to get there, and more and more subject areas in which related projects can and should be started.

As for future developments, I am quite sure that this convergence of disciplinary perspectives will continue, that economists will (have the need to) work on approaches how to combine profit interests with global and local sustainability concerns, that engineers will further embark on a more embedded view on technology development and will develop more holistically oriented design and innovation methods, and that sports and other social scientists will cooperate with both parties to render their expertise more useful to society.

Finally, and hopefully not only wishfully thinking, all these developments will help transfer sports engineering from a quite analytically oriented newcomer discipline into a responsible yardstick for future sports and health related activities. In this arena there is a huge scope for innovations useful to both economy and society, in products and systems as well as in related methods – the contributions that follow will surely only be the beginning.

Custom-Fit: Quality of Life of European Sporting Public through Custom-Fit Products

Anton Gerrits[1], Chris Lewis Jones[2] and Rafael Valero[3]

[1] TNO, Netherlands, anton.gerrits@tno.nl
[2] Delcam plc, UK
[3] AIJU, Technological Institute of Toys, Spain

Abstract. A radical change (Feenstra, Holmer, Tromans, Moos and Mieritz 2003) in manufacturing is starting to occur. The Rapid Manufacturing, which can be defined as "the use of an additive manufacturing process to construct parts that are used directly as finished products or components", is set to supersede many current uses of moulds and dies. Rapid Manufacturing (Wohlers 2003) is based on new additive manufacturing techniques that produce fully functional parts directly from a 3D CAD model without the use of tooling. The ambitious scope of the European Initiative CUSTOM-FIT, a Framework 6 Integrated Project, is to create a fully integrated system for the design, production and supply of individualized custom-products. Within the cases under study there are strong connections to sporting articles. This project is being funded by the European Commission over the next four years and a half and will become central to European research concerning Rapid Manufacturing (RM).

1 Introduction

Custom-Fit is an industry led FP6 Integrated Project co-ordinated by Delcam plc, initiated by TNO Industrial Technology and Loughbourough University. The consortium is made up from broad base of some 30 partners from 12 countries forming a complementary set of roles with their own individual expertise in the field. The social aim of Custom-Fit is to improve the quality of life by providing products to the citizen optimized to their individual geometrical shapes and requirements. This will improve the performance & comfort and at the same time reduce injuries. The following examples have been selected as test cases for the project and will illustrate its goal.

Prosthesis For an amputee traditionally the socket is made using a negative and positive plaster model. The negative reproduces the stump topology and is produced by wrapping a plaster bandage around the stump. The positive is made pouring liquid plaster into the negative model. When the plaster becomes hard, it needs to be skillfully and correctly shaped for the socket's construction. Often several fittings are required before sufficient comfort is achieved. The new Custom-Fit approach will be to scan the stump first, model the socket using new CAD techniques that will model variable graded materials and produce the socket with a Layer Manufacturing Technique (LMT).

Implants This includes cleft pallet patients, trauma and jaw bone fracture of the dental base, hip replacement and bone replacement for cancer patients,

Helmets protect the user against injuries. Custom-Fit will apply RM technology to produce helmets that are custom made in order to improve sport performance and fitted also to reduce the risk of Traumatic Brain Injury (TBI). Furthermore, a customized inner cushioning will allow the possibility of integrating spaces for integrated communication devices.

Back packs and in particular the interface to transmit load through the body will be investigated to improve design for mountaineering and rock climbers.

Seats are another example where prolonged contact will be improved with the design of customized parts. The focus will be directed at motor cycle seats, but the process of data capture, design and manufacture will provide a route for all sporting seats with the benefits of improved posture and performance for the athlete.

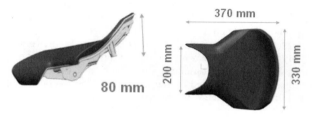

Fig. 1. Motor bike seat

2 Core

Although customization of products (Hague, Campbell and Dickens 2003) has been recognized as having important commercial potential for many years, it has generally been limited to relatively superficial cosmetic variations such as the choice of product color, for example. The additional cost associated with customization has been a major barrier to its wide spread adoption.

During this projects complex customized products will be produced from unique combinations of (nano) materials without the need for dedicated production tooling. Custom-Fit will drastically change how and where products will be designed and made, but there is still a long way to go; RM itself does not exist yet and furthermore a complete new manufacturing and supply system has to be developed (Hague, Dickens, Mansour, Saleh and Sun 2002).

The first step developed in the project has been the definition of the EU citizen's requirements for Custom-Fit products. It has been necessary to research and evaluate the service dimension of this technology to determine the affect on the supply & demand chains. Market studies have been developed in order to predict and guide the new positioning for these individualized products.

Point of customization is also a big issue, studies have been carried out to determine where and how the best place is (retails shop, scanning centre, mobile scanner, home visits, postal service or Internet). Time of data capture, cost of equipment, skill of the sales assistant or practitioner, ease of use of software, transfer of data are all parameters affecting the potential uptake. The resultant experience the client receives as a whole coupled with the advantages of the product satisfaction add more immeasurable uncertainties to the investigations.

A vital new step in this research is the development of a new knowledge base platform, with customer requirement capture and transfer of scan data through a new neutral format, in order to define geometrical and non-geometrical requirements.

This new Neutral Scan Format (NSF) based in XML has been developed to encapsulate more than just geometric topology of scan layers. The new format can store not only the 3D shape but anthropometric measurements embedded as geometrical properties in NSF (width, height, circumference, cutting plane geometries…). In addition non-geometric requirements such as the result of user-product interaction (pressure, velocities, pain, performance and finally satisfaction) are intended to be associated with the design and manufacture to provide a new "dimension" of customization.

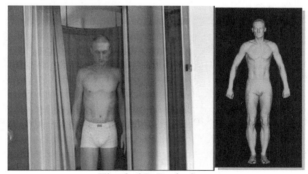

Fig. 2. 3D Body Scan

The Custom-Fit process of design and manufacture for sports equipment will provide the opportunity to build in more performance criteria than ever. For example, the helmet includes customization beyond simply fitting to cranial in 3-diemnsions to the cyclist or skiers head, but it will also consider non-geometric requirements such as eyes, mouth and nose, vision angle, stress, sweat rates, information of previous injuries or special characteristics like temperature of use, humidity, solar radiation, pressure map in the contact surface, inertial forces and other parameters that can affect the athlete's performance.

To re-solve these new design challenges Custom-Fit aims to develop technology which will be able to manage these new characteristics of the part. Presently not only is there limited hardware, currently there is no commercially (Feenstra et al. 2003) available software that allows a designer to incorporate functional grading and incorporated non-geometric data into a new product. This is essential to the whole Custom-Fit process since graded parts with several materials or variable density are the necessary to have sports devices with improved and customized properties.

Several researches are being carried out in order to have a user-friendly system for representing, editing and added graded material structures and also a methodology to transform subjective non-geometrical requirements into objective geometrical and physical shapes.

With parts that are composed of graded or variable density materials new algorithms are required for slicing and representing different material structures. CT-based material assignment enables graded material structures to specific areas on CT scan level. This tool is building onto a FEA analysis, processing it through graded area definitions and enabling to run again a FEA on the graded end result.

The actual production of the parts within Custom-Fit will utilize Rapid Manufacturing processes. Within the project there are three basic methods of production being investigated as potential machining techniques for Rapid Manufacturing purposes. All of them are potentially capable to depositing graded material to create single functionally graded components and all are currently in a development phase.

Metal Printing Process (SINTEF, Norway) is aimed at developing the equivalent of a high-speed photocopier that produces three-dimensional objects from powder material. Layers of powder are generated by attracting the metal or ceramic powder to a charged photoreceptor (PR) under the influence of an electrostatic field. The attracted layer is deposited on a building table where it is consolidated. More photoreceptors will be used to sequential deposit different powders in the same layer. The process is repeated layer-by-layer until the three-dimensional object is formed and consolidated.

Multiple deflection continuous jet process (TNO, The Netherlands) is similar to a standard ink-jet system but it can currently use fluids with viscosities of approximately 300 mPa/s at jetting temperature. With this system a continuous stream of droplets is produced, the separate droplets are then electrostatically charged and deflected to form the desired pattern. Because of the higher achievable viscosities, stronger end products can be produced. Since it is a selective deposition technique it is possible to use several types of materials within the same product.

"Upside down" 3D printing process (De Montfort University, UK) is based in a polymer powder that is positioned precisely where it is required by the selective application of a temporary binder/electrostatic charge. Excess powder is removed and the material that has been deposited is fused together. This process is continuously repeated adding building and supporting materials until the desired functionally graded object has been completed.

RM does not really exist at the moment and RP (rapid prototyping) processes are currently being used in place of Rapid Manufacturing machines. These machines are limited as they have not been designed for these finished applications. Consequently they produce parts with poor or variable surface finish, tolerances that are difficult to maintain and the repeatability of the processes is often poor with very little real time adaptive control. Also, when manufacturing small volumes of prototypes the removal of support material is not a major issue but this will become so if there is a need to produce hundreds or thousands of parts in a short time. The speed of current layer manufacturing processes is a major issue when compared with conventional manufacturing processes with injection moulding being 100 to 1000 times faster in material throughput.

The options to enhance speed and tolerances will depend on the technique developed. MPP process is based on xerographic addition of powder material; so there is a huge research carried on in order to improve the charge stage. The same as the Jetting and "Upside-down" process, there is a research under patent in order to control how material is placed.

The technical objectives are achieve a resolution of 0.03 mm and a roughness of 5 μm. And also take advantage of the chance of manufacturing graded parts with different materials.

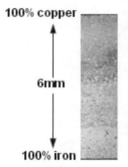

100% copper

6mm

100% iron

Each layer has a progressive change, possible in the 10% or 20% increments, to move from one material to another with different properties.

The gradual change in structure minimize problems with incompatible thermal expansion coefficients or other property differences.

Fig. 3. Graded structure

3 Conclusions

The ability to produce customized products, which are matched to the specific needs of an individual, is expected to have a major benefit on the quality of life and performance of European athletes at all levels from the highest Olympians to the local club member. This integrated project, funded by the European Commission, has been identified as being highly ambitious with corresponding breakthrough and potential impact and therefore requiring an enormous effort and drive to achieve the objectives. The carefully selected consortium, which is going to finish working in March 2009, and the stated research objectives that have been achieved the first year of the project has provided Custom-Fit with a sound basis for success.

Fig. 4. Paralympic athlete

Acknowledgements

The authors, in particular, and Custom-Fit consortium would like to express our gratitude to the European Commission for their financial support and assistance within the Sixth Framework Programme in developing this Integrated Project that would not have been possible without their help.

References

Feenstra F., Holmer B., Tromans G., Moos N. and Mieritz B., RP, RT, RM trends and developments / research survey *Proc. 4th Int. Conference on RP and Virtual Prototyping.* 20 June 2003, London

Wohlers T., "The Wohlers Report 2003", USA 2003

Hague R., Campbell R.I. and Dickens, P.M., Implications on Design of Rapid Manufacturing". *Journal of Mechanical Engineering Science*, January 2003 pp 25-30

Hague R., Dickens P.M., Mansour S., Saleh N. and Sun Z., Design for Rapid Manufacturing", *Rapid Prototyping and manufacturing Connference*, Cincinatti, USA, April 2002, 10 pp

Wimpenny D.I., Hayes, Goodship V., Rapid Manufacturing – It is feasible?. *International Journal of CAD/CAM & Computer Graphics*, Vol15, December 2000, pp 281-293

VeloVent – An Inner City Traffic System for Active People

Maximilian Müller[1], Veit Senner[1], Michael Krohn[2]

[1] Technische Universitaet Muenchen, mueller@sp.tum.de
[2] Hochschule für Gestaltung und Kunst Zuerich

Abstract. The presented concept of a new traffic system for metropolitan areas "VeloVent" suggests a number of innovative developments that may help to solve two major problems of these areas – too much motorized traffic and poor physical health of the citizens. The idea is based on human powered vehicles which run through a tube network with transparent covers and are supported by an externally introduced air stream. Important locations of public interest (i.e. shopping centers, major company facilities, parks, university campus) are connected by the tube. Emphasis was put on the premises and results of human powered VeloVent vehicles for different transportation purposes in the urban environment.

1 Introduction and Motivation

A well developed traffic infrastructure which allows just in time mobility and shipments is a cornerstone of modern societies. In fact the metropolitan areas play a key role in the highly developed technology and service sectors. The degree of urbanization indicates this backbone function quite well. The corresponding numbers from Germany 48.8% (Voith 2005), Korea 79.7% (Deutscher Industrie- und Handelskammertag 2005) and even China with 21% to 30% clearly show urbanization being on a high level in these countries. But apart from their economical power these areas also bear a number of severe problems, i.e. environmental pollution due to a rapidly increasing volume of motorized traffic. Health problems may result from this, not only due to the environmental problems but also due to the lack of physical activity (Steidl 2004).

Cycling is an activity that can help to solve both of them – presuming an adequate infrastructure and intelligent transportation concepts. In Germany only 12.5% of the routes up to a distance of 8 km are covered by bicycle and the overall average distance is 2.9 km per ride (BMVBW 1999). Although these numbers have been increasing since the late 70ies, they show the great potential of bicycles as means of transportation especially in the big cities in Germany. This observation should be true for most metropolitan areas in the world.

Several biomechanical studies have further shown that the cyclic pedalling motion is one of the most efficient human movement patterns making cycling even more appropriate for self-locomotion than walking or running. As a positive side effect it may help to prevent cardiovascular diseases, hypertension and arthritis and to build up the immune-system. In summary everyday cycling should improve the general well-being.

2 The VeloVent Concept and its assumed Side Effects

The idea is to run human powered vehicles (HPVs), regular bicycles or special bicycle like concepts in a network of tubes with transparent covers. The HPVs are supported by airflow in the direction of movement in order to minimize their air resistance.

It is estimated that an 80 kg person riding a conventional bicycle and being supported by a relatively slow constant airflow of 15 km/h will need 90 watt only to maintain a driving velocity of 30 km/h. This moderate power level can even be realized by humans with low physical capacity and be kept over a longer period of time. The tube network is intended to link major cross points, park & ride places, railway stations, big shopping centres or facilities of major companies. It may follow major automobile lanes, subway tracks, rivers or parks. It can be resting on pillars, be attached to buildings above the foot pathways (Fig. 1) or even hung below bridges in order to cross rivers or intersections. For safety reasons and optimal traffic management the system would provide tubes that are wide enough to drive side by side. Different vehicle classes with two, three or four wheels for various transportation purposes are possible. Three positive effects should result from this concept: First VeloVent would be capable to avoid three major disadvantages of today's HPV's that are their dependence on the weather, their lack of safety within the motorized traffic and their relatively low average speed. Therefore and second, VeloVent should have the potential to make HPVs more attractive in supporting everyday transportation demands. Finally and third the use of HPVs requires a modest physical activation and thus should positively contribute to public health.

Fig. 1. VeloVent scenario for metropolitan areas

3 Targets of the current investigations and developments

VeloVent is intended to be integrated in the existing infrastructure of public and individual transportation. Research and development activities regarding VeloVent therefore need to address a variety of different aspects. We distinguish between three different fields of interest:

(1) Technological realization including concept dependent energy cost calculations and production cost estimations.
(2) Social and sociological analysis including cultural and location dependent user acceptance.
(3) Economical analysis including its potentials regarding traffic effects, general innovation effects, additional products and services and effects regarding working market.

This paper addresses the first issue, the technological aspects. It will explain the work steps and methods applied by now and will present first concepts of vehicles as a result from this work.

4 Methods

The concepts presented here are results of an interdisciplinary project which was established between students going for a degree in mechanical engineering at the Technische Universität München and students of industrial design at the Hochschule für Gestaltung und Kunst in Zurich, Switzerland. The timeline was six months and included a systematic approach for the early phases of product development. Corresponding methods (e.g. Brainstorming and Method 6-3-5 to find new solutions) were applied for the mechanical engineering part. Two workshops were held together with the industrial design students to merge form and function of the concepts. Even if it does not answer the major question on the general feasibility of the entire system this step was necessary to get empirical data on the vehicles' boundary conditions (i.e. air resistance coefficient). These values are taken as input for the fundamental energy and material cost calculation of the tube network and the whole system. The main boundary conditions were set after a rough estimation of influence parameters: air-flow velocity 15 km/h, cross section of one tube direction 3.00 m times 2.40 m, air resistance coefficient times surface area (c_W*A, which is crucial to air resistance) from behind as big as possible and frontal as small as possible. The former is important to gain advance in case the vehicle is slower than the tube air stream, the latter in case if it is faster. The interfaces to the system (e.g. wheel to track) were defined and integrated to a function structure which helped to point out trade-offs and led to the requirements list. Three abstract scenarios were extracted from this information which was then enhanced to the following concepts. As for the development, a number of product models were generated. These helped to summarize and categorize requirements (function structure, requirements list) and solutions (morphological matrix), to validate specific questions of package dimensions (anthropological measurements, estimation of part dimensions) or shape (design sketches, wire frame models). They generally contributed to get more information about important product features. Therefore, the models supported the design process from an abstract level of ideas to a more detailed three dimensional CAD-model and a first scaled rapid prototyping model with specific product properties (Lindemann 2004).

5 Results

The "mother child" concept: women tend to connect several activities to errand chains within one bicycle ride (BMVBW 1999). For example, they would bring their children to the kindergarten or school, go to the post office and for the daily shopping. Unfortunately, neither the vehicles offered by the market nor the traffic system in general respond to the questions that arise in this context. These are to take along children in a safe way, carry heavy luggage from door to door and manage to do this on cumbersomely routes (Fig. 2).

Fig. 2. Mother-child concept side view

Regarding this, the "mother-child" concept was developed in a beneficial way by providing a portable trailer that can act as a shopping cart as well and which is capable of carrying children and shopping bags at a time. The trailer and the main part of the bicycle are connected by just pushing the trailer towards the self-locking coupling mechanism of the bicycle and together form a single unit. The trailer comprises a flexible roll-axis in a way that it would bend together with the main part of the bicycle and brings the center of gravity more towards the inside in fast turns. The bicycle's transmission is located at the front wheel to facilitate a simple setup for the trailer. This type of transmission was tested at a manufacturer's site (ZOX bikes, Erlangen, Germany) and revealed no serious shortcomings. Additional bags can be attached to the trailer at the back side which makes it more flexible regarding the storage of shopping or other luggage. The cyclist is sitting in an upright position during cycling which on the one hand permits still good oversight and keeps the point of gravitation low. The back surface is formed in a smooth way to avoid air turbulences and to display as much face as possible.

The "taxi" concept: businessmen or visitors wish to get in and around the cities fast and in time, taking small or middle size luggage with them. The increasing number of congestions not only during the rush hours bears the risk of delays and mental stress for the participants. Common rickshaw models suffer several disadvantages that are low average speed, uncomfortable entry and limited space for luggage. The new concept shows a possible solution to attract more potential passengers for this transportation medium (Fig. 3). Unlike common concepts, the passengers take a seat in the front part of the vehicle by turning the seat from a position bended 90° out-wards to a position in driving direction. The vehicle is a tricycle design with two front wheels and one back wheel (Fig. 4). The design bears a number of characteris-tic conveniences: easy access for the passengers, panoramic view and simple drive setup for one wheel. The driver sits in an upright position and keeps good overlook about the traffic. Additional compartments can be attached to the mainframe on both sides of the driver (e.g. to store suitcases) and an intelligent navigation system eases orientation towards the destination.

Fig. 3. Taxi concept side view

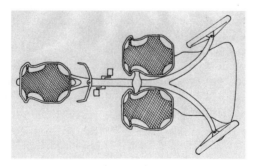

Fig. 4. Taxi concept top view

Finally the "delivery truck" concept was developed upon the fact that the quantity of transportation of small sized goods is an increasing sector, mainly due to the many e-commerce opportunities via internet and more just in time deliveries. For obvious reasons many of the daily delivery services may not able to be done by HPV's. However, some smaller shipments within city borders are already conducted by bicycle couriers. This sector could be served by HPV's in a more extended way and the proposed concept aims to combine advantages of human powered vehicles like flexibility with specific conceptual features like extended and easily accessible com-partments and an intelligent navigation and shipment tracking system (Fig. 5). The concept is a two wheel design and the main compartment is located below the driver seat which can be flapped away for easy access. Additional bags can be attached on both sides of the front wheel and the back wheel. Like the other concepts the driver sits in a lowered, but upright position.

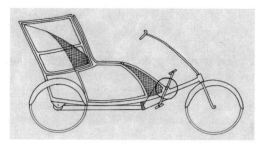

Fig. 5. Delivery-truck concept side view

6 Conclusions and Outlook

The presented concepts combine the potential for a more comfortable and intelligent handling with a cost conscious construction and design features that form a corporate identity of VeloVent vehicles. They point out strategies to solve some major problems of actual inner-city traffic systems and the modern society. However, the concepts remain in an early stage of product development and important steps are necessary to particularize and validate them. The development of VeloVent should figure out side effects which can influence the design of innovative HPV products and systems. Besides, two interdisciplinary research projects have been started together with departments of industrial design, fluid mechanics, product development and constructions. They concern questions in the field of an intelligent design of the tube system and an international survey among potential users of VeloVent to evaluate acceptance and objections as well as social effects.

References

Bundesministerium für Verkehr, Bau- und Wohnungswesen (1999) Erster Bericht der Bundesregierung über die Situation des Fahrradverkehrs in der Bundesrepublik Deutschland. BMVBW, Bonn.
Deutscher Industrie- und Handelskammertag (2005) Korea – Land und Leute. DIHK, Berlin.
Lindemann, U. (2004) Methodische Entwicklung technischer Produkte – Methoden flexibel und situationsgerecht anwenden. Springer, Berlin.
Steidl, J. (2004) Fast jeder zweite Deutsche hat Übergewicht. *Pressemitteilung*, Statistisches Bundesamt, Wiesbaden.
Voit, H. (2005) Rund 15% der Bevölkerung Deutschlands lebt auf dem Land. *Pressemitteilung*, Statistisches Bundesamt, Wiesbaden.

Approach of a Model for the Interaction Between Athlete, Sports Equipment and Environment

Andreas Krueger, Kerstin Witte and Juergen Edelmann-Nusser

Otto-v.-Guericke-University Magdeburg, andreas.krueger@gse-w.uni-magdeburg.de

Abstract. For the interaction between athlete, sports equipment and environment there are general characteristics which are important to consider during the product design process of performance oriented sports equipment. The aim of this study is to describe these aspects of the athlete-sports equipment-system and to get a model for the interaction between athlete, equipment and environment. The model comprises the athlete himself, his activities, the sports equipment and environment. Furthermore the subjective criteria in the model, perception, emotion, status symbol and trend-setting, are of interest. Some examples will illustrate the model. This approach can be used to assist the designer of performance oriented sports equipment and thus to support a systematic product design process in sports.

1 Introduction

High quality sports equipment becomes more and more important in all domains of sport. As defined here the term sports equipment will be used synonymously to sports technology and includes all goods for the use in sports. There are three main reasons for the growing interest in sports technology, which are performance, safety and economics (Gros 2003). Hence, an increase of innovation and complexity, a specialization and differentiation of sports technology as well as a shortening of product life cycles can be observed (Gros 2003; Heinemann 2001; Weber, Schnieder, Kortlueke and Horak 1995). Due to the increasing complexity and specialization of high quality sports equipment different knowledge domains (information technology, mechanical- and electrical engineering) are involved. To ensure the communication and cooperation between these technical disciplines a description of the general characteristics of sports technology is required. Furthermore the requirement lists need to be defined more precisely regarding the interactions between athlete, sports equipment and environment and a differentiated consideration of these requirements in all phases of the design process is necessary (Pahl and Beitz 1996; VDI 2206 2004). Hence, for the design of high quality sports equipment it is important to know the general characteristics, to be familiar with the interaction between athlete, sports equipment and environment and to take this into account in all phases of the design process (Gros 2003). Therefore the general characteristics of sports technology need to be systematised.

The aim of this study is to describe an approach of a model for the interaction between the athlete, his sports equipment and the environment and therefore to systematize the general characteristics of the athlete-sports equipment-system.

2 Description of the Model

In the following the model will be described. Therefore the interaction inside the athlete-sports equipment-system and the interaction of such system with its environment will be explained. Some examples will illustrate the model.

2.1 Interaction Between the Athlete and the Sports Equipment

Figure 1 shows the model for the interaction between athlete and sports equipment modified according to the VDI (Verein Deutscher Ingenieure e.V.) guideline 2242 "Engineering design of products in accordance with ergonomics" (VDI 1986).

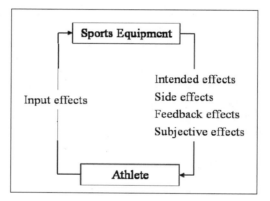

Fig.1. Model for the interaction between athlete and sports equipment (modified, VDI guideline 2242 1986)

To fulfil the function of sports equipment, the shown system requires at least one athlete. Due to input effects, such as exercises, movements and activities the athlete uses the sports equipment. The equipment returns intended, side and feedback effects, which may lead to further actions of the athlete.

The intended effects can be understood as the desired effects, which the sports equipment has been developed for. Depending on the function of the equipment there are four different types of intended effects, which are improvement of the athlete's performance, improvement of the athlete's training, improvement of safety, supply of information or a combination of these intended effects. The intended effect "Improvement of the athlete's performance" is often desired in competitive sports to achieve an advantage and therefore to gain better results in competitions. As an example serves the intended effect of less friction in swimming due to the use of optimized (shark skin) swim suits and by means of that possibly better results. The enhancement of the athlete's condition in the trainings process is also of significance. To improve the athlete's training special sports equipment (e.g. ergometer) is used and thus the intended effect "Improvement of the athlete's training" is realizable. Protection and improvement of the athlete's safety is another main function of sports equipment and therefore often desired, e.g. the use of suspension forks in mountain biking to reduce the acting mechanical loads on the human body. To estimate the athlete's performance it is necessary to obtain information, such as body data (e.g. heart rate) or performance data (e.g. power output). Hence, "Supply of information" is also an important function of sports equipment, especially in competitive sports. Besides the stated types of the intended effect there is often the combination of these types wanted; e.g. a modern helmet in cycling (individual time-trial stages) serves as safety equipment as well as performance equipment at the same time due to the reduction of wind resistance.

Apart from intended, objective effects, subjective ones from the sports equipment can affect the athlete. In contrast to the objective effects, the subjective effects can also act on the athlete without input effects. Perception, emotion, status symbol and trend-setting can be stated as such subjective effects. Since these effects have an enormous influence on the sales volume it is necessary to take these effects into account for the product design process, too.

The feedback effects are defined as a functional relationship due to the action of the sports equipment on the athlete (Pahl et al. 1996). These effects are used by the athlete to control the input effects. The information (feedback) transmission from the sports equipment to the athlete is enabled by means of sensory information uptake, which is separated in visual-, acoustic-, vestibular-, kinaesthetic- and tactile uptake. The sound of a tennis racquet at the moment of the tennis ball impact enables the athlete to gain information concerning the technique and therefore to recontrol his technique.

Besides the intended and feedback effects, side effects might occur as well. Side effects can be defined as undesired and unintended effects of the sports equipment on the athlete (Pahl et al. 1996). The vibration of a tennis racquet is an illustrative example for a side effect. Since the side effect can have an adverse effect on the athlete, it is important to minimize or to eliminate side effects due to the timely consideration in the product design process. If it is not possible to eliminate the side effect of the sports equipment, it is important to estimate the intended and the side effect since the disadvantages of the latter may exceed the advantage of the intended effect. Furthermore the implication of the side effect can vary from athlete to athlete. The constriction of the athlete due to the use of an optimized swim suit may influence one athlete more than another.

2.2 Interaction Between the Athlete-Sports Equipment-System and the Environment

In addition to the mentioned effects between athlete and sports equipment the relationship of the athlete-sports equipment-system to the surrounding environment is also of interest. Figure 2 illustrates the interaction between the system and the environment modified according to the VDI guideline 2242.

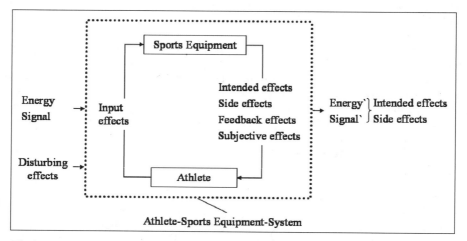

Fig.2. Interaction between the athlete-sports equipment-system and the environment (modified, VDI guideline 2242 1986)

There are desired and undesired effects between the system and the environment. On the one hand undesired effects from outside can influence the system. This can be defined as disturbing effects which can cause adverse, undesired side effects (Pahl et al. 1996). As an example, consider an archer while shooting with an acting crosswind. Due to the crosswind it may become difficult for the athlete to hold the target.

On the other hand the system requires inputs from the environment and produces outputs at the same time. In the engineering science the terms input and output are linked to energy-, signal- and material flow (Pahl et al. 1996). However, in sports no material flow can be found. Thus the described model for the interaction between athlete and sports equipment includes only signal and energy flow. Inside the system energy and signal can be changed or converted and therefore might have different occurrence (Pahl et al. 1996):

-   Energy: mechanical, thermal, chemical, electrical…also force, current…
-   Signal: data, information, control impulse…

There are various ways to convert energy and signal inside the system. A treadmill transforms electrical energy in mechanical energy; the athlete converts chemical energy in mechanical and thermal energy. Signals are received, prepared, compared, combined, displayed and recorded etc. (Pahl et al. 1996). As an example, consider a GPS receiver for the use in sports, which receives signals from satellites, calculates the actual position of the athlete in an earth fixed coordinate system and the actual speed of the athlete and finally saves and displays the calculated data. However, a signal flow is not possible without an energy flow, even if the latter is very small

(Pahl et al. 1996). Subsequent parts of the converted energy and signal are transferred to the environment and for that reason can be understood as outputs of the athlete-sports equipment-system. Converted energy and signal are illustrated in the model through the use of apostrophes. The system's output effects can have desired and undesired effects and thus intended and side effects as defined. An example for an intended effect on the environment is the transfer and converting of mechanical (pressure) into thermal energy of an ice skate, thus it becomes possible to skate. In contrast to such effects occurring vibrations from e.g. a treadmill while working may be an undesired side effect.

2.3 Example for a Practical Application: The Athlete-Shoe-System

The athlete-shoe-system will be used to illustrate and therefore to exemplify the model as whole. Figure 3 illustrates the interaction between the athlete, the shoe and the environment modified according to figure 2. During the interaction between athlete and shoe (e.g. running shoe) different effects can be observed.

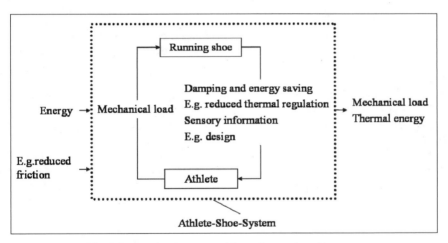

Fig.3. Interaction between athlete, shoe and environment

First of all the athlete uses the shoe for running, which is linked to a mechanical load and can be understood as an input effect. Hence, two different intended effects are caused. On the one hand the mechanical load will be damped at the moment of the heel strike and thus the safety will be improved. In addition to that the shoe might enhance the athlete's performance due to the saving and release of mechanical energy in the forefoot area with every stride. Apart the intended effects the shoe returns feedback effects concerning the running technique, which include mainly tactile- and kinaesthetic sensory information. Possible occurring side effects might be e.g. an insufficient wear comfort due to a poor fit or a reduced thermal regulation. Another example here might be a loss of time in a competition, due to an over damping effect. Besides the technical effects, subjective ones such as shoe design etc. might influence the runner as well. In addition to the mentioned effects the relationships between the athlete-shoe-system to its environment are also of importance. The acting ground reaction force can be interpreted as the system's energy input. If e.g. no ground reaction force in running direction exists caused by a slippery underground, a disturbing effect occurs. Furthermore, the acting force of the system on the underground and the heat caused by the athlete's body and the friction can be understood as the outputs of the system. To be efficient it is important for the designer to consider the effects already in an early stage of the shoe-design process.

3 Summary

This paper shows that there are general characteristics of sports equipment. The interactions between athlete, sports equipment and environment expressed trough input-, indented-, side-, feedback- and disturbing effects were explained and illustrated. All effects have to be considered during the design process of performance orientated sports equipment. In particular the requirement lists need to be defined more precisely in regard to these effects. Furthermore the precise requirements should be considered in all phases of the design process particularly concerning the product planning, the conceptual- and the embodiment design (VDI 2206, 2004). Additionally, the effects are criteria to evaluate the quality of a design proposal. Hence, they can be used to check and evaluate the concept, the embodiment or the prototype regarding the precise requirements. If there are limitations or intolerable side effects it is necessary to redesign the design proposal (VDI 2242 1986). It is essential to keep the side effects to a minimum and thus to maximize the benefit of the sports technology.

It was shown, that the model can be used to assist the designer of performance oriented sports equipment and thus to support a systematic product design process in sports. Due to an application of the model by practitioners the effect concerning the efficiency of the design process of performance oriented sports equipment could be evaluated.

References

Gros, H.J. (2003) Test und Entwicklung von Sportgeräten. Chancen, Risiken und Probleme. In: K. Roemer, J. Edelmann-Nusser, K. Witte und E.F. Moritz (Eds.), *Sporttechnologie zwischen Theorie und Praxis*. Shaker Verlag, Aachen, pp. 11-17.
Heinemann, K. (2001) *Die Technologisierung des Sports*. Karl Hofmann, Schorndorf.
Pahl, G.and Beitz, W. (1996) *Engineering Design*. Springer, Berlin, Heidelberg, New York.
VDI-Richtlinie 2206 (2004) *Design methodology for mechatronic systems*. Beuth Verlag, Berlin.
VDI-Richtlinie 2242 (1986) *Engineering design of products in accordance with ergonomics*. Beuth Verlag, Berlin, Koeln.
Weber, W., Schnieder, C., Kortlueke, N. and Horak, B. (1995) *Die wirtschaftliche Bedeutung des Sports*. Karl Hofmann, Schorndorf.

Strategy Deployment of High-Technology Companies Entering the Consumer Sports Electronic Market

Alexandros Giannakis[1], Claude Stricker[1], Giorgos Kotrotsios[2]

1 AISTS - International Academy of Sports Science and Technology, Lausanne, Switzerland

2 CSEM - Swiss Center for Electronics and Microtechnology, Neuchâtel, Swizerland

Abstract. The consumer sports electronic (CSE) market has been growing at an unprecedented rate, largely due to the advances made in microelectronics, communication technologies, high-level software, optoelectronics, micro-optics, bioengineering, and new materials. This paper proposes a structured process to enter this business sector for innovative, high-technology companies.

1 Brief Industry Depiction

Thriving in the CSE marketplace requires mastery of a diverse set of skills and capabilities. From adroitly reading market trends; investing prudently in future technologies; leveraging the skills and capabilities of technical and marketing personnel in an interactive fashion; understanding sport devices' customers intimately; offering a compelling value proposition; developing astute marketing strategies; pricing with an eye to consumer value even when a company deals with industrial customers only; and harmonizing with the business partners the distribution channels and supply chains.

CSE products' development necessitates the partnering of numerous companies with diverse business core backgrounds; hence network externalities and the need to develop industry standards are key characteristics of commercial success.

The marketing strategy of the products should be tailored to the type of innovation they feature; that being, incremental or radical. This notion of the contingent effects of the type of innovation on the marketing strategy must be carried through the entire lifecycle of the products.

At present, portable instruments' applications have exceeded the boundaries of professional sport, having been endorsed by medical institutes, government agencies, and specialized programs. The market's growth has inevitably attracted other industries, most notably the personal computer (e.g. Intel), telecommunications (e.g. Nokia) and medical sector (e.g. Hitachi or Siemens medical) fields. Therefore, new products or hybrids are being developed either exclusively by these industries or in collaboration with the CSE industry whose leading manufacturers include Suunto, Polar, Casio, and Nike.

2 High-Technology Companies and CSE

In order to analyze and evaluate its potential involvement in the CSE arena, a high-technology company must initially position itself in the scheme of partners that shape the group of companies involved in the development of CSE devices.

The second step would be to perform an assessment of the partnered CSE manu-facturers' status in the echelon of CSE companies. These partners could either be or aspire to be industry leaders, followers, or laggards. Knowing their strategic view is paramount for the high-technology company itself, in order to delineate the scope and the technological boundaries of these partnerships.

2.1 Market Plan and Corporate Strategy

In order to develop the high-technology company's market plan and consequently its strategic actions, a systematic and highly integrated process for evaluating market opportunities and developing strategies must be laid out. Following are the steps in this process:

Market Planning
1. Define goals and mission
2. Choose the arena
3. Identify potentially attractive opportunities
4. Make choices about markets
5. Plan key relationships
6. Develop value proposition and marketing strategy
7. Understand the profit dynamic
8. Implement the chosen strategy

Market Strategy
1. Target existing and new customers
2. Define products and services
3. Market timing
4. Execution

Competitive Advantage
1. Resources and core competencies
2. Tests of competitive advantage for superiority and sustainability
3. Approach to developing resources and competencies

2.2 Positioning in the Revenue Continuum

For a high-tech firm, technology itself is either the product (i.e. licensing proprietary technology) or gives rise to a product (i.e. commercializing products based on a new technology). The company innovates technological know-how and hence faces a unique decision: should it sell the technology itself or license it?

Should it commercialize the idea – marketing, distributing, and selling a full solution including service and support? Or, given that final products can be 'decomposed' into subsystems and components, should the high-tech company manufacture and sell some subsystem or component on an original equipment manufacturer (OEM) basis?

The final decision boils down to the basic issue of how to transform know-how into revenues. The different options are listed below:
-   Sell of license know-how solely
-   Sell 'proof-of-concept'
-   Sell commercial-grade components to OEMs
-   Sell final products or systems with all essential components
-   Sell a complete, end-to-end solution

The factors that affect a firm's decision about where along the continuum to generate revenues are shown in fig.1:

Conditions	Know-How	'Proof of Concept'	Compo-nents	Full Prod-uct	Com-plete Solution
The technology does not fit with the high-tech company's corporate mission					
The high-tech company has insufficient financial resources to exploit the technology					
The window of opportunity is tight and the high-tech company cannot move quickly enough	Sell	Sell	Sell		
The market potential is smaller than expected					
Allowing other firms access to the technology is the most appropriate action					
The range of technologies in a specific market is very diverse					
Market characterized by	License				

demand-side increasing returns					
Components incompatible with industry standards				Commercialize	Commercialize
Offering technology to competitors may encourage industry standardization of the high-tech company's technology	License			Commercialize	
The high-tech company may have skills in some market segments but not others (e.g. industrial but not consumer)					
Major corporate buyers require a second source					

Fig.1. What to Sell

Evaluating the factors leading to a 'sell know-how' approach, high-tech companies should lean towards this end of the continuum. For example a company innovating a probeless method to measure the athlete's biological signs should license this technology to an existing manufacturer avoiding the development and commercialization of an integral product in-house. Therefore it is desirable to compete close to the know-how end of the continuum, pulling towards 'componentization'.

2.3 Product Definition and Development

Once the decision on what type of product the high-tech company can contribute in the CSE market is taken, the next step will be to focalize on defining and developing the product.

CSE manufacturers have become highly savvy regarding their needs as to gain market share, and consequently became more ubiquitous so that their products satisfy customer needs. Consequently, it is essential for the high-tech company to define its offerings in details.

The 12 steps of product definition are:
1. Strategic alignment
2. End user and industrial customer needs
3. Competitive analysis
4. Localization
5. Product positioning
6. Project priorities
7. Risk assessment and project work breakdown
8. Core competencies

9. Strategic dependencies
10. Project leadership
11. Project resources
12. Project and business plan

Product definition includes a data sheet listing what the product will do for the end users and industrial customers. Also it includes an internal specification including detailed information about the mechanical, electrical, firmware, software, and other components of the product as well as their interactions.

The final step in this effort is to analyze which, if any, of the possible scenarios would be best for the high-tech company to invest in. The primary criterion for selection should be a certain measure of profitability, since that is the primary objective of most publicly traded companies.

In summary, product definition can be split into a series of interdependent steps which project teams need to answer. One of the keys to a successfully marketed product is completing each step thoroughly so that the product definition can be based on fact and not supposition. By questioning each step individually, product definition problems can be pinpointed and resolved, or a decision to redeploy the project can be made.

2.4 New Product Testing

Within the product generation concept stage, a significant amount of internal reviews and testing is imperative incorporating a series of further activities. During the testing phase, it often becomes difficult to delineate between the activities of concept testing, prototype development and product testing, because these activities are intimately related and interlinked. It is also necessary to use test centers, such as sport institutes specializing in biological measurements and instrument testing, in order to meet successfully sport-specific constraints like performance, endurance, safety, and rules compliance.

3. Marketing Approach

Marketing of high-tech products is a complex process. New technologies are not always understood properly and as a result partners are reluctant to invest in breakthrough innovations. At the same time, a technological change may not only make a product obsolete, but the whole marketing strategy of a company as well.

The first step is to carry out marketing research using 'early adopters' users who are characterized by their willingness to pay high prices for products, well ahead of the mainstream, but at the same time are savvy and demanding. Their significance is reinforced by the fact that sport practitioners base their choices on recommendations from experienced and elite athletes. Within this context, collaboration with professional teams and sport federations is also important.

The second step is to recognize the importance of the network effect, which implies that a company must aim to establish its technology as the industry standard. But getting the network effect rolling can prove counter-intuitive when, while trying to achieve industry standard status, a company sells its technology at very low prices, or licenses its technology to competitors. Understanding how to design a product that can benefit from network effect, and leveraging the dynamics of such an environment, can greatly strengthen the marketing plan for high-tech markets.

It is well documented that R&D and Marketing departments possess inherently different cultural backgrounds, and subsequently collaboration usually suffers. Nevertheless, it is crucial that these two groups work productively together in order to make sure they are striving towards the same objectives.

Pricing is also a delicate act for high-tech products. Current economic realities dictate that firms must price in such a way that they can both recoup their R&D expenses while also keeping pace with downward pressure on prices.

Even having developed an innovative technology, choosing the target market for launching it, is vital. Selecting the wrong target, the venture may surrender the more profitable segment to the competition; while disseminating the risk against many market segments may spread the firm's resources, proving unsuccessful in any one segment. Overlooking the need to proactively select a target, and let customers self-select into purchasing or not, is both very inefficient in terms of a firm's resource allocation and also very ineffective in terms of the likely response that would be generated from unfocused marketing efforts.

4 Challenges and Recommendations

In the increasingly complex CSE business environment, technology management is at least as important as any other business activity and must encompass: forecasting techniques, technology-change patterns, technology diffusion and spillover mechanisms and trajectories, technology transfer scenarios, linkages with business process reengineering and quality management paradigms, among others.

Traditional management paradigms have limited usefulness in technology management. Newer strategic management thrusts, particularly strategic architecture and its associated elements, possess capabilities for highlighting technological change impacts in management.

Success of high-technology companies in the CSE business environment has both short and long term dimensions. Measuring success by using only one dimension may prove misleading and may not provide a complete picture of the organization's strengths or weaknesses and its future prospects. Profitability is merely a snapshot of a momentary situation, and it may change almost instantaneously. Present short-term business success must be accompanied by satisfactory performance assessment in other areas. New business opportunities, and particularly build up of the future in terms of technology, people, facilities, and other areas, are the high-technology company's royalty for long-term profitability and sustained success.

References

Bruce, M. and Cooper, R.C. (1997) Marketing and design management. Thomson Business Press, London.

Cooper, R.G. (1998) Predevelopment activities determine new product success. Industrial Marketing Management, Vol.17.

Ford, R.C. and Randolph, W.A. (1992) Cross functional structures: a review and integration of matrix organizations and project management. Project Management Journal, Vol.18.

Mohr, J., Sengupta, S., and Slater, S. (2004) Marketing of high-technology products and innovations. Pearson Prentice Hall.

From Innovation to Firm Formation: Contributions by Sports Enthusiasts to the Windsurfing, Snowboarding & Skateboarding Industries

Sonali K. Shah

University of Illinois at Urbana-Champaign, sonali@uiuc.edu

Abstract. Teams of employees at firms innovate. Scientists and engineers at universities and research institutions innovate. Inventors at private labs innovate. Regular people consume. Wrong! Regular people innovate, too. Users have been the source of many large and small innovations across a wide range of product classes, industries, and even scientific disciplines. In this paper I describe the contributions made by user innovators in the windsurfing, skateboarding, and snowboarding industries.

1 Introduction

We are accustomed to thinking of firms as the primary engine of innovative activity and industrial progress. The research and development activities of most firms are based on a proprietary model; exclusive property rights provide the basis for capturing value from innovative investments and managerial control is the basic tool for directing and coordinating innovative efforts. The proprietary model does not, however, stand alone.

The "community-based" model has generated many of the innovations we use on a daily basis. The social structure created by this model has cultivated many entrepreneurial ventures and even seeded new industries and product categories. In stark contrast to the proprietary model, the community based model relies neither on exclusive property rights nor hierarchical managerial control. The model is based upon the open, voluntary, and collaborative efforts of users – a term that describes enthusiasts, tinkerers, amateurs, everyday people, and even firms who derive benefit from a product or service by using it.

Open source software development is perhaps the most prominent example of the community-based model. Although often viewed as an anomaly unique to software production, the community-based model extends well beyond the domain of software. Innovative communities have been influential in product categories as diverse as automobiles, sports equipment, and personal computers (Kline and Pinch 1996; Franz 1999; Freiberger and Swaine 2000; Franke and Shah 2003; Luthje, Herstatt and von Hippel 2005).

In this paper, I describe and discuss three elements of the community-based model. First, users and manufacturers generate different sets of information. This allows users to develop innovations distinct from those typically developed within firms. Specifically, innovations embodying novel product functionality tend to be developed by users. Second, users may choose to share their innovations within user communities. Third, innovations developed by users and freely shared within user communities have provided the basis for successful commercial ventures. Data drawn from the windsurfing, skateboarding, and snowboarding industries illustrate these processes.

2 Sports Equipment Innovation by Users & Their Communities

Both users and manufacturers contributed to the development of equipment innovations in the windsurfing, skateboarding, and snowboarding industries. *Users* are defined as individuals or firms that expect to directly benefit from a product or service by using it (von Hippel 1988). In contrast, *manufacturers* are those who expect to benefit from manufacturing and selling a product, service, or related knowledge; thus, firms, entrepreneurs, and inventors seeking to sell ideas, products, or services are all examples of manufacturers. To illustrate, snowboarders are users of snowboards. Firms such as Burton and Gnu are manufacturers of snowboards. An inventor who hears that there is a market for improved snowboard bindings and develops a new type of binding with the intent of patenting and licensing it is categorized as a manufacturer.

2.1 The User Innovation Process in Three Sports

This section describes the process by which users and their communities develop innovations. I begin with an example that illustrates this process. The following passage describes how Larry Stanley and the community of windsurfing enthusiasts around him innovated in the sport of windsurfing.

Mike Horgan and Larry Stanley began jumping and attempting aerial tricks and turns with their windsurfing boards in 1974. The problem was that they flew off in mid-air because there was no way to keep the board with them. As a result, they hurt their feet and legs, damaged the board, and soon lost interest. In 1978 West German Jurgen Honscheid came to participate in the first Hawaiian World Cup and was introduced to jumping. A renewed enthusiasm for jumping arose and soon a group of windsurfers were all trying to outdo each other. Then Larry Stanley remembered the Chip - a small experimental board that he had equipped with footstraps a year earlier for the purpose of controlling the board at high speeds - and thought:

> *It's dumb not to use this for jumping.*
> *I could go so much faster than I ever thought and when you hit a wave it was like a motorcycle rider hitting a ramp – you just flew into the air. We had been doing that, but had been falling off in mid-air because you couldn't keep*

the board under you. All of a sudden not only could you fly into the air, but you could land the thing. And not only that, you could [also] change direction in the air!

The whole sport of high performance windsurfing really started from that. As soon as I did it, there were about 10 of us who sailed all the time together and within one or two days there were various boards out there that had footstraps of various kinds on them and we were all going fast and jumping waves and stuff. It just kind of snowballed from there.

News of the innovation spread quickly and instructions for how to make and attach footstraps to a windsurf board were shared freely. Later, Larry Stanley, Mike Horgan and a small set of windsurfing friends would begin the commercial production and sale of footstraps (and other innovations). Today the footstrap is considered a standard feature on windsurf boards.

This example illustrates three key components of innovation development by users. First, the act of use itself creates new needs and desires among users that lead to the creation of new equipment and techniques. Second, user cooperation in communities is critical to prototyping, improving, and diffusing solutions to those needs. Working jointly allows rapid development and simultaneous experimentation, however working jointly also requires that users openly reveal their ideas and prototypes to others. Third, user innovations – even after they have been freely revealed - are sometimes commercialized. Each of these three key components is discussed in detail below.

Users generate and accumulate information based on product use in extreme or novel contexts, the creation of new (unintended) uses for the product or service, and accidental discovery - in addition to intended product use. In contrast, marketing teams at firms generally focus on understanding and improving the *intended* use(s) of a product. For example, until the handles of childrens' scooters accidentally fell off and children experimented with the resulting toy, it is unlikely that manufacturers would have identified skateboarding as a fun activity. These differences in usage and search patterns create an information asymmetry between users and manufacturers. Because users and manufacturers hold different stocks of information, they will tend to develop different types of innovations.

Two complementary sets of information are required for product development activity: (1) Information regarding need and the use context. As discussed in the previous paragraph, this information tends to be generated by users. (2) Solution information. This information may be held by both manufacturers who specialize in a particular solution type and by individuals with expertise in specific areas. Not surprisingly, innovators will develop innovations based upon the information they possess. Individual users hold limited stocks of information from which to draw when innovating. Even a user who knows exactly what functionality she desires may be unable to independently create a solution that achieves that functionality, let alone create an efficient or elegant solution. Users frequently overcome this barrier by working together.

Working together provides users with significant benefits.   Working with others allows users to access resources in order to develop their innovations.   Working with others also allows more rapid development due to simultaneous experimentation.   To illustrate, consider the following description given by windsurfing innovator Larry Stanley:

> ...we were all helping each other and giving each other ideas, and we'd brainstorm and go out and do this and the next day the [other] guy would do it a little better, you know, that's how all these things came about...I would say a lot of it stemmed from Mike Horgan because, if something didn't work, he would just rush home and change it or he'd whip the saw out and cut it right there at the beach.

Cooperation between users can take many forms.   Informal one-to-one cooperation between users is frequent.   Semi-structured one-to-many interactions have also been documented (e.g. through publications in newsletters, magazines, and websites).   More structured cooperation within "innovation communities" is also widespread.   Innovation communities provide social structures and, occasionally, tools that facilitate communication and interaction between users and the creation and diffusion of innovations.   Open source software development communities are a good example of this.

Innovation communities are composed of loosely-affiliated users with common interests.   They are characterized by voluntary participation, the relatively free flow of information, and far less hierarchical control and coordination than seen in firms.   These characteristics allow for rich feedback and the potential to match problem with individuals who possess the ideas and means to solve them.   Due to the varied needs and skills of the individuals involved, user communities are often well-equipped to identify and solve a wide range of design problems.

User communities utilize a number of communication channels.   Today, the Internet is one of the most common – and is being used for much more than open source software development.   For example, kite-surfing enthusiasts have created an on-line community where they share innovation-related information on board and sail design.   However, free and open diffusion of ideas and innovations occurred even before the advent of the Internet.   Users have historically and continue to share ideas through word-of-mouth; at club meetings, conferences and competitions; and in newsletters and magazines.   For example, Newman Darby, who is credited with the invention of the windsurfer, published blueprints and instructions for making a windsurfer in *Popular Science* magazine.

The open revelation of information and innovations is a necessary input into cooperative work.   Communities provide several innovation-related benefits that might lead an innovator to develop an innovation within or share a completed innovation with the community.   First, community members work with innovators and provide innovation-related ideas and assistance (Franke and Shah 2003; Harhoff, Henkel and von Hippel 2003).   Second, innovators may share simply because they enjoy the innovation development process and working with others.   Third, user-innovators willing to share their work with others generally want to prevent third-parties from appropriating that work. Communities take a variety of precautions to protect their work and make sure that it will remain available for others to use and modify.

A generally unintended consequence of sharing the innovation in the community is the potential development of a commercial market for the innovation - and the opportunity to build a business to satisfy and further grow this market. Although conventional wisdom argues that the open revelation of innovations and the commercialization of those same innovations for profit are antithetical, here I find that open-revelation can actually set the stage for profitable commercial production. As the innovation diffuses through the community, the reactions of community members to the innovation can be observed. As user-innovators observed interest in their innovations, many chose to commercialize the product. This process is straightforward in some cases, and highly emergent in others. Some user-innovators did not think to produce their innovation for sale to others until after receiving a series of requests from enthusiasts - who had heard of the equipment from other enthusiasts or in newsletters and magazines – interested in purchasing a copy of the innovation. Handmade copies of the equipment were initially constructed for free or at-cost. Eventually, some user-innovators realized that they could sell the equipment at a profit and began to manufacture and market the product.

Firms founded by users in these industries functioned as lifestyle firms for many years. By lifestyle firm, I mean a firm with ten or fewer employees that generates modest revenues for innovating users while they continue to innovate and advance their skills in a sport. These firms were initially operated out of garages or spare rooms. In their early years, these firms generally had no capital equipment beyond portable power tools and produced products one-by-one or in small lots. Over time, many of these firms became leaders in their fields and many were regarded as makers of exceptionally high-quality equipment. Several continue to operate independently, while the brands established by others have been acquired by larger manufacturers. Many of today's well-known brands in the windsurfing, skateboarding, and snowboarding industries – including Windsurfing Hawaii, Gnu, Winterstick, and Dogtown Skates - were created by innovative enthusiasts who later became entrepreneurs.

2.2 How Important is Community-Based Innovation in These Sports?

In 2000, I conducted a longitudinal study of the development and commercialization histories of 57 key equipment innovations in the windsurfing, skateboarding, and snowboarding industries (Shah 2000). The aim of the study was to understand the extent to which users did or did not contribute to innovative and commercial activity in these sports. The study found that users and their communities were critical to the emergence and development of these sports.

Sports equipment users developed the first-of-type innovation in each of the three sports studied, that is, users developed the first skateboard, the first snowboard, and the first windsurfer. Users also developed 57% of all major improvement innovations in the sample, while manufacturers developed 27% of the major improvement innovations. The remaining 16% were developed by other functional sources of innovation, such as joint user-manufacturer teams or professional athletes.

3 Conclusion

Community-based innovation has contributed greatly to technological and industrial advance in consumer sports equipment, as well as in a wide variety of other fields. Users are at the center of this model: they discover new needs and desires, cooperate with other users within innovation communities, and sometimes even commercialize their innovations.

The author can be contacted via email at sonali@alum.mit.edu. The full text version of this paper is titled "Open Beyond Software" and has been published as a book chapter in *Open Sources 2: The Continuing Evolution*, published by O'Reilly Media in 2005.

References

Franke, N. and Shah, S. (2003) How Communities Support Innovative Activities: An Exploration of Assistance and Sharing among End-Users. Research Policy, 32, 157-178.

Franz, K. (1999) *Narrating Automobility: Travelers, Tinkerers, and Technological Authority in the Twentieth Century (Unpublished Doctoral Dissertation)*. Brown University, Providence, Rhode Island.

Freiberger, P. and Swaine, M. (2000) *Fire in the Valley*. McGraw-Hill, New York.

Harhoff, D., Henkel, J. and von Hippel, E. (2003) Profiting from Voluntary Information Spillovers: How Users Benefit by Freely Revealing Their Innovations. Research Policy, 32,(10), 1753-1769.

Kline, R. and Pinch, T. (1996) Users as Agents of Technological Change: The Social Construction of the Automobile in the Rural United States. Technology & Culture, 37, 763-795.

Luthje, C., Herstatt, C. and von Hippel, E. (2005) The Dominant Role of "Local" Information in User Innovation: The Case of Mountain Biking. Research Policy, 34,(6), 951-965.

Shah, S. (2000) Sources and Patterns of Innovation in a Consumer Products Field: Innovations in Sporting Equipment. MIT Sloan School Working Paper #4105, Cambridge, MA.

von Hippel, E. (1988) *The Sources of Innovation*. Oxford University Press, New York.

Some Problems of Pareto-Optimisation in Sports Engineering

Alexander Podgaets[1], Wubbo Ockels[1], Kazuya Seo[2], Valeriy Stolbov[3], Alexey Belyaev[3], Anton Shumihin[4], Sergey Vasilenko[5]

[1] Chair of Aerospace for Sustainable Engineering and Technology,
  Delft University of Technology, Netherlands, sulion@inbox.ru
[2] Faculty of Education, Art and Science, Yamagata University, Japan
[3] Department of Mathematical Modelling, Perm State Technical University, Russia
[4] Department of Theoretical Mechanics, Perm State Technical University, Russia
[5] Department of Computational Fluid Dynamics, JSC Aviadvigatel, Perm Motors, Russia

Abstract. Four problems has been solved: the design engineering of ski, finding several parameters in underdetermined problem of mechanics, optimal flight control of ski-jumper and making a strategic plan for a factory of bicycles. All are solved in the frames of one approach called Pareto-optimisation. Pareto-optimisation is a relatively new method of multi-objective optimisation that makes the job of experts making final decision much more transparent than in other multi-objective approaches involving the dialogue with operator. The problems solved are typical problems when multi-objective optimisation arises in engineering. Method demonstrated in this paper gives a powerful tool of multi-objective optimisation and enables statement and successful solution of such problems in practice of sports engineering.

Introduction

The need for multi-objective optimisation appears often in engineering problems. They are solved with a number of approaches, most often by reducing to single objective optimisation or using genetic algorithms (e.g., Mortensen, 1996).

However there is one method for multi-objective tasks, Pareto-optimisation (Sobol and Statnikov, 1981), which is superior in every aspect. It has strict mathematical grounding, easiness in use and speed in providing solution. It also allows to involve human experts if necessary and makes their job easier than ever.

The purpose of this paper is to demonstrate the use of Pareto-optimisation in several common areas. Typical multi-objective tasks include:

- Experiment planning
- Finding the parameters of mathematical model from experiment
- The optimal design choice that takes into account several considerations
- Solution of underdetermined problems

- Optimal control with several controlling functionals
- Problems of planning with non-linear objectives

Below the method of Pareto-optimisation will be used for solution of following problems: design of the system for fighting ski return, solution of underdetermined problem about chewing, optimal flight control of ski-jumper and making a strategic plan for a factory of bicycles.

The exact problems solved in this paper serve only as examples. They represent simple problems, each of them from different domain of engineering so that superiority of Pareto-optimisation can be shown in all fields of multi-disciplinary optimisation. The chapters are named for the areas to which the examples belong.

1 Design Choice

One of the most typical use of multi-objective optimisation is in choice of design. Some parameters could be discrete choices, others – continuous values. Constraints could be either variables or functions that are found from solution of other problems.

Let us consider the following mechanical system for fighting the return of ski. The bottom of ski is covered with thick film of baked kevlar with viscous teflon plates sticking out of it (see fig. 1). Return of ski forces snow to go between plate and bottom of ski thus increasing ski friction coefficient. During the first moment of forward motion the plates release the snow and ski friction coefficient returns to normal. Geometrical sizes of plates are determined from condition of enduring working load. The number of plates determine the friction coefficient.

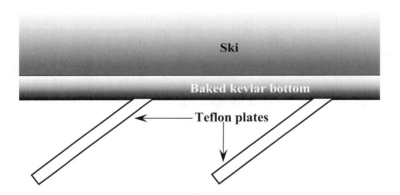

Fig.1. The scheme of ski with reduced return

The objectives are the ski return friction that should be maximised and the mass of such expensive materials as are teflon and kevlar that should be minimised. Two parameters are continuous – the orientation angle and scale of teflon plates – and one is discrete number that determines whether the plates are round, square or triangular. Discrete controlling parameter is introduced because all different options can be taken into account this way.

The purpose of this design choice is to reduce the return of ski without using special gels, oils or waxes, solely by mechanical means for any weather at once. The friction of ski is calculated as friction coefficient multiplied by ground reaction:

$$F_{forw} = f_{forw}N , \qquad F_{back} = f_{back}N \qquad (1)$$

The value of the ski friction coefficient f_{back} for the return movement should be maximised while its value for the forward movement f_{forw} should be minimized.

Each plate on the ski's ground surface has orientation angle α and length l. Parameter s determines the shape of plate's cross-section (1 for circle, 2 for square and 3 for rectangle with longer size equal the width of ski) and r represents the typical size in the cross-section (the radius for circular cross-section, the side for square and the lesser side for rectangular).

The strength of the plate provides maximal elastic stress as a function of shape. It also determines the number of the plates n and serves as a constraint The amount of material in plates is actually their volume which depends on their number, length and area of cross-section:

$$\sigma_{max} = \sigma_{max}(\alpha,l,r,s) \qquad (2)$$
$$n = n(\sigma_{max}) \qquad (3)$$
$$V = nSl \qquad (4)$$

Thus, the following problem should be solved: to find the shape of the plates that less hinders forward motion and reduces return motion as much as possible for minimal amount of material spent for plates. The plates should be also strong enough under normal skiing load:

$$\begin{cases} \Phi_1 \equiv f_{forw} \to \min_{\alpha,l,r,s} \\[1mm] \Phi_2 \equiv f_{back} \to \max_{\alpha,l,r,s} \\[1mm] \Phi_3 \equiv V \to \min_{\alpha,l,r,s} \\[1mm] \sigma \leq \sigma_{max}(\alpha,l,r,s) \end{cases} \qquad (5)$$

This is a multi-objective problem of optimisation with three objectives and constraint. Pareto-optimisation that is used here consists of the following steps:

1. Creating the table of tests. It means finding objectives Φ_i for all possible parameters α, l, r and s. If objectives do not include exponential functions which is a very rare case then it is possible to limit ourselves to checking some finite number of points which is determined from condition of convergence of results. The worst approach is building a grid with even intervals between points over every coordinate (Sobol, 1957). Monte-Carlo methods are usually used in planning of experiment for similar purposes of finding unknown value that depends on several parameters.

Special sequences such as LP_τ (Sobol and Statnikov, 1981) produce slightly better results for fewer test points. If particular point is near the boundary of the area D then adjacent points of the boundary should be included into the table, too.

2. Choosing constraints on objectives. $\Phi^* = \left(\Phi_1^*, \Phi_2^*, \Phi_3^*\right)$ are values chosen by engineer after looking at the table of tests. They can be also chosen automatically for a fully automated version of method. The points that comply with these constraints are called set of Pareto. They have better values of every objective than the rest of test points:

$$x \in D^* : \left\{\Phi_1(x) < \Phi^*, \Phi_2(x) < \Phi^*, ..., \Phi_M(x) < \Phi^*\right\} \tag{6}$$

The strong sides of such approach are two-fold: at first, this optimisation method has a strict mathematical grounding, theorems exists that guarantee finding global optimal solution. At second, if human-PC version of method is used, experts will be asked how bad could be friction of ski and volume of material spent on plates, that is the actual design parameters. With seeing results in an interactive manner while changing demands it makes procedure much simpler than answering questions about some abstract weights that many methods use.

Again, this simple problem is just an example for demonstration of Pareto-optimisation. The actual design problems can be much more complicated but their complexity will not affect the procedure of optimisation. It has been successfully used by (Sobol and Statnikov, 1981) in mechanical, civil and aerospace engineering for problems with up to several dozens of objectives and up to several hundreds of parameters.

2 Planning

Every planning problem is the problem of multi-objective optimisation. It is usually possible to solve it using linear or dynamic programming. But sometimes target objectives are significantly nonlinear which makes the error of linearization too big so methods of multi-objective optimisation become a natural choice. More complex case of master schedule plan requires additional use of several heuristics because the problem has exponential number of variants. More simple case of strategic planning allows direct solution. The problem of strategic planning for a bicycle factory Stefivelo (Perm, Russia) is considered. There are several types of bicycles and we need to find which types should be produced and when, considering results of marketing and mission of factory. The schedule of delivery of supplies represents constraints. The schedule of production gives parameters. Maximal profit and maximal satisfaction of market demands are objectives. Resulting tree of variants is viewed as a table of test points for Pareto-optimisation.

3 Finding Missing Values

Probably, the most amazing use of multi-objective optimisation is that for solution of undertermined problems of mechanics where it helps to fill the huge gap in available data by introducing dramatically less additional suppositions than any other approach.

Let us consider the problem of finding forces in face muscles during a bite. This problem can be useful for prediction of damage during a punch in martial arts and other contact sports because it is well known that muscles are more vulnerable to damage when they are not relaxed. Thirty muscles participate in closing our mouth and if we assume that directions of their forces are similar to directions of muscles then we shall have only 30 variables F_i, $i=1$, 2, ..., 30. Left and right temporomandibular joints give us three reactions each: X_L, Y_L, Z_L and X_R, Y_R, Z_R. The moment of biting makes a problem of statics with 6 equations of equilibrium. Thus, we have 6 equations with 36 variables.

Such underdetermined problems could be solved by methods of structural mechanics:

$$\sum_{i=1}^{30} F_{ix} + X_L + X_R = 0, \quad \sum_{i=1}^{30} F_{iy} + Y_L + Y_R = 0, \quad \sum_{i=1}^{30} F_{iz} + Z_L + Z_R = 0,$$

$$\sum_{i=1}^{30} M_x(\mathbf{F}_i) + M_x(\mathbf{L}) + M_x(\mathbf{R}) = 0, \quad \sum_{i=1}^{30} M_y(\mathbf{F}_i) + M_y(\mathbf{L}) + M_y(\mathbf{R}) = 0, \qquad (7)$$

$$\sum_{i=1}^{30} M_z(\mathbf{F}_i) + M_z(\mathbf{L}) + M_z(\mathbf{R}) = 0.$$

But there is another way. Let us assume that we bite in such a way that the reactions in both temporomandibular joints are minimal and equal. It is no longer problem of mechanics – it is a problem of optimisation:

$$x = (F_1, F_2, ..., F_{30}) \in D \subset R^{30},$$
$$\Phi_1(x) = X_L \to \min_x, \quad \Phi_2(x) = Y_L \to \min_x, \quad \Phi_3(x) = Z_L \to \min_x,$$
$$\Phi_4(x) = X_R \to \min_x, \quad \Phi_5(x) = Y_R \to \min_x, \quad \Phi_6(x) = Z_R \to \min_x, \qquad (8)$$
$$\Phi_7 = |X_L - X_R| + |Y_L - Y_R| + |Z_L - Z_R| \to \min_x.$$

Here joint reactions and their discrepancy are objectives to be minimised and forces in face muscles are parameters to be chosen.

4 Multi-Objective Optimal Control

Let us consider the flight of ski jumper. The mathematical model of ski jump is taken from Podgaets *et al*, 2004 and is completed with the shape of Okurayama jumping hill in Sapporo, Japan (Seo *et al*, 2004a) and aerodynamic data of Seo *et al*, 2004b. Three objectives are jump distance, safety of flight represented by maximal torque with regard to possible wind gusts and landing speed.

The principle of maximum of Pontryagin (Pontryagin *et al*, 1961) is used to find solution. Three controlling functions are forward leaning angle, attack angle and ski opening angle (Seo *et al*, 2004a). Method of Galerkin (Galerkin *et al*, 1952; Fletcher *et al*, 1984) is used for decomposition of controlling functions. The number of functions in series is determined from the condition of convergence of results.

Decomposition of controlling functions into series reduces optimal control into optimisation with several hundreds of controlling parameters. Algorithms for solution of such problems are well known and work quick enough on modern personal computers, but it is always hard to find good starting points and not to get stuck in a local minimum. The answer for both difficulties is creating the table of test points that will present the whole area of possible controlling functions to the view of engineer in an intelligible and clear way.

Conclusion

Multi-objective optimisation often arises in engineering problems. We hope that the examples given here will help in formulating such problems so that multi-objectiveness is not avoided or reduced to a single objective. Direct statement and solution of such problems is quick enough for practical needs.

References

Fletcher C.A.J. (1984) Computational Galerkin methods. Springer-Verlag
Mortensen U.K. (1996) Evolution in a microchip: A brief study into genetic algorithms. ESA: ESTEC Repro
Podgaets A.R., Seo K., Jošt B., Štuhec S. (2004) Multicriterion optimisation of a ski jump. *The Engineering of Sport 5: Proceedings of International Conference* (Editors: M. Hubbard, R.D. Mehta, J.M. Pallis). **2**, 440-446
Pontryagin L.S., Boltyanskii B.G., Gamkrelidze R.V., Mischenko E.F. (1961). Mathematical theory of optimal processes. Moscow: PhysMathLit *(in Russian)*
Seo K., Watanabe I., Murakami M. (2004a) Optimal flight technique for V-style ski jumping. *Sport Engineering.* **7**, 1, 31-40
Seo K., Murakami M., Yoshida K. (2004b). Aerodynamic force data for a V-style ski jumping flight. *Sport Engineering.* **7**, 2, 97-104
Sobol I.M. and Statnikov R.B. (1981) Choosing optimal parameters in problems with many criteria. Moscow: Science *(in Russian)*
Sobol I.M. (1957) Multidimensional integrals and method of Monte-Karlo. *Reports of Academy of Sciences of USSR.* **114**, 4, 706-709 *(in Russian)*

Life Cycle Assessment and Evaluation of Environmental Impact of Sports Equipment

Aleksandar Subic and Niall Paterson

RMIT University, Mechanical and Automotive Engineering,
Aleksandar.Subic@rmit.edu.au

Abstract. Design of sports equipment is primarily performance driven. As a consequence of this approach the sports equipment industry has eagerly adopted new materials and processes over the years that provide the competitive edge but which have unintentionally placed additional burdens on the environment. With the introduction of more rigorous environmental regulations in the developed countries there has been a further shift of manufacturing operations to areas of the world where such policies are more relaxed. There is much evidence that such practices have caused further global environmental degradation. With around 80% of the environmental burden of a product determined during the design stage it is clear that new design practices and tools are required in order to address this problem in a more sustainable way. Environmental concerns need to become a design objective rather than a constraint. This paper discusses the principles and strategies for environmentally sound design of sports equipment and in particular the eco-design approach based on life cycle assessment (LCA). A case study is presented involving comparative environmental life cycle assessment of carbon fibre and glass fibre composite tennis racquets using software EcoScan.

1 Introduction

Materials and processes used for sports equipment carry with them significant environmental risks. For example ski boots use PVC (poly vinyl chloride) based materials, athletic footwear uses petroleum-based solvents and other potentially damaging compounds such as sulphur hexafluoride in air bladders for cushioning and impact shock protection. Also, composites, such as carbon fibre reinforced polymers that are increasingly used in tennis racquets and other sports equipment provide greater strength to weight performance but cannot be readily recycled. Advances in sports equipment have unintentionally placed additional burdens on the environment.

It is estimated that around 80% of the environmental burden of a product is determined during the design stage. Hence, in modern design, environmental issues are given high priority, which has resulted in the development and application of new design tools and practices encompassed by the design for the environment or eco-design approach (Subic 2005). Eco-design is about developing more environmentally benign products and processes based on detailed understanding of the environmental hazards, risks and impacts of products and processes over their entire life cycle including production, usage and disposal stages.

Governments, particularly in Europe, have been providing incentives to companies willing to adopt "greener" product design and manufacturing approaches. For example, Germany's product take-back laws have prompted European and US industries to reduce packaging and start designing products with disassembly and recycling in mind. France, Netherlands and Australia have special government agencies to foster clean technologies. In the US the Federal government has initiated a number of energy efficiency programs and has prescribed the use of recycled and/or recyclable materials in product design. Clearly there is a growing genuine concern world wide for the environment whereby it is not acceptable any longer for products to be incinerated or to end up in landfill after their useful life. Eco-design implies lower social cost of pollution control and environmental protection through more efficient use of resources, reduced emissions and waste.

In order to assess the environmental impact of a product it has become necessary for informed, science based weighting of the wide variety of environmental damages to give a single criterion. This allows the different products and phases of the lifecycle to be compared on the same scale. The method used to weight the damages is defined in the scope of each individual investigation. This flexible approach to defining the criterion to compare products allows for numerous methods, but there are few examples of a standardized, broadly accepted criterion in the literature with the exception of the EI-99 and its predecessor EI-95. This paper presents a case study on comparative environmental life cycle assessment (LCA) of carbon fibre and glass fibre tennis racquets using software EcoScan based on the EI-99 indicator. The main objective of this study is to provide scope for a more general discussion about the role and importance of LCA in sports equipment design.

2 Overview of Environmental Life Cycle Assessment Methods

The LCA methodology allows for the environmental impact of a product to be calculated across the entire lifecycle. The quantification of the environmental impact can be performed using several different methods, including unique customized methods. All techniques share the same basis, namely the characterization of a process (e.g. material production) based on several environmental effects, each of which can be quantified in terms of reference substances e.g. global warming in terms of CO_2 kg produced, or acidification in terms of SO_2 kg.

One common indicator of an environmental effect is a Global Warming Potential (GWP). GWPs were developed and are regularly updated by the Intergovernmental Panel for Climate Change (IPCC) to determine the effect of gases on global warming. They are used to compare the abilities of different greenhouse gases to trap heat in the atmosphere over a set time considering the decay rate of the gas. They are often found in standard LCA software. One such indicator is GWP100-CO_2-eq (kg), which provides a measure of the time-integrated radiative forcing (heating over time calculated at the surface troposphere) of the emissions of a process relative to that of

CO_2 over 100 years (20 and 500 year indicators are also available) (Houghton, Ding, Griggs, Noguer, van der Linden and Xiaosu 2001).

Although GWPs are useful for comparing the environmental effects in LCA, another indicator is found more commonly in the literature, the Eco-Indicator (EI). The EI is a dimensionless figure used to compare the relative difference between products, processes or components. In determining indicators for the various materials, processes, transport and disposal scenarios all relevant aspects are taken into account, such as extraction of raw materials, transport processes of raw materials, precursor processes such as the production of granules required for certain plastics, and the emissions and energy requirements of machines and average recycling plants etc.

The EI scale is chosen so that the value of 1 pt is representative of one thousandth of the yearly load of a European citizen (Goedkoop and Spriensma 2001). The most up to date version of this indicator is the EI-99, which is defined by weighting the damage that all resulting emissions and extractions can cause to 3 areas: human health, eco system quality and resources. The damage is normally expressed in 10 or more categories such as acidification, ozone layer depletion, ecotoxicity and resource extraction (Goedkoop et al. 2001). The panel that developed the indicator found damage to human health and ecosystem quality to be approximately twice as important as damage to resources, which is included in the EI score. Although the weighting terms given to damage categories (human health, ecosystem quality and resources) are objective, numerous data available to the environmental scientists that defined the indicator has allowed the EI-99 to become widely accepted as a good indicator of environmental impact. For further information on the definitions of terminology, modelling techniques, uncertainties and assumptions employed in the development of EI-99 refer to the developers' reports (Goedkoop et al. 2001).

3 Comparative Environmental Life Cycle Assessment of Composite Tennis Racquets

The importance of environmentally sound design is becoming more prominent. This includes both product design issues and broader socio-economic concerns. Companies that are able and willing to address these issues are predicted to become market leaders of their respective 'green' products (Hundal 2002). Using life cycle assessment (LCA) methods is one way of implementing good environmental design practice and assessing the environmental impact of alternative products.

Modern sports equipment is often made from advanced composite materials because they can be used to manufacture structurally and mechanically effective components while reducing weight. For example, in the production phase of the product life cycle it has been noted that fibre reinforced composites compare favourably to aluminium (Lee, Jonas and Disalvo 1991). However, they are much more difficult to recycle because of the complex and longer recycling chains (process from collection to disposal) resulting in a lower value material that is rarely used again in comparable products.

In the remainder of this section two types of fibre reinforced composite racquets, namely carbon fibre (CF) and glass fibre (GF) reinforced, will be compared using a brief, simplified LCA with EcoScan life 3.1 (TNO, Netherlands). This provides scope for a general discussion about the importance and role of LCA in design of composite sports equipment. Although all LCA phases can be easily investigated using software such as EcoScan, the default database does not include specific data on materials, production or disposal scenarios for composites, thus providing an accurate LCA for a composite product is difficult. Future work will use specific data for more detailed analysis of selected tennis racquet designs, for now a brief example is given using the default EcoScan database.

The transport phase of LCA is not included in this analysis because it is quite difficult to influence logistical procedures from a design perspective (Abele, Anderl and Birkhoefer 2005), and the usage phase is not included because the use of a racquet requires no energy source other than that of the participant.

Both racquets under consideration possess the same overall weight while fibre reinforcement is varied from 5 to 50 %. The functional unit is assumed to be 10 years which requires 4 replacement grips and sets of strings. Usually a process tree would be included in LCA, but for the sake of brevity it is omitted here. Estimations of the weight of materials used for construction and packaging (assumed to be the same for both racquet types) were made as listed in table 1. The materials and component manufacturing processes were then entered into EcoScan.

Material	Weight
Carbon black or Glass fibre	5% (16 g) to 50% (160 g)
PA 6.6 (Nylon) resin	95% (304 g) to 50% (160 g)
PUR rigid foam	40 g
Nylon strings (x5)	100 g
EPDM rubber grip (x5)	100 g
Aluminium for case zip	20 g
LDPE case cover	120 g
Packaging paper	20 g
Packaging cardboard	160 g
PET bottle grade packaging plastic	60 g

Table 1 Estimates of materials used in racquet construction and packaging

Figure 1 shows the EI-99 data for production of the two racquet types. Both racquets show a decrease in impact for an increase in reinforcing content. Figure 2 shows the EI-99 data for disposal of the two racquets, with the CF racquet showing an increase in disposal impact for and increase in reinforcing content, whereas the GF racquet shows a constant EI score for all variations.

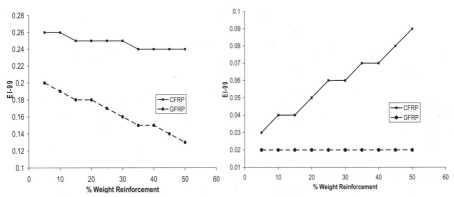

Fig.1 Comparison of environmental impact of the production phase for different racquet frames

Fig.2 Comparison of environmental impact of the disposal phase for different racquet frames

In figure 1 it is evident that the environmental effects of nylon resin production are higher than those of both fibre types. Furthermore, the glass fibres have a lower impact than the carbon fibres, principally because of the types of materials chosen to represent GF and CF in the analysis. To reduce the environmental impact of the production phase of these racquets a high proportion of resin should be included. It should be possible to decrease the amount of reinforcement by increasing the quality of the fibres. However, higher grade materials generally require more refinement resulting in greater energy requirements and a higher environmental impact. Figure 2 shows the environmental impact for the default waste disposal scenario, which predicts municipal waste disposal (24% disposal by incineration and 76% by landfill) for all materials except the CF. Because carbon black (the CF material in this analysis) is considered a hazardous chemical the only disposal data available is for 100% incineration, which means the CF racquet increases the environmental impact with more reinforcement.

From the comparison of figures 1 and 2 it is clear that the production phase produces a much higher environmental impact than the disposal phase. For 25% fibre reinforcement, the impact associated with the production phase of the CF racquet is approximately 4 times higher than that of the disposal phase, and for the GF racquet it is approximately 8 times higher. This suggests that the majority of environmental improvements can be made in the production phase of both racquets. However, the use of novel composite recycling techniques may also reduce the impact of the disposal phase. Furthermore, based on these results, a 50% GF racquet would be the most environmentally sound.

4 Conclusion

a) In this paper we have presented a simplified life cycle assessment (LCA) approach using the eco-indicator EI-99 that can be applied in sports equipment design to avoid and/or reduce environmental degradation associated with the manufacturing and disposal of such products.

b) A case study involving comparative environmental life cycle assessment of carbon fibre and glass fibre reinforced tennis racquets showed how this methodology can be used to identify and predict the environmental impacts of alternative designs.

c) From the study we may draw the conclusion that the environmental impact of sports equipment is typically determined by the manufacturing and disposal phases. In the case considered, the production phase produced a much higher environmental impact than the disposal phase.

d) The environmental impact of the composite tennis racquets during their entire life-cycle can be influenced significantly by the weight and level of fibre reinforcement and the longevity of the composite.

e) Further work will focus on more detailed life cycle assessment of selected tennis racquets using industrial composite data bases and process specifications as a basis.

References

Abele, E., Anderl, R. and Birkhoefer, H. (Eds) (2005) *Environmentally Friendly Product Development. Methods and Tools*. Springer.

Goedkoop, M. and Spriensma, R. (2000) The Eco-Indicator 99. A damage orientated method for life cycle impact assessment. Methodology report 3rd Edition. http://www.pre.nl/eco-indicator99/ei99-reports.htm

Houghton, J. T., Ding, Y., Griggs D.J., Noguer, M., van der Linden P. J. and Xiaosu D. (Eds) (2001) Climate Change 2001: The Scientific Basis Contribution of Working Group I to the Third Assessment Report of the Intergovernmental Panel on Climate Change (IPCC) Cambridge University Press, UK.

Hundal, M. S. (2002) Introduction to Design for Environment and Life Cycle Engineering. In: M. S. Hundal (Ed) *Mechanical Life Cycle Handbook*. Marcel Dekker, pp 1-25.

Lee, S. M., Jonas, T. and DiSalvo, G. (1991) The beneficial and environmental impact of composite materials – an unexpected bonus. SAMPE 27, pp 19-25.

Subic, A. (2005) A life cycle approach to sustainable design of sports equipment. In: Subic, A. and Ujihashi, S. (Eds) *The Impact of Technology on Sport*. ASTA, pp 12-19.

Sports Engineering and Sustainability

R. Keith Hanna[1], Eckehard Fozzy Moritz[2]

[1] Fluent Europe Ltd, Sheffield, UK, *rkh@fluent.co.uk*
[2] SportKreativWerkstatt GmbH, München, Germany, efm@sportkreativwerkstatt.de

Abstract. The last 20 years has witnessed an explosion of professional sporting and leisure activities, driven by consumers having more disposable income to spend, economic globalization, multi-channel 24 hour TV, the WWWeb and mobile telecommunications all hungry for content and information. It comes as no surprise then, that science and engineering have already embraced these opportunities with the advent of new disciplines like sports nutrition, sports psychology and the latest one onto the scene, sports engineering. This paper analyses some of the recent and likely future developments, and on this basis puts forward the view that sports engineering will be a transforming agent for the Sports & Leisure Industry well into the 21st Century. The yardstick for these developments will be considerations of sustainability, briefly introduced here in analytical categories, and their resulting innovation potentials.

1 Sustainability: A Road Less Traveled?

In a brief synopsis of recent activity in the sports industry, we can pinpoint the following developments:

We are seeing new sporting and leisure activities proliferate and old sports are even rein-venting themselves in the rush to grab both our attention and a slice of the financial pie that results from our hunger for entertainment and our disposable incomes.

The quest for monetary benefits now dominates sports activities. In the last decade the amount of money poured into elite sport has hit staggering heights. The German seven times Formula 1 World Champion, Michael Schumacher, has been estimated to have earned $1,000,000,000 throughout his 15 year career (British Airways Business Magazine, 2001), and the American golfer, Tiger Woods, is not that far behind. Big Sporting Events are now linked with big business opportunities, and the worldwide Sports & Leisure Industry is estimated to be worth something like US$500bn per year today while growing at 3% per annum (Sports Industry Federation, 2001).

It is still difficult to define what activities should be considered inside or outside this industry sector especially since tenuous things like "Gambling" and "Toys & Games" are usually considered part of the sector. Figure 1 shows a breakdown of the Sports Business Sector in the US in 2003. However we split this pie, it is interesting to note that the Sports & Leisure Industry globally is as big as the Aerospace Industry. In the US alone, the Sports Industry is seven times bigger that the Movie Industry and twice as big as the US Automotive Industry.

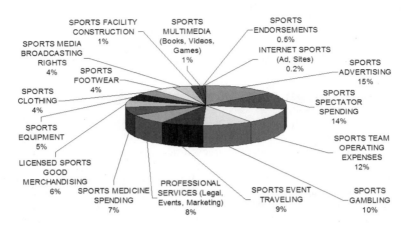

Fig. 1. US Sports & Leisure Industry Subsector Spend 2003 ($188.4 Bn) – Street & Smith's **www.sportsbusinessjournal.com**

When the word "Sustainability" is mentioned today, it is hard not to conjure up images of aging hippies in open-toed sandals declaring "love and peace" to all and sundry. However, this concept has become central to both political thought and politician's rhetoric today…. if not action. Still, it is not an easy subject to define. A United Nations Report in 1987 has a definition that is probably our best starting point: "*Sustainable Development is the development that meets the need of the present without compromising the ability of future generations to meet their own needs*" (Brandtland, 1987). But how can we get from this definition to specific action; how can we make sports engineering sustainable? First of all, it makes sense to try to reconcile the different aspects of sustainability. In many publications in this field, sustainability is divided into at least three aspects: social sustainability, ecological sustainability, and economic sustainability. Social sustainability, according to Kopfmüller et al (2001), can further be distinguished into categories of justice of distribution (of goods, wealth, happiness), cultural integration, climate of cooperation, social cohesion and individual freedom. Ecological sustainability means the preservation of the "functions" of nature – which not only includes self-regulation and protection of the bio-system, but also the production and reproduction of resources, the provision of space and an atmosphere of sensibility. Economic sustainability, finally, is that which is demanded by all industries, that is to make clear that their interests have to be taken care of as well, since industrial activities provide the basis for realizing social and ecological sustainability.

Obviously, the Sports & Leisure Industry today is doing well in economic sustainability (at least most big sports companies are quite profitable) but poorly on the social and ecological scales. This is all the more a pity because sport itself offers a huge potential to contribute to social sustainability; for example, the Undersecretary of the UN for Sports, Ogi, has pointed out (Ogi 2004, p.51) that, "*Many of the core values inherent to sport reflect essential prerequisites for development and peace, such as fair play, cooperation, sharing, and respect.*" Moreover, regarding ecological sustainability a support of non-motorized ways to travel (e.g. by bicycle, inline-skates) may help to reduce the ever-growing environmental burden of car traffic, especially in urban areas. In the following, we will therefore not only discuss about economic sustainability, but also focus on social and environmental sustainability, understanding that only a balanced consideration of all three aspects may lead to a truly sustainable future.

2 Landscape of Societal and Technology Trends in the 21st Century

Considerations on sustainability in sports engineering must start with an analysis of general developments, since the world today is experiencing social, economical, and technological changes at a rate never seen before. Discerning these trends and anticipating future opportunities is a hazardous exercise. However, planning for their likely impact might be prudent. Engineering will play a major part in solving environmental problems that our descendants will face. Given our present consumption rates, all known petroleum resources should be depleted by 2030 (Kirby, 2004). Even if more eco-friendly energy sources emerge, it is certain that the Sports & Leisure Industry of 20 years from now will have to adapt to a world where sustainable ecological development will be an everyday imperative rather than a pipe dream. One of the sad realities we face today is the epidemic in overweight people in the world. It is estimated that over 1 billion people in the world today are overweight, around 300 million of which are clinically obese - all due to sedentary lifestyles, fast food and limited physical activity. If present trends continue, by 2025 over 50% of Britons & Germans, and 75% of Americans will be clinically obese. Countries like China and India have also seen enormous jumps of outsourcing of IT and manufacturing from Western Europe, North America and Japan in the last 5 years. Friedman (2005) refers to this as the *"playing field is now level"* or *"the commercial world is flat"*. Salaries of trained engineers in the developed world are up to 20 times those of engineers in China or India, and even if we neglect efficiency issues, it is clear why many commercial businesses are now moving operations to these nations.

99 LIVES
We are living our lives at too fast a pace. We look for ways to take back our time to avoid over-scheduling

ANCHORING
We are reaching back to the past and our spiritual roots to prepare for and understand the future

ATMOSFEAR
We read about polluted air and contaminated water and fear for our safety

BEING ALIVE
We are aware that good health means a longer life

CASHING OUT
We want to opt out, in search of a simpler, more fulfilling life

CLANNING
We all want to belong to a group that represents common feelings or shared ideals

COCOONING
We need to protect ourselves from the unpredictable realities of the outside world

DOWN-AGING
We are nostalgic for youth, and find comfort in the things we did as children

EGONOMICS
We live in a depersonalised society, but want businesses to recognise our individuality

EVE-OLUTION
We know that women are different from men. So why aren't

companies marketing to target each sex separately?

FANTASY ADVENTURE
We are increasingly excited by the unknown and the unexplored – in the comfort of our own homes

FUTURETENSE
We are anxious about the speed of change in society and find it difficult to cope with the broken promises of today (computer software crashing, drugs having new side effects and e-businesses that can't deliver in time for Christmas) or to imagine tomorrow

ICON TOPPLING
We question and reject the pillars of

society – whether it's government, big business, medicine or religion

PLEASURE REVENGE
We are sick of being politically correct. We are finally letting loose and pursuing pleasure again

SOS (SAVE OUR SOCIETY)
We rediscover a social conscience

SMALL INDULGENCES
We are so stressed out that we indulge in affordable little luxuries

VIGILANTE CONSUMER
We increasingly manipulate the marketplace through protest and politics

Fig. 2. Several Predicted Trends in Developed World Societies Today

Looking at societal and cultural trends from the consumer's perspective (see Fig. 2), it is instructive to examine what one influential "futurologist" in America (Telegraph Magazine,

2004) is saying. By tapping into these trends, sports companies could tailor their products and services to get a competitive marketing advantage, and new start-up companies could exploit these opportunities. In conclusion, **customization** of generic products offers huge potentials for the sports industry this century, as will be products for **demographic groups**. Western consumers, who are living ever longer, and facing the problems of a sedentary lifestyle will also pursue **health and leisure** related activities more and more leading to new products.

3 Whither Sports Engineering?

So how will all these societal and technological trends feed into the Sports Industry of tomorrow and how can sports engineering benefit? Defying all outsourcing by the big industrial players, the authors believe that in the industrialized world Small & Medium Sized Enterprises (SME's) in particular will have the opportunity to use modern sports engineering approaches and technologies in their R&D efforts to come up with innovative products to stay competitive. Today companies like Reebok & Adidas typically invest only 1.4% of Turnover in R&D. If high-tech added-value sports products and services can be offered through SME's who engage in high levels of constant technological innovation via sports engineering, there is the potential for our discipline to be a transforming agent in the Sports Industry of tomorrow. One big problem that still faces the Sports & Leisure Industry today is the cost of doing innovative or "blue sky" research. Sports Industry Multinationals tend to push their R&D towards university research groups who may not have the resources to do other beyond-the-envelope research. Anyone who has had a creative idea and then tried to commercialize it can testify to the overwhelming hurdles that are faced in this industry. In most instances, getting the money to move through prototyping and demonstration phases can be a huge hurdle. Then when you have a product, how do you sell it and through what channels? This "funding gap" for risky innovative ideas has been a big stumbling-block for sports engineering SME's for some time.

It is clear from the above discussion that several opportunities are likely to emerge over the coming decades that the discipline of sports engineering, and sports engineers in general, should seek to exploit. In what follows, we outline likely developments that we believe will result from extrapolating current trends. Taking social and ecological sustainability demands as innovation challenges too will open up yet more opportunities, like cooperative sports toys and nature-integrated installations motivating health-sustaining physical activities (for more of these see Moritz 2004):

Great strides towards the **"Virtual Athlete"** should be possible in the next 20 years. The ability to produce physically and biologically realistic computer models of elite athletes which will be able to virtually test any proposed new sports surface or piece of equipment or injury scenario must be a compelling goal for the engineering profession. Realistic scientifically accurate moving athletes plus cheaper supercomputers should make holistic computer simulations of sportsmen and women a reality in the next 20 years. We should also see **Real-Time virtual modeling** of athletes and equipment at competitive events to gain competitive advantage on the day of the competition.

Miniaturization of computer chips should lead to the possibility of in-body monitoring of elite athletes as well as realistic data acquisition during competition. This will come out of a twin drive for more realistic sports media coverage and better telemetry in Sport. Sports engineering opportunities for new sensors, monitors, and advanced non-intrusive cameras will emerge.

We will see increasing usage of **"Smart" materials in sport** within the constraints of a given set of rules and the demands of governing sports bodies. Advanced intelligent materials for training purposes, if not for competition, will provide engineering opportunities. Advances in artificial intelligence will also lead to new products in this area such as customized intelligent clothing and footwear.

There is a need to continue to **build on the fundamental foundations of our understanding of the science underlying most competitive sports**. Increasing sports engineering research on "breathing" and "sweating" athletes will result in new pieces of clothing and footwear. Elucidatory research on for instance the precise aerodynamics of sports balls and equipment will continue for quite some time to come before we fully understand what's happening in sport. Engineering research should also lead to a deeper understanding of the effects of weather on sport and spin off products will result.

The **Internet will allow us to have more lifetime monitoring** and tracking of athletes' performance metrics both during training regimes and in competition.

The future will almost certainly see more **globally distributed innovation in sports engineering** with international collaboration between centers of excellence. Indeed, physical and virtual clusters of research will become the norm as will more **interdisciplinary research** across engineering and non-engineering disciplines to provide holistic sports solutions.

Personal and custom fabrication of sports equipment and "lifestyle" products will become a growing industry sector. Custom sports products for demographic groups in society will be commonplace – children, men, women, the disabled, the obese, the elderly etc. and these will be markets in their own right.

The authors foresee an explosion of **start-up companies in the Sports & Leisure Industry** with high R&D technology spending relative to turnover especially with lower niche manufacturing thresholds being readily available through globalization.

Opportunities will exist for socially conscious, environmentally friendly, sustainable, high-tech sports engineering products and companies.

Real time "decision-by-technology" will happen in the near future to augment the decisions of umpires and referees in various sports with product opportunities.

Biomedical engineering is gathering pace in the healthcare industry and it will be **applied to elite athletes** to better understand their performance.

Politicians are turning increasingly to the Sports & Leisure Industry to solve **pressing health and fitness problems**, to promote social inclusion for different demographic groups, and to be a generator of economic development.

The **Toy Industry** today, which is worth about $100 Billion, does not really use modern sports engineering techniques extensively. However, with more adults trying to recapture childhood passion, opportunities for new solutions will occur.

The academic community in sports engineering will have to **teach entrepreneurial skills** as well as basic engineering principles. How do we **nurture creativity and innovation?** Professional engineers who can manage a SME by overseeing R&D, acquire venture capital, do sales & marketing, lodge patents, make aesthetic design choices, manufacture remotely and possibly via virtual international collaboration, will require new sets of skills few engineers currently have.

4 Conclusions

The 21st Century should see sustainability becoming an overriding priority in the global economy across all industry sectors. Social, environmental and global economic trends in the world are driving this sustainability agenda, and when it is coupled with emerging technology streams it can be used to fashion a new landscape for the Sports & Leisure Industry in the decades ahead. This marriage will produce opportunities for sports engineers to ride the new sustainability waves to develop customized products using high tech (virtual, smart, miniaturized) solutions, leading to improved performance and novel types of experiences. In this new climate, many start-up enterprises will be empowered to enter the market at a lower entry point than ever before, and more niche innovative products will appear that require technology input. Sports engineering will therefore help to drive economic regeneration in the developed and developing world. Politicians will also drive the industry to produce new solutions to help tackle societal health and fitness problems in order to improve people's quality of life and social cohesion, and to find ways to cope with ecological burdens. The need for more holistic, interdisciplinary, transnational sports engineering solutions to these problems will become more obvious and opportunities will appear. Finally, the ISEA itself needs to cater for the emerging landscape being described here, in both the education courses being developed, the research going on in the world, and in its remit of equipping sports engineering professionals to be prepared for a fulfilling career in a brave new sports world.

References

British Airways "Business Life" Mag., Feb 2002, *Michael Schumacher: Racing to Riches*, p46
Brundtland, G.H. (1987) World Commission on Environment and Development, *Our Common Future*. United Nations Report A/42/427
Friedman, T. (2005) The World is Flat – A Brief History of the Globalized World in the 21st Century. Allen Lane/Penguin Publishing, London, ISBN 0-713-99878-4
International Union of Nutritional Sciences Website, 2006, http://www.iuns.org/features/obesity/obesity.htm
Kirby, A., BBC Website Article, April 2004 http://news.bbc.co.uk/1/hi/sci/tech/3623549.stm
Kopfmüller et al (2001) Nachhaltige Entwicklung integrativ betrachtet, edition sigma, Berlin
Mercer Human Resource Consulting, 2006, http://www.mercerhr.com/pressrelease/details.jhtml/dynamic/idContent/1175865
Moritz, E.F. (2004). Nachhaltige Innovation im Sport. *In: Gros et al (Eds.). Sporttechnologie zwischen Theorie und Praxis 3*. Shaker Verlag Aachen
Ogi, A. (2004). Promoting Development and Peace through Sport. *World Federation of the Sporting Goods Industry Official International Handbook*, Verbier p51-54
Sports Industries Federation (2001) *Sports Goods – A Survey of the UK, the EU and 7 Selected EU Markets,* SEARCE, Jan 2001, www.sports-life.com

2 Design

Synopsis of Current Developments: Design

Dr Mike Caine

Sports Technology Research Group, Mechanical and Manufacturing Engineering, Loughborough University, M.P.Caine@lboro.ac.uk

Introduction

It is difficult to imagine a more interesting time to be involved in the design of sports and fitness equipment. The growing sporting goods industry is characterised by innovation, and new product development initiatives abound as sports brands vie for increased sales and profitability. From recreational to elite sport the role of design in new product development has never been more high profile.

Design defined

Good design should span the entire new product development process - from concept to customer. However, the core design phase starts with concept communication and evaluation and finishes with a detailed embodiment from which a mass manufactured product can be made. It is within this phase where the ultimate success of a product is most likely to be determined.

The role of design within the sporting goods industry

Sporting goods designers face variable challenges depending upon the intended end user. Elite sports people are preoccupied with performance enhancement and are often prepared to sacrifice non functional features in the products that they use provided a performance gain is attained. However, the role of sports psychology is interesting in this regard, as to perform optimally athletes and players often report the need to look and feel good. They desire products that inspire and instil trust; however, enhanced functionality alone is seldom enough to ensure this confidence is achieved. The role of the designer is thus to couple great performance with stunning aesthetic and perfect fit - cost and ease of manufacture are seldom key considerations. Its worth reflecting on the fact that the designer often plays an important ethical role, often being required to deliberate over the balance between improved performance and increased risk of injury. Furthermore, in some sports the potential legality of the new concept or product also needs to be considered.

For recreational sports people the demands are different, cost of manufacture becomes a key concern to the designer, as does product durability. Users are almost always customers and as such typically make purchasing decisions based upon in-store appeal and perceived value. Branding is important and product endorsement is key, thus the role of the marketer is significant. However, the product design team also plays an important role. Imaginative design can reduce the bill of materials, improve ease of manufacture and assembly and enable the product to be packaged and shipped in a cost effective manner. Designers of mass market goods also need to consider sustainability issues, whereby choice of renewable and recyclable materials should be a priority.

Recent developments and trends

The sporting goods industry often sees the introduction of new materials and manufacturing processes before other sectors due to the relative lack of regulation combined with the willingness of sports participants to adopt new innovations. As novel materials, new manufacturing processes and advanced information communication technologies continue to emerge, sporting goods designers have more freedom than ever. Several major sporting goods brands have looked to pass on these design freedom to their customers, offering on-line customers the option to customise products that are subsequently made to order. Current examples are limited to aesthetic details such as choice of colourways and logos within athletic footwear. However, more functional choices will be permitted in the near future. There are already a small number of specialist providers designing and making customised products, such as football boots, by utilizing rapid manufacturing technologies to replace components that would traditionally have been made using injection mould tooling. Historically sporting goods designers have sought to combine functionality with style at a price point that represents best value. However, as the industry matures, premium brands need to offer more utility in their product offerings. This can be achieved by considering how to increase the pleasure that products impart to their users. By incorporating features that go beyond the purely functional, designers seek to capture this often intangible but crucial element within their new products. Understanding how to measure end user perceptions, and then determining how to incorporate this information within the design phase is thus an important consideration.

Future direction

Areas that are currently emerging and look set to develop further include embedded technologies e.g. sensors in apparel, active vibration dampening systems in footwear, tracking systems in skis etc. Other trends worth watching include designs influenced by biomimetics i.e. functionality engineering using 'designs' encountered in nature for inspiration. Products designed to be 'intelligent' are also set to proliferate, these will adapt to accommodate user preferences often using sophisticated feedback systems. However, whatever the latest design trend you can guarantee that sport will continue to be a showcase for human ingenuity and creativity in design.

The Use of System Analysis and Design Methodology in the Development of a Novel Cricket Bowling System

Laura Justham, Andrew West

Loughborough University, L.Justham@Lboro.ac.uk

Abstract. System analysis and design is an engineering technique which is widely used in industrial projects to ensure a solution is reached which fulfills the requirements of the user. The same technique has been used in this research for the design and implementation of a cricket bowling system which has been designed with both existing technology and the requirements of players and coaches in mind. The structured analysis and design methodology has been teamed with an object oriented approach to provide a component based solution which includes a bowling machine able to recreate any common bowling delivery, a visualization to provide temporal and spatial information about the delivery and an independent user interface to ensure robustness. This papers looks in more detail at the design and analysis methodologies used to create this solution.

1 Introduction to System Analysis and Design

System analysis and design is an interdisciplinary development technique to understand what the user needs and how to provide it. Having fully defined the scope of the system an iterative approach is adopted which allows analysts to modify and reassess on the strength of results obtained during the design process. A number of different methods are currently used in industry. The document led waterfall approach has historically been advocated as the best practice in system design. A linearly iterative process is used to work through the requirements, analysis, design and implementation stages with feedback loops to re-assess progress throughout the system development life cycle (SDLC). An example of this method is Structured System Analysis and Design Methodology (SSADM) (Bowman 2004).

The principles of the structured methodology are well defined using diagrams and modeling techniques. On a small scale these structured methodologies are very successful. However as the size of the system increases they are less easy to manage and over the past 15 years Object Oriented Analysis and Design (OOAD) methods have become more widely used in industry for medium to large scale systems. OOAD deals with the operations carried out by the system as well as the data on which these operations are performed which allows the analysis and design phases to occur concurrently rather than independently. The Unified Modeling Language (UML) is used as the standard language for documenting OOAD (Schach 2004).

2 Development of the Novel Cricket Bowling System

Cricket bowling machines are widely used by players of all standards and are a highly respected training tool. They do not usually provide the uncertainty and variability of a human bowler, but are used for numerous repeatable deliveries which may be used to improve batting technique, learn new skills and carry out specific training drills. The batsman can find himself playing each shot in the same way or pre-determining the delivery before the ball is launched due to the position of the machine and the use of polyurethane balls rather than real cricket balls. In the case of a real bowler this is not possible so the batsman will watch the run-up and anticipate the delivery using the information available to him (Abernethy 1981; Abernethy 1984; Penrose 1995).

2.1 Capturing the Requirements of a New System

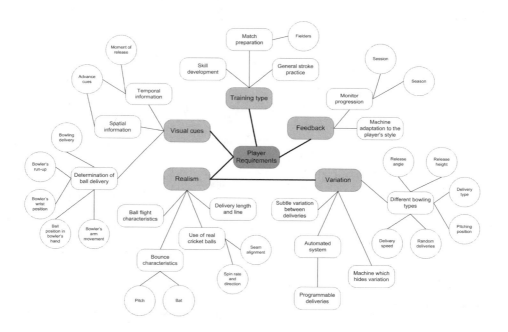

Fig. 1. An overview of the player and coach requirements for the novel bowling system, captured from interviews at the England and Wales Cricket Board National Cricket Center.

Quality Function Deployment (QFD) methods were used to capture the requirements of elite level players and coaches at the England and Wales Cricket Board (ECB) and to compare how existing machines measure up to these specifications. Figure 1 is a scatter diagram detailing the results obtained from interviews with players when they were asked about functionality required in an ideal bowling machine. Three main requirements were apparent, namely a realistic training environment which would allow for training drills and pre-season training as well as match preparation, the inclusion of visual cues to provide temporal and spatial information regarding the details of the upcoming delivery and controlled variability which models how a bowler would systematically alter each delivery.

2.2 The Finalized System Components

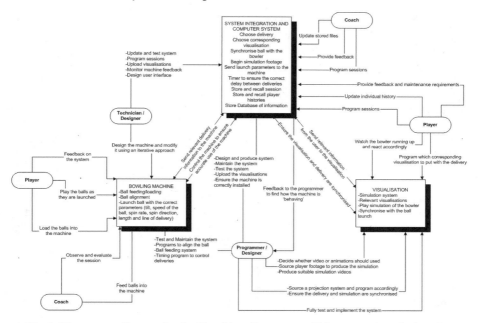

Fig. 2. The context diagram for the Novel bowling system. This summarizes the key data flows, process and entities acting within the system.

The system has been defined as a complete training environment which fulfills the requirements set out in Fig. 1. It is made up of three sub-systems; the bowling machine, a visualization and a user interface. The bowling machine has been designed such that any common bowling delivery may be re-created using a proper cricket ball with a correctly oriented seam. The launch characteristics of the ball from the machine and a bowler are identical and liable to the same external environmental factors during flight. The use of a cricket ball ensures similar impact characteristics with the pitch and bat to provide realism, unpredictability and variation. The visualization is used to provide advance information and temporal cues regarding the upcoming delivery. The user interface assures robustness and reusability of the system components within other applications and allows complete user control. A common database which contains all the information to re-create deliveries and training sessions has been directly coupled with the user interface to allow information flow from the database to the visualization and machine. The system relies upon the interface to ensure accurate recreation and synchronization of the delivery parameters. The flow of data and processes within the system components were conceptualized in Fig. 2 using a context diagram which provides an overview of the complete system.

3 Analysis of the System Requirements

A combination of SSADM and OOAD has been used as the basis for the development of the novel bowling system to ensure a generic structured approach which may be adapted to fit any specific application. The ECB National Cricket Center (NCC) is based at Loughborough University and is one of the best indoor cricket training facilities in the world. Three bowling machines are used for training on a regular basis and the ball launch characteristics from each of these machines were investigated to understand current equipment capabilities using a high speed camera with a frame rate of 10,000 fps. This corresponds to a pixel resolution of 512 x 256, and a field of view of 1m x 0.5m. From this filming the speed of the ball at release, spin rate, direction of spin and a 2-dimensional ball trajectory were calculated. The Hawk-Eye ball tracking system was used to simultaneously plot the complete trajectory of each ball release to measure the characteristics of a selection of deliveries which considered the full operating range of the machine.

Similar experimental results were obtained from bowlers and compared with the results obtained from the three machines. This highlighted the shortcomings of existing technology and aided the definition of the novel system requirements. The major issues associated with existing machines are that they do not recreate technically correct bowling deliveries and are not able to simulate every common delivery. Results have shown that the speed of the ball and the pitching position may be controlled to be realistic, yet the flight of the ball and the impact characteristics of the ball with the pitch are unrealistic. These quantitative requirements were compared to the qualitative ideas gathered from interviews and show a strong agreement between the needs of players and the functionality which is not yet available from existing machines.

4 New System Design Process

The complete system was designed as a number of modular components. The machine was developed around the accurate recreation of a ball delivery, a fully synchronized visualization, an automated ball feeding device and a miscellaneous selection of additional requirements. The visualization and independent user interface were developed in parallel with the machine. The processes and entities acting on the system components were investigated using functional decomposition based around data flow diagrams (DFD), as shown in Fig. 3.

The new machine had to be able to recreate technically correct deliveries with a properly oriented cricket ball which could be bowled by any type of bowler. Therefore the machine capabilities had to include controllable spin about any axis on the ball at speeds over 40rps, ball release speeds from 13.4 m/s (30 mph) to over 44.7 m/s (100 mph) and variable ball release positions to alter the line, length and flight characteristics of the delivery.

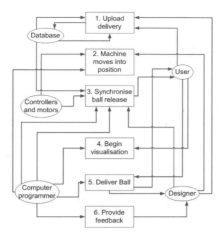

Fig. 3. The top level data flow diagram depicting how the processes and entities interact with one another through data flows around the system.

Fig. 4. The laboratory set-up used for testing the first prototype bowling machine system at Loughborough University.

The design process took place over a number of weeks through brainstorming using both SSADM and OOAD in an iterative process of testing and adapting ideas. The machine was designed to be a development on existing technology rather than a totally new design. Existing machines have been used for over twenty years and have been proven to be an effective training tool so it was decided that a development on this technology to incorporate the additional functionality required by elite players was more straightforward and would allow for more time to be concentrated on the novel and additional requirements. The capability to create spin around any axis of the ball has been based upon rifle theory which stabilizes the ball's flight and incorporates a rifling style spin. Any combination of spin may be applied to the ball in a controllable and repeatable manner. Automation and repositioning of the machine between deliveries is possible using pre-programmed settings from the database.

Testing of the prototype machine has been carried out in a laboratory setting, as shown in Fig. 4, and at the indoor training facility at the ECB-NCC. Ball release speeds over 100mph (44.7 m/s) and spin rates over 50 rps have been observed using high speed camera analysis and the ball tracking system which is based at the NCC. Cricket balls were used throughout testing and the orientation of the seam was controlled such that a realistic ball release was observed.

5 Conclusions

System analysis and design, using either the SSADM or OOAD, is a valuable tool when considering the design and development of any system. This research has shown that the techniques provide a structured basis for enhancing existing tools and redesigning equipment to improve technology and training facilities. The design and manufacture of the novel bowling machine featured in this research has been a rapid process resulting in a complete working prototype which exceeds current machine capabilities whilst providing better facilities for the end user. The use of system analysis and design has aided all stages of the design and manufacturing process enabling the user's requirements and the actual machine prototype to be monitored continually to ensure an accurate solution which fulfills every aspect of the original specification documentation.

The value of systems analysis engineering tools for the end user are vast; the design, development and manufacture is quicker than using traditional methods and results in a machine which more accurately models the exact requirements laid out in the initial planning stages. Modifications may also be made at various stages in the design and development process resulting in less costly post manufacture maintenance and service requirements.

Acknowledgements

The authors would like to acknowledge the financial support of the Engineering and Physical Sciences Research Council of Great Britain (EPSRC) and the IMCRC at Loughborough University. They would also like to thank the players and coaches at the ECB-NCC and the technical staff at Loughborough University for their ongoing support.

References

Abernethy, B. (1981). "Mechanisms of skill in cricket batting." Australian Journal of Science and Medicine in Sport 13(1): 3-10.
Abernethy, B., Russell, D.G. (1984). "Advance cue utilisation by skilled cricket batsmen." Australian Journal of Science and Medicine in Sport 16(2): 2-10.
Bowman, K. (2004). Systems Analysis: A Beginner's Guide, Palgrave Macmillan.
Penrose, J. M. T., Roach, N.K. (1995). "Decision making and advanced cue utilisation by cricket batsmen." Journal of Human Movement Studies 29(5): 199-218.
Schach, S. R. (2004). Introduction to object oriented analysis and design with UML and the unified process, McGraw-Hill.

Degradation of Tennis Balls and their Recovery

Aimee C Cubitt[1], Alan N Bramley[2]

[1] University of Bath, UK, Department of Mechanical Engineering, A.C.Cubitt@bath.ac.uk
[2] University of Bath, UK, Department of Mechanical Engineering, A.N.Bramley@bath.ac.uk

Abstract. The internal pressure in a tennis ball controls its rebound characteristics but it is known that there can be a pressure loss from the ball resulting in them becoming unusable. This is annoying to many players who are thus continuously sustaining the costs of ball replacement. This paper describes a series of experiments conducted to evaluate this pressure loss. These included taken samples of balls from long-term storage, from actual use in tennis and also subjecting balls to short-term high frequency bouncing. A standard rebound test was used to evaluate the characteristics of the balls. It was concluded that the period of storage is probably the major factor in causing pressure loss. It is assumed the loss is a result of the permeability of the rubber used for the shell of the ball. An experimental pressure vessel was constructed that would enable balls to be externally pressurised for varying periods of time at varying levels of pressure to recover their internal pressure. The paper describes the results of these tests and sets out conclusions that can be used for the design of a player friendly device to rejuvenate tennis balls.

1 Introduction

The tennis ball must provide consistent playing properties while being sufficiently durable to withstand repeated high-speed impacts with the racket and court surface. On average, a tube of four tennis balls will cost a player around £6 and to sustain this cost by replacing them as regularly as after each play is an expense that tennis players would prefer to avoid. Playing with flat balls not only leads to less predictable play but it can also lead to serious injury such as tennis elbow.

The leakage occurs because the polymers used in construction of the balls is of low density and thus relatively permeable by gases and liquids This loss of pressure occurs by combinations of adsorption, diffusion and desorption. Adsorption and desorption exacerabate complete the process of pressure loss from the cores of tennis balls and both play an important role in the degradation of mechanical properties of the rubber cores which in turn results in more rapid pressure loss. Due to the low density of polymers, every surrounding medium that has moveable molecules will infiltrate or permeate its molecular structure. This subjects the material to large amounts of stress which may cause cracks to appear and eventually propagate, leading to failure of the rubber core of tennis balls eventually making them unable to contain any pressure whatsoever.

This paper describes a series of tests used to evaluate the pressure loss occurring in balls arising from use and from just elapsed time together with an investigation into a means of re-pressurising the balls. The experimental data so obtained was to be used to inform the design of a tennis ball re-pressuring device

2 Experimental Programme

2.1 The bounce test

Throughout the set of experiments described in this paper, the standard 'Bounce Test' proposed by the ITF (x) was used and adapted to suit the available laboratory conditions. This enabled the effect of differing scenarios of pressure loss and pressure re-introduction to be evaluated. The test was always carried out in the same laboratory for each trial where the recognised ambient temperatures and pressures were kept constant throughout. A video camera was used to record the bounce data.

Each ball was dropped three times for each individual test in order that an average could be calculated. The scale against which the bounce heights were measured was adjusted so that it could easily be read from the video recording.

2.2 Pressure loss from storage

The effect on bounce height of storing new and used tennis balls in ambient pressure was examined experimentally. The factors affecting the rate of pressure loss are; the induced pressure gradient and the time they are left in ambient pressure. The initial bounce height of three used and three new tennis balls was measured and over a time period of two weeks, the resulting bounce height was examined. There was negligible difference in these tests. It can therefore be concluded that the tennis balls need to be left in ambient air pressure for a longer period of time than 14 days to demonstrate a noticeable (more than 2%) pressure loss.

2.3 Pressure loss from usage

It would be a fair assumption that the compressions a tennis ball experiences during play will increase the rate and quantity of pressure loss from the cores of tennis balls. This assumption ignores the effects the compressions have on changing the properties of the rubber cores of tennis balls. This is an issue that is not explored in this paper, yet it is thought that these effects could be the reason for some of the experimental results. Tennis balls were examined for the effect of both artificial and natural play conditions where the frequency and force of impacts were varied independently. The results from natural usage where the tests were carried out immediately after usage are shown in Fig. 1. It can be seen that the effects are negligible.

Artificial Accelerated Usage was achieved by using a ball launching machine to launch tennis balls directly onto a hard surface. The number of repetitions and speed of impact were varied independently and their effects noted. The results are shown in Fig.2.

Increasing the speed of the ball and therefore rate of compression had the effect of increasing the reduction in bounce height. The ball launched at 90mph showed the largest percentage decrease in bounce height of 6.7%.

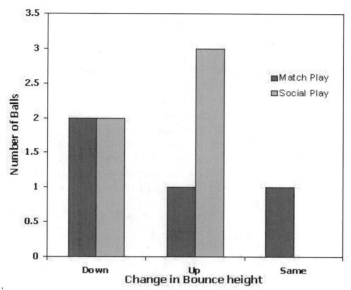

Fig. 1. Effect of natural usage on bounce characteristics

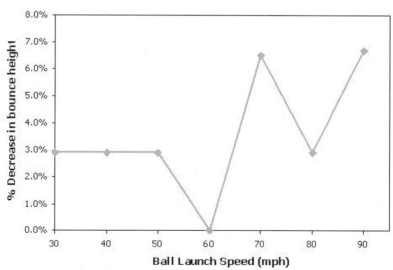

Fig. 2. Effect of accelerated usage on bounce characteristics

Subjecting tennis balls to more impacts at the same speed, results in very little change to the percentage decrease in bounce height. The results show that after 20 repetitions the percentage decrease is approximately 2% and after 40 repetitions, the percentage decrease is approximately 4%.   These results were obtained when launching the ball at 60 mph.

2.4 Re-inflation of balls

In order to reverse the effects of pressure loss, the ability to use the permeability of the rubber cores to allow them to be re-inflated was explored. The test rig consisted of a metal tube with an adapter valve to connect to mains air supply. An ncorporated pressure gauge enabled the pressure to be varied controlled and monitored. It was a very simple testing rig that could contain seven balls in a sealed environment. vidence was gathered from various sources to define a suitable range of testing pressures and time periods for the series of experiments. Long term (more than one week) tests could not be performed due to the limitations of the project timescales.

Six experiments were designed to investigate the required ranges of each variable as shown in Table 1

Table 1. Test Conditions

Test Number	Time in test-rig (Days)	Pressure in test-rig (lb/in^2)
1	1	25
2	3	25
3	1	35
4	1	15
5	4	15
6	8	15

The balls that were used in the testing procedures were initially examined in order that their amount of prior usage could be quantified using the Bounce Test. Balls J and H were unused and the shortfall from regulation was due to pressure loss under ambient conditions. This data is given in Table 2.

Table 2. Test results

Ball	Initial Rebound Height (m)	Percentage of drop height (1.83m)	Shortfall from regulation (70%)
7	0.92	50%	20%
8	0.85	46%	24%
9	0.79	43%	27%
10	0.88	48%	22%
14	0.91	50%	20%
J	1.05	57%	13%
H	1.04	57%	13%

The results are shown in Fig. 3.

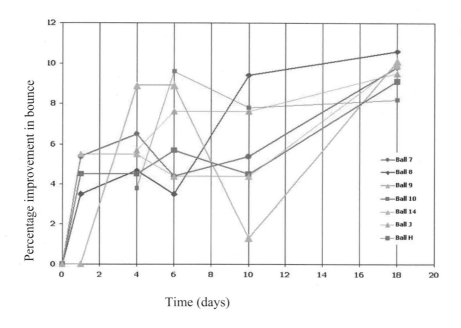

Time (days)

Fig.3. Effect of external pressure on bounce characteristics

The results indicated that a minimum pressure of 10 lb/in^2 was required to achieve any noticeable effect. Every ball demonstrated a percentage increase in bounce height above 9% after Test 6 was carried out and the variance in final percentage increase is minimal. Three out of five used balls had a decreased bounce height after Test 3. The remaining two used balls demonstrated only a 2% increase in bounce height. Both of the new balls showed improvement after Test 3 due their increased ability to withstand the effect of the collapsing pressure. The longer the balls are contained under any pressure, the greater their percentage change in bounce height. The re-pressurisation rate slows down after the first day that the balls are stored in the pressurised container, regardless of the value of pressure. The rate of re-pressurisation is dependent on the pressure in the container where the higher the value of pressure, the more rapid the rate of re-inflation. Above a pressure of 35 lb/in^2 the balls experience a reverse re-pressurisation and the rubber spheres are deflated by the external force. The final percentage increase in bounce height is independent of the amount of usage that the ball has been subjected to prior to re-inflation.

Conclusions

Rebound height reduction is marginally affected by the amount of wear that the ball has experience.

The effects of pressure loss caused by the pressure gradient to ambient pressure can only be noticed after a period of more than 14 days.

The rate of re-pressurisation is dependent on the pressure in the container.

There is an upper limit of pressure for this increased rate, after which the bounce height is decreased.

The final percentage increase in bounce height is independent of the amount of usage that the ball has been subjected to.

Development of a Highly Adjustable Cushioned Treadmill

Philip Hodgkins, Steve Rothberg and Mike Caine

Loughborough University, Sports Technology Research Group, p.p.hodgkins@lboro.ac.uk

Abstract. The treadmill is favoured by many due to the reduction of impact forces on joints in comparison to overground locomotion. The treadmill allows users to set the distance, speed and gradient, however, currently surface properties can not be adjusted. At impact with the ground the foot undergoes a rapid change in velocity creating high forces to be transmitted throughout the body. Repeat exposure to such shocks is thought to be responsible for the widespread overuse-type injuries (Cavanagh 1990). Literature is available on the kinematics and energy expenditure of various overground surfaces, however, not specifically for treadmills providing surface stiffness and damping adjustments. A cushioned treadmill has been conceived and designed which permits the running/walking surface to be adjusted for stiffness and damping. The concept has been designed to be implemented on home/commercial treadmills unlike Kerdok *et al.* (2002) who created a bespoke laboratory treadmill rig to evaluate the energetics and mechanics of running on surfaces of different stiffnesses. The new concept is thought to provide enhanced user functionality which to date has yet to be achieved in a mass market version. The locomotion kinematics and associated energy cost over a range of surface settings is to be evaluated, along with the rating of perceived exertion. The design is presented and discussed, along with the methodology for evaluation.

1 Introduction

Favoured by many due to familiarity with movement pattern and high calorific expenditure, the treadmill is used to enhance general fitness and performance levels. As a training aid it allows distance, speed, and gradient to be precisely set and monitored during use, benefits favouring treadmill usage as opposed to overground locomotion. There is however, one important variable not included on the majority of treadmills - surface replication. People choose to run or walk on a variety of different surfaces (e.g. all-weather polyurethane tracks, dirt trails, grass and sand) for training benefit or satisfaction, in spite of this very little research or product development has been conducted into replicating surfaces in treadmills.

2 Investigation

At impact with the ground surface, the foot experiences a rapid change in velocity. The rearfoot typically strikes the ground with velocities exceeding 1.0 ms^{-1} and this velocity decreases to zero in a short period of time. The hardness of the interacting surfaces alters the time period taken for the velocity to decrease to zero. Harder surfaces are displaced less and so the velocity reduces to zero in a shorter time (Cavanagh 1990). The amplitude of vertical impact forces in heel-toe running range from to 2.2 to 3.2 times body weight depending on running velocity, surface and individual style (Cavanagh and Lafortune 1980). Speed has a direct effect on the magnitude of the vertical ground reaction forces, so as speed increases the magnitude of vertical force increases, as does the rate of loading. This foot-surface impact creates a transmission of a vertical shock through the body and brings with it the potential for injury (Cavanagh 1990).

2.1 Injury Prevention

Injuries are far more prevalent during running than walking mainly due to the higher ground reaction forces involved. The loads generated from running have many beneficial effects on the human body, such as maintaining bone mineral density. Unfortunately high levels of repetitive loading are cited as causing 'overuse' injuries (Cavanagh 1990). A common type of injury resulting from continual training is stress fractures of the tibial bone. These fractures result from repetitive sub-threshold loading that, over time, exceeds the bone's intrinsic ability to repair itself. A study by Milgrom, Finestone, Segev, Olin, Arndt and Ekenman (2003) investigated overground versus treadmill running and found that axial compression strains, tension strains, compression strain rates, and tension strain rates were 48–285% higher during overground running than during treadmill running. Therefore the treadmill runner is at a lower risk of developing tibial stress fractures, but less likely to achieve tibial bone strengthening, than overground runners.

2.2 Surfaces

The compliance of the surface can serve to lessen the stresses on the lower extremities of the runner as generally cushioning works to decrease the forces between colliding bodies by increasing the time of collision. Several studies show that surfaces play a significant role in the dynamic loading experienced by a subject (Kim & Voloshin 1992; Stussi, Denoth, Muller, and Stacoff 1997). Most natural surfaces do not return all of the energy they absorb. Less is known about the kinematics of locomotion on energy dissipating, damped surfaces. However, studies have been conducted on snow and sand surfaces. It was found that, to transverse these terrains extra mechanical work must be performed to replace the energy dissipated by the surface (Lejeune, Willems and Heglund 1998) and therefore consume extra metabolic energy (Pandolf, Haisman and Goldman 1976; Zamparo, Perini, Orizio, Sacher, and Ferretti 1992). An indoor track surface was developed by McMahon & Greene (1979) to enhance performance by attempting to match the spring characteristics of the track with natural stiffness of the lower extremity of the runner. McMahon and Greene (1979) showed that a range of surface stiffness values exist over which a runner's performance was enhanced. The tracks built within this performance range were shown to increase running speeds by 2-3% and decrease running injuries by 50%.

A study by Kerdok, Biewener, McMahon, Weyand and Herr (2002) of the energetics and mechanics of running on surfaces of different stiffnesses examined how leg stiffness and metabolic cost are affected by changes in surface stiffness. They built a treadmill with adjustable platforms to alter the surface stiffness and measured eight subjects running over five different stiffnesses ranging from 75–946 kN/m. The 12.5-fold decrease in surface stiffness resulted in a 12% decrease in the runner's metabolic rate and a 29% increase in their leg stiffness. The relationship between the reductions in metabolic rates and the stiffness of the track were believed to be due to a greater proportion of the elastic rebound energy of the surface elevating the centre of mass in the latter portion of the contact phase.

3 Design Methodology

A systematic approach to the development of a new highly adjustable cushioned treadmill was implemented. The five core phases were:

(1) Concept Generation (2) Initial Modelling (3) Embodiment (4) Detail Design (5) Evaluation

In addition, a fundamentally different design philosophy was followed than that typically implemented by the fitness industry. The new approach focused upon a user-centred design process whereby contemporary knowledge pertaining to motivation to exercise and factors determining maintenance of exercise were prioritised at the concept development stage. It is believed that new fitness improving technologies can be created or modified to enable users to derive additional benefits exceeding those achievable using existing designs providing a more fulfilling user experience.

3.2 Concept Generation

To emulate overground surfaces on a treadmill the stiffness and rebound characteristics need to be closely matched. Other surface properties exist such as traction and surface roughness etc but, due to the constraints of existing treadmill design, the ability to adjust the compression stiffness and rebound damping are paramount. The proposed concept was to incorporate a spring and damper unit into a home/commercial treadmill. This enables the compression stiffness to be preset via different stiffness rated springs as well as the ability to set the pre-load by altering the position of the upper collar. The speed of compression can be adjusted precisely by altering the compression damping and also the speed of rebound can be altered in the same manner by altering the rebound damping. Some shock absorbers allow the high and low speed damping for compression and rebound to be individually adjusted. High speed rebound damping affects the response of the shock when returning from deep in the travel. Low speed rebound damping controls the response to small travel returns. High speed compression damping refers to how fast the shock compresses and so affects the bottoming resistance. Low speed damping controls the compression of the shock to small travel movements.

3.3 Initial Modelling

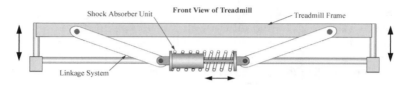

Fig. 1. Concept 1, front view of horizontally mounted single shock absorber system

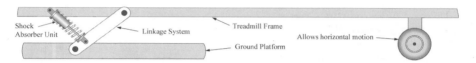

Fig. 2. Concept 2, side view of dual angular mounted shock absorber and linkage.

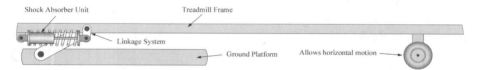

Fig. 3. Concept 3, side view of horizontally mounted shock absorber and linkage.

Initial modelling was conducted to investigate the feasibility of three different concepts. The advantage of concept 1 is that it utilises linear slides to restrict motion to only vertical movement and by using twin levers this motion is transmitted to a single shock absorber. The main disadvantage of this system is the complexity of the linkage mechanism. The advantage of concept 2 is the simplicity of mounting the shock absorbers either side of the treadmill frame. However, it is an unnecessary additional cost to have two shock absorbers and also this arrangement raises the deck of the treadmill rather far off the ground. Concept 3 has a single shock absorber mounted horizontally underneath the frame. The advantage of horizontally mounting the shock absorber is that space saving is greatly enhanced, an issue of great importance in the home consumer market. Evaluating the concepts by means of a matrix it was decided to embody concept 3.

3.4 Embodiment

The embodiment design phase involved 3D Cad modelling in SolidWorks 2006 (SolidWorks Corporation, Concord, MA) of the optimised linkage design (see Fig. 4).

Fig. 4. Model of linkage and shock absorber incorporated on a home/commercial treadmill

The linkage system has been fitted to a popular home treadmill along with a suitable shock absorber. The treadmill is raised slightly to accommodate the new cushioning system and to allow displacement of the treadmill deck during foot impact. To isolate the movement of the shock unit and reduce running deck flex, the frame will be stiffened by the addition of cross bracing.

3.5 Detail Design

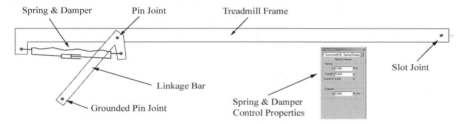

Fig. 5. Working Model simulation

The proposed design has been modelled in Working Model 2D 2004 (Knowledge Revolution, San Mateo CA) to examine the required range of spring stiffness and also the length of the linkage bar to suit the length of the shock absorber. Working Model is a motion simulation software package enabling basic models to be created to evaluate the magnitude of the forces. The model (see Fig. 5) incorporates a spring and damper software feature attached to the linkage system whereby the stiffness and damping rate can be adjusted. An impact force, i.e. three times the body weight of a runner, can then be applied to gain an understanding of the displacement and motion of the treadmill. The data gained from Working Model meant that a 240mm length shock absorber, providing 80mm of travel was required and as the maximum user mass for the treadmill was 125 kg, a spring stiffness of 110 N/mm was necessary.

3.6 User Evaluation

Evaluation of the treadmill will be conducted once manufactured to gain an understanding of the kinematic and energy cost effects on the user when altering the surface stiffness and damping properties. Subjects will be enlisted to perform running tests at predetermined speeds on different surface stiffnesses and damping properties.

To evaluate the kinematics differences, gait analysis will be conducted using CODA 3D motion tracking equipment (Charnwood Dynamics, Leicestershire). Small infra-red light emitting diodes (markers) placed on anatomical features of the subject are tracked by CODA sensor units which record and display the resulting motion. To evaluate the energy cost of each subject running on different surface properties a BIOPAC gas analysis system (BIOPAC Systems, Goleta, CA) is to be used. The system includes a non rebreathing mouthpiece connected to flow transducer to measure breath inflow rate and a mixing chamber to collect and sample CO_2 and O_2 concentrations. The CO_2 and O_2 percentage is analysed in conjunction with the inflow rate to calculate energy cost. The results will be evaluated along with the kinematic analysis to highlight any systematic variance in response to surface alterations. In addition, the rating of perceived exertion of each subject during each test will be taken to evaluate the user response to each surface alteration.

3 Conclusions

The design process has culminated in the advanced CAD model presented. The design has the potential to provide an enhanced user experience over existing home/commercial treadmills, along with the ability to reduce impact forces thus reducing the occurrence of running related injuries. Empirical data is required to fully validate the design.

References

Cavanagh P.R. (1990) *Biomechanics of distance running,* Human Kinetics Books, Champaign, IL.

Cavanagh P.R. and Lafortune M.A. (1980). Ground reaction forces in distance running. J Biomech 13, 397-406.

Kerdok, A.E., Biewener, A.A., McMahon, T.A., Weyand, P.G. and Herr, H.M. (2002). Energetics and mechanics of human running on surfaces of different stiffnesses. J Appl Physiol, 92, 469-478.

Lejeune, T.M., Willems, P.A. and Heglund, N.C. (1998). Mechanics and energetics of human locomotion on sand. J Exp Biol, 201, 2071-2080.

McMahon, T.A. and Greene, P.R. (1979). The influence of track compliance on running. J Biomech, 12, 893-904.

Milgrom C., Finestone A., Segev S., Olin C., Arndt T. and Ekenman I (2003). Are overground or treadmill runners more likely to sustain tibial stress fracture? British journal of sports medicine 37, 160-163.

Pandolf, K.B., Haisman, M.F. and Goldman, R.F. (1976). Metabolic energy expenditure and terrain coefficients for walking on snow. Ergonomics, 19, 683-690.

Stussi, E., Denoth, J., Muller, R. and Stacoff, A. (1997). Sports medicine and rehabilitation. Surface and footwear. Orthopade, 26: 993-8.

Zamparo, P., Perini, R., Orizio, C., Sacher, M. and Ferretti, G. (1992). The energy cost of walking or running on sand. Eur J Appl Physiol Occup Physiol, 65, 183-187.

A New Piece of Equipment for Skiing and Fitness – A Comparison with Existing Models

Peter R. Fischer and Christian Nolte

University Augsburg, peter.fischer@sport.uni-augsburg.de

Abstract. The difference between SKI-GYM and existing devices for fitness and skiing is that this model has a lot of functions to simulate situations on a slope (see fig. 2-4). Different target groups can use this multi-functional machine, world class athletes as well as beginners.

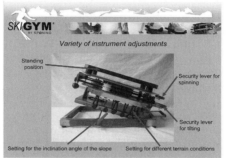

Fig. 1. SKI-GYM - multifunctional tool

1 Introduction

SKI-GYM is a multi-functional training machine for all sports(wo)men to improve their motor skills and their strength and stamina. It will be especially valuable for alpine skiers who can use SKI-GYM to practise during the summer for the winter season to come.

It can also help you to recover faster from injuries. Additional advantages of SKI-GYM are that:
- it is easy to manage
- it is of small and compact size
- its special adjustments offer great flexibility and a range of demanding challenges
- it is suitable for different sports (fitness, skiing) and for different locations (gyms, hotels, sports shops, rental services, clubs, home)

- it has a highly developed security system to avoid injuries

This training machine is the first one which simulates the situation on a ski-slope realistically. There are others which offer coordination training, some which train stamina and, of course, also some which train strength.

But there has not been a machine which offers fitness training for skiing and simulates the situation on snow at the same time. (e.g. DE 44 08 179 A1; US 49 46 160 A; FR 27 02 667 A1; JP 2001 / 11 57 726 A; DE 195 33 757 A1; US 2003/0060338 A1). None of them (except POSTUROMED) has got a comparable security level: involuntary tilting and spinning can be stopped on SKI-GYM, whereas this is not possible with other machines.

Because of the compact and special construction, it is possible to lock and unlock separate directions of movement. This allows for an individual training that can be easily adapted for developing motor skills.

2 Details (construction of the SKI-GYM)

A wealth of adjustments allows a broad range of training possibilities.

2.1 Variety of instrument adjustments

There are four basic settings:
- the inclined plane can be adjusted in three positions
- the adjustment of the angle of inclination of the board allows for different re-sistances (1-5 elastic bands for each side)
- the adjustment of the resistance of the wheel has three levels (0-2 elastic bands for each side)
- with or without plates for the binding

There are special adjustments for a variety of additional exercises:
- exercises with locked wheel in place
- exercises with locked board
- exercises with complete locking mechanism
- exercises with broad and narrow distance between the skis

SKI-GYM has three levels of height that can be adjusted by throwing the lever (fig. 2), thus simulating different angles of inclination (blue, red and black slopes - fig. 3).

Fig. 2. SKI-GYM - multifunctional tool

Effect: the higher the angle of inclination the higher the exercise load for the legs and the feet.

Different terrain conditions (moguls, powder, etc.) can be simulated by using elastic bands on the side of SKI-GYM (fig. 4).

    easy slope (blue)   medium slope (red)   difficult slope (black)

Fig. 3. Variety of positions (angle of inclination)

Fig. 4. Possibilities of setting (elastic bands)

The more elastic bands are used, the more stable the board becomes, guaranteeing a stable position simulating a smooth slope.

The fewer elastic bands are used, the less stable the board becomes, simulating moguls.

Using different numbers of elastic bands on each side simulates a slightly hanging slope imitating the traversing of a slope or is suitable for special training e.g. for one leg after a knee injury.

Effect: Fixing the board to be less stable requires better coordination.

As well as these adjustments, the wheel and the use of a ski binding will result in different effects.

2.2 Workout

The variety of workouts is illustrated using two examples:

    ex. 1:
- Reasonably steep slope: red lever (fig. 3 – white arrow) on medium position
- Relatively smooth slope, using two or more elastic bands (depending on the weight) on both sides for enhanced board stability.

- Training a weak left leg by using more elastic bands on the right side (e.g. one on the left side - none on the right side or two on the left side – one on the right side)
- Skiing in spring when snow is relatively wet and heavy, that means two elastic bands on both sides of the board. The resistance is relatively high.

ex. 2:
- Steep slope with moguls: red lever (fig. 3 – white arrow) on the highest level
- Moguls, only one or two elastic bands (depending on the athlete's weight). Additionally a weak left leg is to be trained: one additional band on the right side that makes the left leg work more.
- Moguls relatively smooth, so resistance relatively low, meaning only one elastic band on the side of the board.

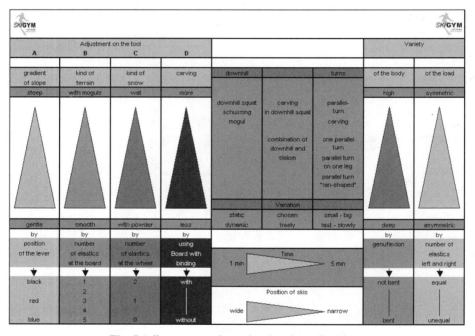

Fig. 5 Adjustment on the tool and variety of workout

3 Methodology

Safety conditions are very important with this machine: none of the other machines can reduce or stop involuntary tilting and spinning movements. This is a key issue for elderly or for anxious people and for people whose coordination skills are not good.

The design of the machine allows a methodical and systematic training with different degrees of difficulty (fig 5). That means:

- setting of different inclination angles (the steeper the more difficult the exercises become);
- setting of the tilting inclination (the board allows the simulation of different slope conditions by using more or fewer elastic bands on both sides of the board);
- imitation of different snow conditions by using more or fewer elastic bands at the fly wheel (more elastic bands on both sides increase the variable resistor which is easier and suitable for beginners. On the other hand more strength is necessary to rotate the fly wheel but of course that's easier to control.);
- imitation of a more realistic situation on snow is possible by using ski bindings which can be mounted on the board (e.g. to train carving technique by using the edges more efficiently; at the beginning of rehabilitation it's better not to use the boards, later it's possible to use them in order to train in a more complex way).

The aim of this project was the creation of a training machine which really allows simulating the situation on a slope.

Many people are interested in training possibilities for the skiing season, especially those who live far away from ski resorts. They don't want to train just strength and stamina in general, they also want to be well prepared for their one or two weeks on snow and to be well trained in order to avoid injuries. Classical ski training does not offer this. Therefore this special machine meets an urgent need.

Another aspect emphasises the necessity of this training equipment. After injuries people can't often start skiing again, not because of problems caused by the injury, but because of their anxiety. Training equipment, which simulates the frightening situation, can help to overcome the problem. In this context it is important to note that it is possible to train one leg or both legs on different levels, which can be useful in recovering after injuries.

Conclusion:

After studying all the information of existing machines, none manages to unite all these advantages in one single piece of equipment.

4 Scientific research

First tests have shown the high efficiency of this equipment. In a pilot study, three people who trained on SKI-GYM after knee injuries were back on snow much faster than people with similar problems who didn't train on the SKI-GYM.

Training on SKI-GYM is not time consuming, yet very efficient (two to three training units per week). The physical abilities necessary for skiing can be improved very well: strength and stamina can be clearly improved.

Different elements of skiing technique can be trained successfully, as the first tests with experimentees have shown. Adjustments enable individual training at an individual level; training is kept entertaining through a variety of exercises. The invention is still new and research is still going on, more scientific data are needed.

By July more tests will have been completed and the results will be presented, hopefully confirming the advantages of this training equipment.

5 Conclusions

The first tests show some important aspects:

This device offers an excellent variety of training, more than any other already existing.

The intensity can be varied from low to very high. That means it is effective for both beginners and professionals.

It also seems to be useful training equipment for people who are recovering from an injury.

More than any other training device this tool models closely the different snow situations (smooth, hard, moguls…).

6 Outlook

The first data will be presented during the congress. As the machine is really brand new (the first five SKI-GYMs were built two months ago), it was only possible to start some tests which are promising, but which could not be completed in this short period of time. Nevertheless this device stands a good chance to win the Innovation Award on the FIBO in Germany this year.

To monitor an individual's performance, SKI-GYM can be used with a special plate that stores a tester's data for transfer to a computer.

References

Fischer, P. R. (2000/2001) (Ed.) TIMaX Sport. *Themen, Hintergrund-Informationen und viele neue Materialien auf X-Medien (topics, background information and many new materials on different media). Sport for pupils in secondary schools. Basic material with eight topics (228 pages), one video und one CD and three additional supplements with ten topics (278 pages).* WEKA Kissing et al.

Fischer, P. R. / Wieser, R. (2001) Kenntnisse und Einsichten, Übergreifende Themen: „Gym for two" – Fitnesspartnergymnastik bringt Spaß zu zweit. *Topic with 46 pages.* In: FISCHER, Peter R. (Ed.) TIMaX Sport. *First additional supplement with 3 topics (108 pages).* WEKA Kissing et al.

Fischer, P. R., Wieser, R. (2001) Kenntnisse und Einsichten, Übergreifende Themen: Fit durch motivierendes Konditionstraining. *Topic with 38 pages.* In: FISCHER, P. R. (Ed.) TIMaX Sport. *Second additional supplement with 3 topics (74 pages).* WEKA Kissing et al.

Fischer P. R. & Nolte Ch. (2005) SKI-GYM – new equipment for skiing and fitness. Augsburg.

A Fresh Approach to Sports Equipment Design: Evolving Hockey Sticks Using Genetic Algorithms

Mark McHutchon, Graeme Manson and Matt Carré

University of Sheffield, Sports Engineering Research Group, m.mchutchon@shef.ac.uk,

Abstract. In this paper the authors describe the first stages of developing a genetic algorithm for optimising hockey stick design. The final iteration of this algorithm optimises the mass distribution and total mass of the hockey stick, and ranks the performance of each design based on its modelled ability in hitting and perceived performance in dribbling. The fitness function used draws on player performance testing results, validated models and implied human perception data.

1 Introduction

This paper covers the early development of a genetic algorithm (GA) based design tool which optimises the physical properties of a hockey stick to meet a customisable design brief. Several versions of the algorithm are implemented, driven at least in part by a previously validated model (McHutchon, Curtis and Carré 2004).

Sports equipment is generally being designed in an increasingly technical and scientific manner. Manufacturers are recognising the need to invest in research and development and remain abreast of new materials and technologies. Marketing in some sports is becoming more focused on genuine technical innovation and improvement rather than resorting to buzz words and flashy graphics. Tennis and golf provide valid examples of mass-participation sports where technology has been employed on a large scale, although some sports and areas of performance remain unaffected by such research. Hockey and cricket are two sports where development has previously been very much based on past experience, with a much hand-crafting being involved in the manufacture of wooden sticks and bats. Figure 1 shows how the shape of sticks has changed over the years (images are not in proportion).

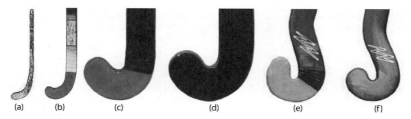

(a)   (b)   (c)   (d)   (e)   (f)

Fig. 1. – The changing face of hockey sticks over the last 100 years, (a) early 20th Century, (b) 1950s, (c) late 1990s wooden, (d) late 1990s composite, (e) & (f) modern goal keepers sticks.

It may be said that the craftsmanship involved in manufacturing from mulberry wood has hindered hockey stick development, as the physical limits of wood present obstacles to innovation. However, as the use of composites continues to increase the scope for dramatic changes in shape, mass distribution and vibration characteristics will be taken advantage of along with more consistent manufacture.

GAs have several advantages, such as the ability to efficiently search a wide design space without any prior knowledge of the search area. The fitness function, which is the function to be optimised, can easily be adapted to include new knowledge from research whether it is a model, experimental results or conclusions based on human perception feedback.

1.1 Literature Review

Genetic Algorithms were first realised in the 1960s by John Holland. Early developments are well documented (Holland 1975 and Goldberg 1988). Michalewicz (1992) provides a comprehensive guide to GA theory and implementation from which the author's code was created. Evolutionary computing is a growing discipline with rapidly diversifying applications including economics, design, robotics and networking.

Parametric design is a field of increasing scope as the boundary between CAD software and Finite Element packages becomes less distinct. Some research has been done into simplifying the design process in sports, reducing it to a problem with only a handful of variables to optimise. Grant and Nixon (1996) successfully isolate certain parameters in cricket bats and their effect on performance, whilst Lee and Kim (2004) and Kim *et al.* (2005) link an ABAQUS finite element model with GA code.

McHutchon *et al.* (2004) validated a basic Rigid Body Model derived from Newtonian mechanics of the form in Eq. 1,

$$Power = \frac{V_s(1+e)}{\dfrac{m_b}{M} + \dfrac{m_b z^2}{I} + 1} \qquad (1)$$

where m, I and z are the inertial properties of the stick, V_s is the velocity of the stick head and m_b and e are the physical properties of the ball. In spite of all this understanding of how parameters affect performance, there is a significant lack of research devoted to pursuing fully optimised designs.

1.2 Objective

To develop a genetic algorithm based design tool which will accurately and rapidly cover all available design space in pursuit of the ideal hockey stick for a particular situation (e.g.: playing style, player ability).

2 Overview of Genetic Algorithms

A genetic algorithm works by iterating a loop which carries out a set of evolutionary functions on an array of potential solutions to a problem. This array of solutions, often referred to as a population, is first evaluated to see which of the solutions performs well. The best performing, or most-fit solutions are carried into the next iteration. Two specific modifier functions make the genetic algorithm distinct; by performing a 'genetic cross-over' and 'mutation' upon designs selected for the next generation the algorithm emulates the processes of evolution and natural selection. Figure 2 shows a typical GA structure.

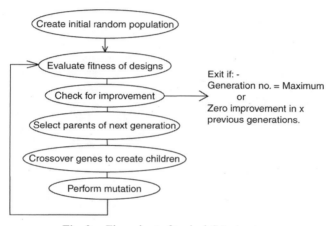

Fig. 2. – Flow chart of typical GA structure

There are other modifier functions available, and variants upon standard mutation, some of which will be covered here as the binary and real number variants of the author's code are discussed.

GAs were originally coded using binary representations of each candidate solution. The set of decimal numbers making up each solution was multiplied up to an integer, converted to binary and the set of binary numbers concatenated into one string. Crossover (Fig. 3) is simple for a pair of binary strings, mutation simply consists of 'flipping' any randomly selected bit to its counterpart.

```
                      Crossover point

Parent A  1 0 1 1 0 0|1 0 1 1          1 0 1 1 0 0  0 0 0 0   Child A
Parent B  0 1 1 1 0 1|0 0 0 0  ───▶    0 1 1 1 0 1  1 0 1 1   Child B
```

Fig. 3. –An example of binary crossover

Whilst binary encoded GAs are simple to code and understand in operation, for a problem which demands real numbers accurate to several significant figures the computational burden of working in binary is significant. Not only are the genetic operations more cumbersome with arrays of larger dimensions, but the encoding and decoding from binary takes a lot of time. With simple problems computational expense is not of great concern, but for a complex fitness function with many input variables and a population of 100 solutions, several thousand iterations may be needed to ensure convergence making any streamlining of code worthwhile.

Mutation for a real number based GA can produce much more dramatic shifts in results than for a binary GA. For a 2 variable problem encoded in binary and accurate to 5 significant figures each variable must be encoded by at least 14 bits as $2^{14}=16384$, the smallest 5 digit number found by raising 2^n. Thus any mutation of a digit in the whole 28 bit string will have an effect of random magnitude, and so there remains some opportunity for fine tuning of variables as the algorithm progresses. For mutating a real number the most straightforward method is to randomise the selected variable within its legal range. However, this can result in large shifts in the solution at any stage of algorithm execution. Michalewicz (1992) proposes a method of attenuating mutation according to Eq. 2.

$$\Delta(t, y) = y \left(1 - r^{\left(1 - \frac{t}{T} \right)^b} \right) \tag{2}$$

where delta is the amount of mutation applied, r is a random number from 0 to 1, and t and T are the current and final generation numbers. Thus the amount of mutation applied to a variable decreases as the algorithm progresses. It has been found through extensive testing that this addition to the code consistently increases the speed of convergence. It should be noted that there is no mathematical way of proving convergence. However, provided good judgement is exercised by the user to ensure adequate execution time a high level of confidence can be placed in results.

3 Algorithm Development

The initial version *1.0Bin* of the author's algorithm solved a simple rigid body impact model of a stick and ball (McHutchon *et al.* 2004), using the apparent COR as the fitness of each design. Several limits were applied to each version of the model. The maximum mass was constrained to 0.737 kg in accordance with FIH regulations (FIH 2004) and an arbitrary minimum mass of 0.2 kg set. A similar range was imposed on the other parameters so no time was lost calculating the performance of invalid designs.

Later versions solved the same impact model but optimised an n-segment distributed mass representation of the stick. Real numbers were also adopted for the sake of execution speed and precision. The final version included Excel results output and the generation-decreasing mutation function in Eq. 2.

Whilst power is a performance attribute that all players can associate with, it is not the only consideration. Early indications from human perception orientated player testing carried out by the authors indicates that a lighter stick with centre of mass nearer the hands is preferable for control i.e. dribbling the ball. In light of this Eq. 3 was formulated to work in tension with Eq. 1.

$$Control = \frac{1}{M + |\lambda - z|} \tag{3}$$

Where λ is the distance from the centre of mass to the lower hand grip.

4 Discussion of Results

Each version of the algorithm was tested in turn to evaluate the run time, speed of convergence and quality of results. The early single objective versions of the algorithm produce a very predictable result in a short time. For a 300 generation run with population of 100 and a 3 variable problem, results within 1% of the theoretical maximum were obtained in run times of under 60 seconds on a 2.8 GHz PC. The real number GA was consistently more than twice as fast as its binary counterpart.

Figure 4 gives an indication of what the GA output looks like when collated for the full range of 'power' and 'control' values.

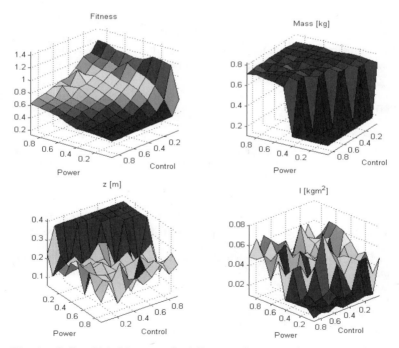

Fig. 4. – Collated Matlab output for full range of *power* and *control* variables.

One hundred executions of the GA were required to formulate Fig. 4, which clearly shows the optimum mass (M), centre of mass distance from the impact location (z) and moment of inertia (I) for any combination of 'power' and 'control'. A surface such as this can give insight into the performance of designs which deviate a lot from current trends.

Whilst the basic performance model behind the algorithm is known to be accurate it only describes a portion of the overall performance of a hockey stick. The *control* performance indicator introduced is somewhat contrived at present and some results given by the algorithm are likely to be impossible to manufacture such as those with large mass variations along the length of the stick. However, it is a trivial matter to incorporate new research knowledge into the algorithm and so the potential remains for a very powerful and accurate design tool.

5 Conclusions and Future Development

In this paper the authors have demonstrated the development and initial application of a Genetic Algorithm (GA) based design tool which optimises the 'power' and 'control' of a hockey stick based on a player's preferences. Whilst the problem being solved by the GA in this paper is relatively straightforward the emphasis here is on the applicability of GAs to much more complex design problems. There is clear potential for a design tool which uses an n-segment finite element beam model. This would combine conventional hitting performance models with player perception of vibrations response and control. Particularly for a problem where the relationship between parameters is not understood, but the problem can be modelled accurately, the GA would be an efficient method of optimising a set of parameters.

References

FIH Rulebook 2004.
Goldberg, D. (1988). Genetic Algorithms, Addison Wesley.
Grant, C. and Nixon, S. A. (1996). Parametric modelling of the dynamic performance of a cricket bat. The Engineering of Sport, Sheffield, Balkema, Rotterdam.
Holland, J. (1975). Adaptation in natural and artificial systems, The University of Michigan Press.
Kim, M. S., Han, D. C., Lee, W. I., and Kim, S. J. (2005). Optimization of golf club shaft characteristics using genetic algorithm. The Impact of Technology on Sport, Tokyo, Australasian Sports Technology Alliance Pty Ltd.
Lee, M. and C. Kim (2004). Design optimisation of graphite golf shafts based on weight and dynamics of swing. International Engineering of Sport 5, Davis, California.
McHutchon, M. A., D. Curtis, et al. (2004). Parametric design of field hockey sticks. IES 2004, Davis, California, ISEA.
Michalewicz, Z. (1992). Genetic Algorithms + Data Structures = Evolution Programs. Berlin Heidelberg, Springer-Verlag.

Development of a New Nordic Walking Equipment and a New Sporting Technique

Anton Sabo[1], Martin Reichel[1], Markus Eckelt[1], Hannes Kogler[2]

[1] Technikum Wien, Sports-Equipment Technology, anton.sabo@technikum-wien.at
[2] Fischer GmbH

Abstract: As a consequence of our intensive Nordic Walking research from the last three years, it was found out that the use of special shoes is essential. The pole length, pole construction and the construction of the handle have a biomechanical influence. Furthermore there should be a difference in Nordic Walking techniques.

There is one fundamental technique from Finland, which is more or less a sporting one. But in Austria Nordic Walking is a health-movement and so we have developed, with the knowledge of the physio-therapeutics, a new, matching technique, which also requires new equipment (handle, strap...).

Our last research covered the influence of Nordic Walking on the prevention of muscular tensions in the upper range of the back (M. Trapezius...). The subjects had an office job and had to work almost all day long in a sitting position in front of a computer. The subject's muscular activity was measured with EMG before and after a training cycle, in which they went Nordic Walking with the finish technique. The results have shown that the subjects trained muscles which are parts of the flexorloop. The muscular tensions became lower but they did not disappear. So a new Nordic Walking technique was developed. This new technique trains the muscles of the extensorloop, which is not possible with conventional Nordic Walking equipment, in particular with the conventional handle and strap. Therefore we developed a new "strap system". This "strap system" does not only differ from the conventional handles and straps in form and look, there is also a difference in the transmission of force and the performance of the technique.

With that new "strap system" and the new technique it is possible to train the muscles of the extensorloop effectively, which results in a prevention of muscular tensions in the upper range of the back and a prevention of muscular dysbalances.

1 Introduction

In the last few years Nordic Walking has become more and more popular and many people are trying out this new sport. But like many other sports it is very important to differentiate between correct shoes, poles and techniques. When we talk about Nordic Walking it has to be clear that this sport can not be compared with other common sports like biking, running, basketball and so on. The primary target group for Nordic Walking are people, who would not do any other sports or can not do any other sports because of the strains. But as a consequence of our intensive research we discovered that for these people and people with muscular dysbalances there should be a difference in techniques. One technique should be a sportive one, which trains the cardiovascular system, and the other should be a kind of physio-therapeutic one, which trains the back muscles effectively. The main difference in these two techniques are the working muscles, which we distinguished in the flexorloop for the sportive technique and the extensorloop for the therapeutics one. This paper describes the research about the difference in these techniques and the new equipment, which is needed for the therapeutic one.

2   Methods

In the fundamental research from the year 2004 – 2005 in which we did a study on the possible prevention of muscular tensions in the upper range of the back by exercising Nordic Walking. The subjects in this research had an office job and had to work almost all day long in a sitting position in front of a computer. The subject's muscular activity was measured by electromyography (EMG) at the following positions:

- the upper range of the M. Trapezius
- the lower range of the M. Trapezius
- M. Deltoideus
- M. Pectoralis, as an antagonist

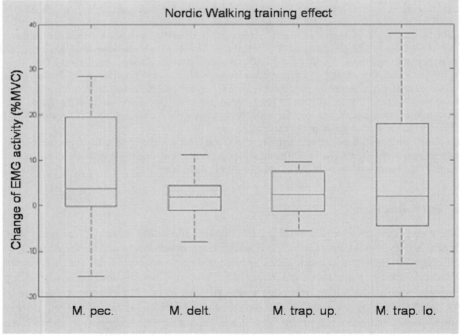

Fig. 1. Change of EMG activity after the Nordic Walking training

The results from this study showed (fig. 1.), that after the training there was higher muscular activity in the lower ranger of the M. Trapezius, which was desired. But we found out, that there was also higher muscular activity in the M. Pectoralis, which was undesirable. So we analyzed the entire movement and so we found out that there should be a new Nordic Walking technique. The fundamental technique from Finland particularly trains the flexorloop (M. Biceps – M. Pectoralis -.....). Our aim was to reduce muscular tensions in the upper range of the back and that Nordic Walking is a prevention for muscular dysbalances. So we talked to physiotherapists and with our combined knowledge we developed a new Nordic Walking technique, which trains the extensorloop, and new Nordic Walking equipment, which is needed to perform this technique.

Our new equipment differs from the conventional equipment in the construction of the handle and the strap.

The new handle:

- The shape supports the stabilisation of the wrist ($20 - 25°$ dorsal flexion of the hand; $15 - 20°$ radial duktion of the hand). So the exposure at the moment when the pole contacts the ground should permit an insurance of the wrist.
- The point for the transmission of the force in combination with the shape of the strap supports the stabilisation and is the crucial factor for the activation of the extensorloop.

The new strap:

- The new strap has additional padding on the edge of the hand.
- The shape of the new strap supports the correct performance of the new technique and the activation of the muscles of the extensorloop.

The subjects in the study with our new Nordic Walking equipment were Nordic Walking instructors (NWO) and people who got an insertion in Nordic Walking. The subject's muscular activity was measured with EMG at the following positions:

- the upper range of the M. Trapezius
- the lower range of the M. Trapezius
- M. Deltoideus
- M. Pectoralis
- M. Triceps
- M. Biceps
- M. extensor carpi
- M. flexor carpi

In additional to the measurements during Nordic Walking we measured the maximum voluntary contraction (MVC) for every muscle at the beginning, the end and when the subject changed the equipment.

The relevant range of the EMG data for the study is the phase of pushing with the pole from the beginning until the pushing is completed. These two points are obvious in the videos and in the data of the acceleration sensor (fig. 2.), which was placed on the Caput ulnae.

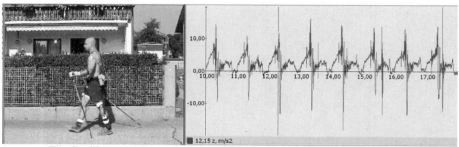

Fig. 2. video recording synchronously to the data of the acceleration sensor

The processing of the EMG raw data effected with Matlab 7.0.4 (fig. 3.) in three steps (Stallkamp, 1998). The first step consists of signal processing by a Butterworth filter. This step is very important because otherwise the motional artefacts would be parts of the further steps and would have an effect on the final results. These motional artefacts were filtered out with a high pass filter and the cut-off-frequency for the filter was found out with the analysis of the data from MVC measurement. The second step was the demodulation of the filtered signal. The third step consists of placing an envelope on the signal. The face under the envelope was calculated and represents the muscle work, which is fundamental for the comparison of the signals from the other subjects with different Nordic Walking equipment.

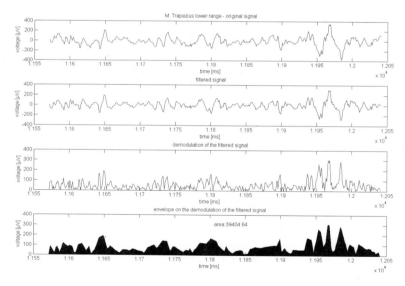

Fig. 3. processing of the EMG raw data with Matlab

3 Results

In the years 2004 and 2005 we worked out a letter of recommendation for the sale. This letter results from co operation with many companies, which produce Nordic Walking equipment (like Leki, Exel, Fischer, Lowa, Salomon ...), and the results of our Nordic Walking studies from the last three years. But this research was based on the Finnish technique.

For our new technique, it would possibly be necessary that the poles are longer , than it is described in our recommendation for the sale (FachWeltSport, VSSÖ, 2005).

Because of longer poles and the new technique, there is a longer and better training of the muscles from the extensorloop. The longer poles result in a displacement of the contact between the pole and the ground (the beginning of the phase of pushing) backwards. This displacement results again in a lower vibration of the material, which reduces strain in the wrist.

There is also another fundamental EMG activity in the muscle groups.

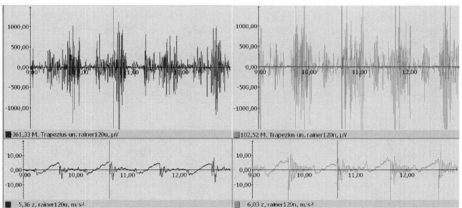

Fig. 4. Comparison of the EMG raw signal of the new Nordic Walking equipment (left side) and common Nordic Walking equipment (right side) in the beginning of the pushing phase

In fig. 4. can be seen, that there is a difference between the EMG signal of the subject's muscular activity during Nordic Walking with the Finnish technique and the new technique.

4 Conclusion

New equipment for a new technique was developed. These two things result in a reduction of strain in the wrist, a benefit of the natural course of motions during Nordic Walking and specialized training of the muscles, which are part of the extensorloop, in particular the muscles of the upper range of the M. Trapezius. This training effect should result in prevention of muscular tensions in the upper range of the back and prevention of muscular dysbalances. So Nordic Walking would be an effective sport in the health movement in Austria.

References

B. Schmölzer, (2003) Nordic Walking – Überprüfung der Funktion von Stöcken und der Effektivität des Bewegungsablaufes, Salzburg

F. Stallkamp, (1998) Dreidimensionale Bewegungsanalyse und elektromyographische Untersuchung beim Inline-Skating unter Berücksichtigung eines Weichschalen- und eines Hartschalenschuhs, Klinik und Poliklinik für Allgemeine Orthopädie der Westfälischen Wilhelms-Universität Münster

Horst de Marées und Joachim Mester, 2.Auflage 1991, Sportphysiologie I, Diesterweg Sauerländer

Rolf Wirhed, Sport- Anatomie und Bewegungslehre, 2. Auflage, Schattauer

Werner Platzer, . vollständige überarbeitete Auflage, Taschenatlas der Anatomie, 7Thieme

Benninghoff, 15. völlig neu bearbeitete Auflage, Anatomie, Band 1, Urban und Schwarzenberg

H.-J. Dobner, G. Perry; Biomechanik für Physiotherapeuten, Hippokrates

B. Segesser, W. Pförringer; Der Schuh im Sport, perimed Fachbuch – Vertragsges.m.b.H

Ibrahim A. Kapandji Funktionelle Anatomie der Gelenke, Band.1, Obere Extremität

H. Lippert, P. Herbold, W. Lippert-Burmester, Anatomie, Text und Atlas 7. Auflage

Zintl, F: Ausdauertraining, Grundlagen, Methoden, Trainingssteuerung

Martin, D./Carl,K./Lehnertz K.: Handbuch Trainingslehre

Badtke, G.: Lehrbuch der Sportmedizin

3 Materials

Synopsis of Current Developments: Materials

Dr.-Ing. Reiner Lützeler

Institut für Kunststoffverarbeitung an der RWTH Aachen, luetzeler@ikv.rwth-aachen.de

Materials are not only in technical application a very important aspect but also in designing functional sports equipment. In high performance as well as in mass sports the materials for equipment, sports wear and supports are responsible for the function and the quality. More and more the advertising uses the developments in designing materials for the marketing of sports products, to convince of buying or to restore the athletes for new sportive options.

Materials not only influence the weight, the design or the price, but also the sportive movements. The best example for this is the pole vault, which could be revolutionized by the use of glasfibre reinforced thermoplastics. But also in other sports (e.g. Tennis, Golf, etc…) the materials are of very high importance.

The following contributions address several material topics concerning sports equipment. There is no limitation to a specific material, because the papers deal with metallic as well as with plastics materials.

The contribution "The Effect of Pressure on Friction of Steel and Ice and Implementation to Bobsleigh Runners" shows that there are several influencing factors on the friction between steel and ice. Very important to notice seems to be the result that there is an optimum that can be influenced by a heat treatment of the runner material. The contribution "Development of Equipment to Compare Novel Ice Skate Blade Materials" deals with a similar topic. This paper presents the development of a testing system, with which the behaviour of new materials on ice can be tested. This field of interest is rounded with a paper that presents the manufacturing of bobsleigh runners, which because of their slender and complex geometry are difficult to produce. Besides the choice of a manufacturing process the used material and the optimisation of the design is discussed in the contribution.

A nano-reinforced metall is the main focus of the contribution "Sanwick Nanoflex – designed for Ultimate Performance". Here a new material is presented, that shows an interesting strength/density-ratio and therefore is suitable for a weight reduction with simultaneous good mechanical properties of sports equipment.

The weight reduction is also a characteristic of carbon fibre reinforced plastics, which are often used for bicycle frames. The contribution "Carbon Fibre Reinforced Plastics – Trendsetting Material for High Performance Racing Bike Classics" characterizes the material properties and screens the material capabilities. The results approve that the material properties are very suitable for bike components because of their weight advantages in proportion to the material properties.

Reinforced thermoplastics are very important for the production of kajak paddles, so that the contribution "Prediction of the Flexural Modulus of Fibre Reinforced Thermoplastics for the use as Kajak Paddle Blades" deals with the adjustment of mechanical properties to the paddle movement. In this paper a modelling is presented with which the stiffness of paddles can be predicted, so that with a calculation a design optimisation can be realised.

Finally the contribution "Recent Advances in Understanding the Behaviour of Shockpads for Outdoor Synthetic Sports Pitches" shows the influence of different design variables on the behaviour of shockpads. Here is shown that the playing performance, the quality and the user comfort as well as the safety are affected, when using different material compositions.

Looking on all the contributions it can be stated, that the contributions deal with interesting topics and provide valuable results. Moreover it is clearly visible that the material properties should be regarded and used more intensively. Thus the top performing of sports results and a development of sports equipment for high performance as well as for mass sports can be realised.

Recent Advances in Understanding the Behavior of Shockpads for Outdoor Synthetic Sports Pitches

Lauren Anderson, Paul Fleming and Ali Ansarifar

Loughborough University, L.J.Anderson@lboro.ac.uk

Abstract. Shock-absorbing layers perform vital roles in the user comfort, safety and ball interaction characteristics of synthetic sports pitches. The layer typically comprises a porous composite of granulated recycled rubber bound in a polyurethane resin, compacted to form a flat continuous pad upon which the carpet is laid. A lack of published information regarding sports shockpads has prompted research at Loughborough University that aims to investigate the fundamental aspects of shockpad layers, namely their design, construction, characteristic behavior, and test methods. This paper outlines the findings of a detailed study investigating the effect of mix design variables on shockpad properties. Primary mix design variables, binder content, bulk density and rubber size distribution, were varied individually in industry standard shockpads produced using a reproducible hand-construction method. A comparison of tensile strength, ball rebound measurements and Clegg Hammer impact behaivior, showed marked influence of these variables over shockpad performance. The dominance of smaller sized rubber particles in a 2-6mm rubber size produced a softer shockpad (lower Clegg Impact Values), higher tensile strength and slightly decreased ball rebound resilience (more energy absorbed). Higher binder contents increased shockpad strength but had no effect on ball rebound or Clegg Impact Values. Increasing the bulk density of shockpads increased the shockpad tensile strength and decreased Clegg Impact Values (i.e. softer). Further work is ongoing to assess the effect of other design and construction variables. Testing to assess the effect of various carpets placed above the shockpad is also ongoing to assess the whole pitch system's performance.

1 Introduction

Synthetic sports pitches are composite structures; generally comprising a foundation of layers of crushed rock and bound macadam, and a surface system of a shockpad and carpet (either filled or unfilled). Each layer is required to interact simultaneously to produce ball rebound, user comfort and safety behavior characteristic of the sport for which the pitch is intended.

Design Variables	Construction Variables
Rubber size	Mixing time
Rubber particle size distribution	Compaction plant
Rubber shape	Cure environment
Binder type	
Binder Content	
Layer thickness	
Bulk Density	

Table 1. Design and construction variables in shockpads.

A previous review focusing on the shockpad layer of synthetic pitches (Anderson, Fleming & Ansarifar 2004) reported a dearth of published investigation and few specified limits stipulated by sporting authorities to ensure consistent layer quality. Recommendations were made for further research into the aspects of constituent material properties and consistency, effect of design and construction variables and understanding shockpad behavior.

Bound in-situ (also called wetpour) shockpads comprise recycled rubber granule bound in a thin coating of polyurethane, which is compacted to produce a porous and continuous layer of specified thickness. Limited research has observed the effects of altering rubber size and binder content on the tensile properties of polyurethane bound rubber crumb. Kim (1997) found larger rubber sizes with a binder content of 13% produced optimal pad strength. Conversely, Sobral et al. (2003) found smaller rubber sizes produced higher strengths. The rubber sizes used by both these researchers were much smaller than those used to construct shockpads in the UK (0.1-3mm compared with 2-6mm) and therefore cannot directly infer the effect of rubber size and binder content. The tensile test is only useful as a 'quality' indicator and gives no indication of the effect of these variables on the ball and athlete interaction characteristics of a synthetic sports pitch.

This paper reports on progress in understanding the effect of selected design variables. Variables identified from the literature survey, site visits and industrialists comments are listed in Table 1, and are separated into their respective stages of the construction process. It is expected that all variables listed will affect the behavior of shockpads, however, this paper will only focus on three; binder content, bulk density and rubber particle size distribution. Characteristic behavior will be assessed through three test methods intended to indicate ball interaction, athlete interaction and quality aspects for the shockpad layer alone.

2 Effect of Design Variables on Shockpad Behavior

2.1 Test Method

A standard shockpad was created to assess the extent changes in design would have on behavior. The range for the design variables chosen are given in Table 2, and were derived from characterization measurements of shockpad samples taken during the construction of two water-based hockey pitches. Three shockpads with standard specifications were produced and the average test results recorded. Variations in tests results for standard shockpads were low and are attributed to constituent material and construction method variability.

A separate shockpad was produced as each variable was altered from the standard value through a range of three to four values that were considered realistic for shockpad construction. All other variables remained with standard values. This method produced ten shockpads in total (three standard shockpads and seven with variations). Particle size distributions for the standard, small sized and large sized variations fitted to a 2-6mm rubber particle size. The method used to construct shockpads is discussed in Section 2.2.

A range of tests to indicate the effects of altering design variables on the playing performance, construction quality, user comfort and safety were selected. The first test involved dropping a hockey ball from a height of 1.5m onto a shockpad (FIH 1999) and measuring the rebound height. The drop height (Hi) and rebound height (Hr) were used in Eq. 1 to determine rebound resilience (R.R). Rebound resilience is a measure of the energy lost during a ball impact and would indicate a difference in shockpad playing performance during ball interactions.

$$R.R = \left(\frac{Hr}{Hi}\right) \times 100 \tag{1}$$

A small portable device called a Clegg Hammer was used to simulate the effect of athlete interaction. The device uses a 2.25kg weight that is dropped from a height of 45mm through a plastic tube to impact the shockpad below, measuring the maximum deceleration of the weight during the impact. Peak forces generated by the have been shown to be similar to that of an athlete heal strike over a much faster contact time(Fleming et al., 2004). Higher Clegg Impact Values generally indicate a harder surface and the greater potential, therefore, for injury during a fall or from repeated use. Tensile strength measurements were used as a third test to indirectly measure the quality and service life of the shockpad produced. Testing involved cutting shockpad samples according to BS EN 12230, then strained to failure at a rate of 50mm/min.

	Standard Shockpad	Variation Range
Binder Content	9 %	5, 12, 15 %
Bulk Density	550 kg/m³	500, 600 kg/m³
Rubber Size	2-6 mm	-
Rubber PSD	Standard	Small, Large
Rubber Shape	Crumb	-
Thickness	12mm	-

Table 2. Standard shockpad specifications and variable ranges.

2.2 Shockpad Construction and Cure

Recycled rubber granules were sieved into separate particle size ranges using mesh sieves. The rubber was washed and dried to remove residual dust ensure adhesion of the binder was not affected. The total mass of rubber required for each shockpad was determined by the product of multiplying bulk density and mould volume.

Rubber and binder were mixed in the required proportions using a small bench-top mixer for three minutes. The mix was placed in wooden moulds 30mm x 90mm x 12mm and compacted by rolling a cylindrical steel bar along the mould edges in perpendicular and diagonal directions until a flat surface was achieved.

Compacted shockpads were placed in an environmental chamber where temperature and humidity were controlled during cure at 22±3°C and 65±5%RH respectively. A previous investigation into shockpad cure times revealed 14 days was sufficient time for a shockpad to reach a plateau in mechanical properties. Upon reaching 14 days curing time, the shockpads were removed from the moulds for testing.

2.3 Test Results

Test results for the ten shockpads constructed to determine the effect of design variables are given in Table 3. Coefficient of variance (CoV) for each result is given as a bracketed number, and was determined by dividing the standard deviation by the mean and multiplying by 100. It should be noted that values from the standard shockpad (9% binder content, 550kg/m^3 bulk density and standard particle size distribution) are the average of the values from the three standard shockpads.

3 Discussion

The ball interaction characteristics of shockpads (rebound resilience) shows little change for variations in binder content and bulk density. A small reduction in resilience is seen for shockpads with small and large particle sizes. The effect of changing particle size on overall pitch properties will be to be further reduced by the carpet layer above.

Clegg Hammer impacts show a greater effect of shockpad design variables. The most significant changes are shown by bulk density and particle size distribution. Clegg Impact Values decrease with increasing bulk density and rubber particle size, indicating a more compliant shockpad is produced in both cases. These effects are considered due to the shockpad having less influence from the rigid substrate below through an increased number of particle interfaces to dissipate the impact energy

Variable		C.I.V [g's]	R.R [%]	T.S. [kPa]
Binder Content	5	250.8 (2.6)	36.9 (1.9)	20.7 (42.0)
[%]	9	257.4 (1.3)	36.7 (1.1)	33.9 (25.0)
	12	227.6 (0.3)	36.4 (1.7)	45.1 (21.5)
	15	252.1 (7.7)	36.6 (1.9)	70.9 (12.0)
Bulk Density	500	290.3 (3.7)	36.2 (1.8)	26.8 (39.9)
[kg/m^3]	550	257.4 (1.3)	36.7 (1.1)	33.9 (25.1)
	600	246.8 (23.3)	36.5 (0.4)	49.3 (20.5)
Rubber PSD	Std.	257.4 (3.4)	36.7 (0.4)	33.9 (8.5)
	Small	229.7 (7.4)	35.7 (0.5)	38.9 (10.5)
	Large	225.3 (2.3)	35.9 (0.4)	34.0 (8.9)

Table 3. Mechanical test results for shockpads with design variations. CoV given in brackets.

Although ball rebound and Clegg hammer impacts are similar in their nature, Clegg Impact Values are more influenced by design variables. Clegg Hammer impacts are much higher energy (similar to a Berlin Artificial Athlete) and therefore produce higher strains and deformation. Ball impacts strain only the upper shockpad surface, and are not influenced by the bulk of the shockpad. Rubber particle size and its packing density determines how the shockpad will behave. As the binder only acts to hold the rubber particles together, under compressive loads its influence over shockpad behavior is not significant compared to physical characteristics of the rubber, such as shape, packing density and intrinsic compressional behavior.

All three variables had an influence over the tensile strength and therefore may influence shockpad life. Binder content exhibits the greatest effect on tensile strength of the three variables studied, with the strength increasing linearly with content. Strength increases to a lesser degree with increased bulk density and decreased particle size. Tensile strength is reliant on sufficient binder being available at contact points to adhere rubber particles together, a result of binder content and the number of contact points in the cross section of the shockpad. Contact points are increased by increasing bulk density and using smaller rubber granules. It should be noted that by reducing particle sizes within a distribution creates higher surface area, therefore increasing the amount of binder required to sufficiently coat the rubber surface. A shockpad with optimum quality would therefore, from these data, be dominant in small rubber particle sizes and have relatively high binder content and bulk density.

Shockpads are rarely used as the surface layer in pitch constructions, so the effect of the carpet and infill materials requires careful consideration. Third generation football and rugby pitches with a long carpet pile filled with rubber are not considered to be greatly affected by the shockpad until infill compaction has occurred (several years). Water-based hockey carpets (no infill) are more affected by the shockpad, especially after wear has occurred. It must be noted that the results of this testing do not indicate changes to the actual whole pitch performance. They only act to indicate that pitch performance may be affected by altering these variables and magnitude will depend on the carpet and infill materials used.

4 Conclusions & Further Work

Altering design variables in shockpads has been shown to affect its playing perform-ance, quality and user comfort. Ball rebound resilience was shown to be slightly influenced by the rubber particle size, however was not thought to be significant for pitch playing characteristics when considering the combined effect of the carpet. Clegg Impact Values were seen to decrease with increasing bulk density and de-creasing rubber particles sizes within the distribution, indicating that a more compli-ant pitch would be achieved. The quality of shockpads produced was influenced by all three variables, with optimum strength being produced by high binder content, high bulk density and small-dominated rubber particle sizes.

Further work is required to assess the effect of other design and construction variables on shockpad behavior. Preliminary testing of shockpads of varying thick-ness has shown significant effects on behavior and warrants further investigation. Repeating tests with different carpets and infill materials will be useful in showing the effect of these variables on whole pitch performance.

References

Anderson, L.J., Fleming P.R. and Ansarifar, A. (2004) Shock Absorbing Layers for Synthetic Sports Pitches. In: M. Hubbard, R.D. Mehta and J.M.Pallis (Eds), *The Engineering of Sport 5*. California, USA, pp. 509-516

FIH (1999) Handbook of Performance Requirements for Outdoor Synthetic Hockey Pitches, International Hockey Federation. Brussels, Belgium

Fleming P, Young C, Dixon N, (2004) Performance Measurements on Synthetic Turf Hockey Pitches, The Engineering of Sport 5, ISEA, M. Hubbard, R. D. Mehta, and J.M Pallis (Eds), Vol 2, pp524-531. ISBN 0-9547861-1-4.

Kim, J.K. (1997) Experimental and Theoretical Studies on Crumb Ruber/Polyurethane Blend System, *Korea Polymer Journal*, 5(4), pp. 241-247

Sobral, M., Samagaio, A.J.B., Ferreira, J.M.F. and Labrincha, J.A. (2003) Mechanical and Acoustical Characteristics of Bound Rubber Granulate, *Journal of Materials Processing Technology*, 142(2), pp. 427-433

The Effect of Pressure on Friction of Steel and Ice and Implementation to Bobsleigh Runners

Melanie Dumm[1], Christian Hainzlmaier[2], Stephan Boerboom[3], Erich Wintermantel[4]

[1] TU Munich, Chair of Biomedical Engineering, Biocompatible Materials and Process Engineering, dumm@medtech.mw.tum.de
[2] TU Munich Chair of Biomedical Engineering, Biocompatible Materials and Process Engineering, hainzlmaier@medtech.mw.tum.de
[3] TU Munich, Chair of Biomedical Engineering, Biocompatible Materials and Process Engineering, boerboom@medtech.mw.tum.de
[4] TU Munich, Chair of Biomedical Engineering, Biocompatible Materials and Process Engineering, wintermantel@medtech.mw.tum.de

Abstract. Bobsleigh runners are a critical factor for success in bobsleigh races. It is very important to use runners which generate a minimum frictional loss in contact with ice. As other studies show, the pressure greatly influences the coefficient of friction between steel and ice. In literature, there are indications for a decreasing coefficient of friction with increasing pressure as well as for an increasing coefficient of friction.

This paper presents a method to measure the coefficient of friction μ between steel and ice. The normal pressure p on the sample was increased continuously. It was shown, that there is a minimum coefficient of friction for an optimum pressure value popt. Results were transferred to bobsleigh engineering.

A molding method was developed to examine the surface of the ice, in particular the trace a runner generates in the bobsleigh track. A highly precise impression material was used and impressions were analyzed by scanning electron microscopy (SEM).

The actual pressure between runner and ice was estimated according to SEM analysis. Comparing the actual pressure with laboratory results, design recommendations for a new bobsleigh runner were determined.

1 Materials and Methods

1.1 Test Bench

The TriboDisc test bench was used for the tribological studies on friction of steel and ice. It operates in a climatic chamber which can be regulated between $-20°C$ and $+15°C$. The test bench is based on the "pin-on-disc" principle in which a fixed sample is set on a rotating disc. The sample is moved on a spiral track over the ice so that a distance of 240m of fresh ice can be used. The relative speed of the sample is adjustable between 0,1m/s and 30m/s. The relative speed keeps constant during one test operation. The normal force on the sample is varied by weights on top of its receptacle. The surface of the ice gets prepared by an ice cutter comparable to that used in the bobsleigh track.

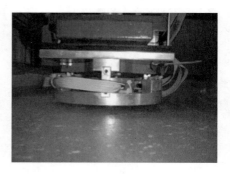

Fig. 1. Test bench TriboDisc   **Fig. 2.** Receptacle of the sample

Fig. 3. Sample

1.2 Sample

The sample itself has a standardized geometry and an area of 32mm² contacting the ice surface. The sliding surface gets polished to a roughness of 1µm. For that reason the influence of the sliding surface structure is neglectable.

1.3 Measuring Unit

To measure the frictional and the normal force strain gages are used. The measuring unit is placed in a way that both forces can be measured separately and simultaneously. The generated data is transmitted to a readings recorder which can log all the values at the same time stamp. The logging frequency is 50 Hz.

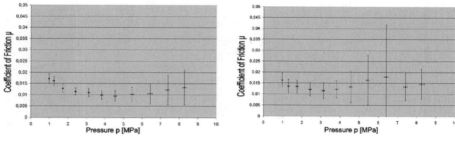

Fig. 4. μ-p-Graph for steel at -2°C   **Fig. 5.** μ-p-Graph hardened steel at -2°C

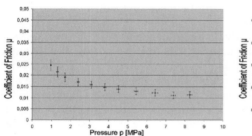
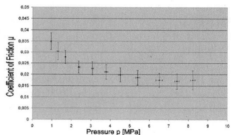

Fig. 6. μ-p-Graph for steel at -15°C   **Fig. 7.** μ-p-Graph hardened steel at -15°C

2 Results

Test series were made with samples of steel and hardened steel at temperatures of
–2°C and –15°C. To get a continuous description of the μ-p-graph the pressure was
increased step by step during the tests.

In general we can see at both samples a decreasing μ with increasing pressure.
Furthermore it is obvious that μ depends on the ice temperature: Tests at –15°C
result in higher friction coefficients than tests at –2°C. But there is an evident difference in the characteristics of the graphs for –2°C and –15°C: At –2°C there is an
observable optimum of the friction coefficient for both samples. Whereas at –15°C
the tendency of μ is still decreasing at the end of our pressure range.

Another interesting aspect is the different behavior of the steel sample compared
to the hardened sample.

3 Discussion

As shown in former studies (Hainzlmaier 2005, Boerboom Vienna 2005, Boerboom
Belgrade 2005) there is a decreasing friction coefficient with increasing pressure and

increasing friction coefficient with decreasing temperature. Both effects could be confirmed in the actual study, too. But in this case a continuous increasing pressure gave the possibility to describe the characteristical behavior of μ for a comparative wide pressure range within one testing procedure. Therewith an optimum pressure for the lowest frictional loss could be found. In contradiction to former studies of Hainzlmaier (Hainzlmaier 2005), who assumes the optimum pressure for lower temperatures decreases because the ice is getting more brittle, this study shows a tendency of a higher optimum pressure for decreasing temperatures. The ploughing effect of the sample and therewith the destruction of the ice surface occurs later with decreasing temperature.

Responsible for this oppositional results could be the improvement of the ice surface by more experience in the ice preparation. Today there is less trapped air in the ice surface so the sliding of the sample is more evenly. Moreover the measure unit was improved and a higher accuracy of the measurement values could be achieved.

The other important result of this study is the influence of heat treatment on the frictional behavior of the material. In this case the heat treatment of the steel led to a degradation of the friction coefficient.

To implement the results of this study to bobsleigh runners it was necessary to know the actual pressure on the runner during the race. Therefore a highly precise impression material was used to cast the traces a runner generates in the bobsleigh track (Hainzlmeier Ontario 2005). Impressions were analyzed by scanning electron microscopy (SEM) and the actual pressure between runner and ice was estimated according to SEM results. This indicates that the bobsleigh runs at a very low pressure in the straight lines and at a very high pressure in the curves, but never at the optimum pressure for lowest frictional loss. With this knowledge geometric optimizations of the runner concerning the pressure distribution over the contact surface can be made.

Regarding the influence of heat treatment to the friction coefficient there is a high potential for optimization of the runner. This is limited by a rigorous rules catalogue of the international bobsleigh organization. Since 01.01.2006 every kind of heat treatment of the runner is forbidden. Nevertheless, with regard to other implementations, this issue should be further investigated.

References

Hainzlmaier, C. (2005) *A new tribologically optimized bobsleigh runner*. Zentralinstitut für Medizintechnik, Technische Universität München.

Boerboom, S., Hainzlmaier, C., Kraus, F., Wintermantel, E. (2005) *Tribologie im Stahl-Eis-Kontakt: Einfluss des Druckes*. Divers Workshop 2005, Vienna.

Boerboom, S., Hainzlmaier, C., Kraus, F., Wintermantel, E. (2005) *Effect of runner material on ice friction in bobsleigh*. 10th Annual Congress of the European College of Sports Science 2005, Belgrade.

Hainzlmaier, C. (2005) *A method to study the surface of ice using an impression material*. 62nd Estern Snow Conference 2005, Ontario.

Prediction of the Flexural Modulus of Fibre Reinforced Thermoplastics for use as Kayak Paddle Blades

Paul.D. Ewart and C.J.R. Verbeek

The University of Waikato, p.ewart@waikato.ac.nz

Abstract. It can be shown that there is a strong link between paddler performance and paddle stiffness. Despite the fact that composite materials are often chosen to enable design of more rigid sporting equipment, readily available flexural modulus values for composites other than proprietary blends are not easily obtained. Therefore, when composition deviates from proprietary blends, accurate prediction of material properties becomes necessary. A mathematical model for predicting the flexural modulus of short fibre reinforced composite materials is developed based on the application of simple beam theory. The flexural modulus can be modelled by considering a small section of the composite comprising a finite number of polymer and reinforcement layers. Simple beam theory assumes that there is perfect bonding between these layers, but it is well known that interfacial adhesion plays a significant role in composite properties. To account for the interfacial layer and interfacial bonding the second moment of area of the composite beam element was modified by assuming that the contribution to the second moment of area from the matrix layer is reduced by an amount representing the non-bonded interfacial layer. The flexural modulus values obtained from the model are compared to experimental values for glass-fibre reinforced linear low density polyethylene. It has been found that the use of non-contact regions in the model resulted in improved accuracy over the model with perfect bonding for short fibre reinforced LLDPE.

1 Introduction

It has been previously shown that there is a strong relationship between athlete performance and equipment stiffness (Carreira, Ly et al. 2002). In an effort to increase paddle efficiency for elite level kayak athletes, improvement to paddle materials, especially their selection and design should be considered. This paper involves modelling mechanical properties of composites in order to assist in improved design.

Previous modelling work into the use of fibre reinforced composite materials for sporting equipment highlighted the inaccuracies inherent in simple beam theory due to the assumptions of a homogeneous microstructure, isotropic strength characteristics and ideal bonding between components (Ewart 2004). The model is based on the concept where by the discrete layers found within a composite material can be used to predict the flexural modulus of the material when treated as a laminate structure

(Fig 1). In order to improve on the predictions gained from the earlier model it is considered necessary to account for the interface between composite components (Ewart and Verbeek 2005). Current research is focused on the composite interface and the interaction between materials with no ideal or dominant theory accounting for the phenomena (Huffington 1960; Sharpe 1971; Lutz and Zimmerman 1996; Hong-Yuan and Yiu-Wing 1998; Haboussi, Helene et al. 2001; Lauke, Schuller et al. 2003). In this paper an improved model is proposed based on the laminate structure of earlier work, but non-ideal bonding is accounted for by considering a non-bonded layer at the interface.

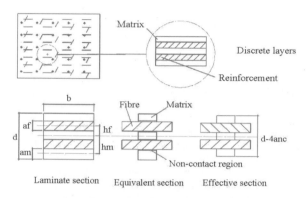

Fig 1. Top, Laminate structure of short fibre reinforced composite. Bottom, Equivalent section and effective section accounting for non-contact region.

2 Composite materials and the interface

It is well known that the mechanical properties of composite materials are significantly influenced by the interface between components (Higgins 1995). Perfect bonding is often assumed in modelling composite mechanical properties and is the basis of fundamental iso-stress and iso-strain models.

What might be considered the most simplistic account for component bonding is that of the interface being the contact surface between two components. Interfacial adhesion can therefore be modelled in terms the stress transfer between matrix and reinforcement. Adhesion can also be considered in terms of a third adhesive layer between the composite components. The contact surfaces between component and adhesive are considered perfectly bonded. The adhesive layer can also be referred to as an interphase between the matrix and the reinforcement (Haboussi, Helene et al. 2001). Leading on from the concept of an interphase, the concept of interfacial adhesion has also been modelled in terms of three interphases mainly due to the presence of surface modification. The fibre interphase, adhesive interphase and matrix interphase with a fibre-adhesive interface and an adhesive-matrix interface, all considered affecting stress transfer mechanisms (Holmes, Feresenbet et al. 2003). In a study by Grozdanov and Bogoeva-Gaceva it was shown that the crystallinity can increase at the reinforcement-matrix interface which leads to improvements in mechanical properties. The formation of an elastic modulus gradient can also be shown as a product of the increased crystallinity as the differential between the component moduli or due to differences of thermal expansion between components (Grozdanov and Bogoeva-Gaceva 2002).

Fig 2. SEM micrograph showing non-contact region between matrix and fibres.

Interfacial adhesion typically occurs as a result of one or more of the following mechanisms, although it has been shown that the bonding mechanism can change during manufacturing processes (cited in Ewart 2002):
(a)   mechanical interlocking
(b)   physio-chemical interactions
(c)   chemical bonding
(d)   mechanical deformation of the fibre–matrix interphase region

3 The model accounting for the interfacial layer

The non-contact area between the fibre and matrix seen in Fig 2 is effectively reducing the overall modulus of the matrix layer in much the same way a polymer is affected by addition of foaming agents to produce structural foam. To account for the additional non-contact region the equivalent section is reduced to the effective section for calculation as shown in Fig. 1 (Throne 1972; Moore and Ironman 1973; Wu and Tsong-Ming 1994).

Maintaining the simplified approach to this model the non-contact region is considered to reduce the second moment of area of the equivalent section. The matrix thickness is reduced by the value of $2\mu m$, obtained from an estimate of the average non-contact region thickness (a_{nc}) in Fig 2.

A five layer model, the optimum number reported from previous work, was developed to determine the overall second moment of area for the laminate element (Ewart and Verbeek 2005). The fibre and matrix layers are considered separately using the parallel axis theorem as Eq. 1-2.

$$I_{fibre} = \sum_{i=1}^{k}\left(I_{xxi} + Ah_{fi}^2\right) \qquad (1)$$

$$I_{matrix} = \sum_{j=1}^{k}\left(I_{xxj} + Ah_{fj}^2\right) \qquad (2)$$

The second moment of area for the laminate manipulated from Eq. 1-2 becomes:

$$I_L = \frac{I_{matrix}}{R} + I_{fibre} = I_C\left[\frac{\left(\Lambda_m + \Gamma_m\right)}{R} + \left(\Lambda_f + \Gamma_f\right)\right] \tag{3}$$

I_C is second moment of area for the sample cross-section; R is the ratio between the reinforcement modulus and matrix modulus, Λ and Γ for the matrix and reinforcement phases, where d is the section depth, were determined to be:

$$\Lambda_m = \frac{V_m^{\,3}}{9} - \frac{2V_m^{\,2}a_{nc}}{d} + \frac{12Vma_{nc}^{\,2}}{d^2} - \frac{24a_{nc}^{\,3}}{d^3}$$

$$\Gamma_m = \frac{8V_m^{\,3}}{9} - \frac{64V_m^{\,2}a_{nc}}{3d} - \frac{48V_m^{\,2}a_{nc}}{9} + \frac{8V_m^{\,2}V_f}{3} - \frac{40V_mV_fa_{nc}}{d}$$

$$+ \frac{112V_ma_{nc}^{\,2}}{d^2} + 2V_mV_f^{\,2} + \frac{144V_fa_{nc}^{\,2}}{d^2} + \frac{96a_{nc}^{\,3}}{d^3} - \frac{12V_f^{\,2}a_{nc}}{d}$$

$$\Lambda_f = \frac{V_f^{\,3}}{4}$$

$$\Gamma_f = V_f\left(\frac{V_m^{\,2}}{3} - \frac{8V_ma_{nc}}{d} + V_mV_f - \frac{9V_fa_{nc}}{d} - \frac{6a_{nc}^{\,2}}{d^2} + \frac{3V_f^{\,2}}{4}\right)$$

The composite bending modulus is then calculated as Eq. 4:

$$E_C = E_m\left(\Lambda_m + \Gamma_m\right) + E_f\left(\Lambda_f + \Gamma_f\right) \tag{4}$$

4 Experimental

Samples used to validate the model were injection moulded using glass-fibre and linear low density polyethylene (LLDPE) varying from 0-0.6 volume fraction of fibre. Elastic modulus values for the components were taken from their respective material data sheets to be glass fibre 78.5 GPa, 13μm diameter, 6mm nominal length and LLDPE 0.19 GPa, non-contact region depth of 2.0μm. The paddle blade test samples were made using proprietary composite blends. Aplax, 30% glass fibre shorts in a polypropylene (PP) matrix modelled using a modulus range for glass fibre of 50-80GPa, 12μm diameter and PP at 1.4 GPa, non-contact region depth of 1.5μm. Duralon, 33% carbon fibre shorts in a polyamide (PA) matrix using carbon fibre modulus range of 180-260 GPa, 8μm diameter and PA at 2.0 GPa, non-contact region depth of 1.0μm. The non-contact region depths were estimated from SEM micrographs taken along the fibre length and end on to the fibre axis. Average values for all of the tests were taken from 5-8 specimens 12 X 3.2 X 118 mm (bxdxl) using the ASTM D790-03 standard.

5 Results and discussion

Figure 3 show comparisons between the isostrain model, the 5 layer perfect adhesion model and the improved interfacial model using 1.5 and 2.0 μm. The improved model shows a similar trend to the experimental values with excellent approximation for the flexural modulus values using the estimated non-contact region depth of 2µm. From $0.5V_f$ the modulus values are seen to reduce. The reduction in modulus values would appear to be due to insufficient matrix material to encapsulate the fibre and create efficient load transfer between the components.

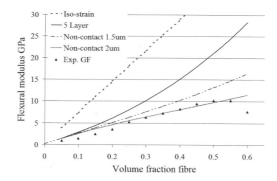

Fig 3. Experimental and predicted values for flexural modulus of glass fibre reinforced LLDPE composites.

The predicted values for the glass fibre and carbon fibre reinforced paddle material (Table 1) also show good agreement. Without being able to ascertain accurate material values for the components used in the proprietary paddle blade material the predicted values needed to be calculated from a range of generic materials data. A lack of precise material data will not allow for precise predictions although it is seen from the values obtained by this method that the predicted values are within an acceptable tolerance.

Table 1. Composite paddle blade material modulus values (σ = standard deviation).

Sample	Experimental modulus	Predicted modulus
Glass/ PP (Aplax)	5.8 GPa ($\sigma = 0.02$)	5.3-8.0 GPa
Carbon/ PA (Duralon)	20.2 GPa ($\sigma = 1.35$)	19.1-27.1 GPa

6 Conclusion

The assumption of perfect bonding for short fibre composites using a laminate approach has been shown to be greatly improved by accounting for the matrix/reinforcement interaction. The ability of the model to predict flexural modulus is however limited to those composites that have a single matrix and a single reinforcement phase. This research highlights the sensitivity of the model in regard to input data such as non-contact region depth, and to some extent component elastic modulus values. It is also likely that non-contact region depth is proportional to interfacial adhesion. Further work could be done to extend the understanding of the non-contact regions that resulted in improved accuracy of the model. This could involve further sensitivity analysis and experimentation with weakened interfacial regions. Further validation of the model could also take the form of paddle blade design analysis using values obtained from the model to form 3D solid models suitable for finite element simulation using first pass engineering tools such as CosmosWorks.

References

Carreira, R. P., Q. H. Ly, et al. (2002). A bicycle frame Finite Element Analysis: standard tests and common cycling situations simulation. The 4th International Conference on the Engineering of Sport, Blackwell Science.

Ewart, P. D. (2002). High Performance Composite Materials for High Performance Sporting Equipment. Report 0777.512Y. Hamilton, University of Waikato.

Ewart, P. D. (2004). Predictive modelling of the flexural properties of composite materials. Report ENMP.590W. Hamilton, University of Waikato.

Ewart, P. D. and C. J. R. Verbeek (2005). Prediction of the Flexural Modulus of Composite Materials for Sporting Equipment. Asia-Pacific Conference on Sports Technology., Tokyo, Japan, Australian Sports Technology Alliance Pty Ltd.

Grozdanov, A. and G. Bogoeva-Gaceva (2002). "Correlations among micro- and macromechanical properties of composite materials based on the interphase concept." Composite Interfaces Vol. 10, (No. 2-3,): pp. 297–317.

Haboussi, D., D. Helene, et al. (2001). "Proposal of refined interface models and their application for free-edge effect." Composite Interfaces, Vol. 8, (No. 1,): pp. 93–107.

Higgins, R. A. (1995). Properties of Engineering Materials. London, Edward Arnold.

Holmes, G. A., E. Feresenbet, et al. (2003). "Using self-assembled monolayer technology to probe the mechanical response of the fiber interphase-matrix interphase interface." Composite Interfaces Vol. 10, (No. 6,): pp. 515–546.

Hong-Yuan, L. and M. Yiu-Wing (1998). "An appraisal of composite interface mechanics models and some challenging problems." Composite interfaces 6(4): 343-362.

Huffington, J. D. (1960). "Internal friction in fibre assemblies." British journal of Applied Physics VOL. 12: 99-102.

Lauke, B., T. Schuller, et al. (2003). "Determination of interface strength between two polymer materials by a new curved interface tensile test." Composite Interfaces Vol. 10 (No. 1): 1–15.

Lutz, M. P. and R. W. Zimmerman (1996). "Effect of the Interphase Zone on a Bulk Modulus of a Particulate Composite." Journal of Applied Mechanics. 63: 855-861.

Moore, D. R. and M. J. Ironman (1973). "The prediction of the flexural rigidity of sandwich foam mouldings." Journal of Cellular Plastics 10 (5).

Sharpe, L. H. (1971). "The Interphase of Adhesion." Journal of Adhesion 4: 51-64.

Throne, J. L. (1972). "A note on the mechanical strength of self-skinning foams." Journal of Cellular Plastics 8(4).

Wu, J.-S. and Y. Tsong-Ming (1994). "Studies on the Flexural Modulus of Structural Foams." Journal of Polymer Research (1): 61-68.

Sandvik Nanoflex® - Designed for Ultimate Performance

Carl Johan Irander and Göran Berglund

Sandvik Materials Technology, Sweden, johan.irander@sandvik.com

Abstract. Sandvik Nanoflex® is a revolutionary metal alloy developed by Sandvik Materials Technology, Sweden. Exploiting the latest advances in nanotechnology, Sandvik Nanoflex® offers an exceptional strength-to-weight ratio for the most demanding conditions in high-performance sports. By combining ductility with an extremely high tensile strength and excellent joining and forming properties, Sandvik Nanoflex® opens for very interesting new design solutions in sports equipment.

The paper outlines the characteristics of the material and describes cases of application within different fields of sports, each item optimized for low-weight and robustness. It is demonstrated that traditional light-weight materials, such as aluminum and titanium, are regularly outperformed by Sandvik Nanoflex®.

1 Introduction

Sandvik Nanoflex® is a new generation of precipitation hardenable stainless steel, originally developed to meet extreme mechanical demands for medical applications. Thanks to a unique combination of key properties the materials has proved to be very fit for use in high-performance sports equipment, such as bicycle components, climbing gear, racket sports, golf equipment, etc.

The excellent formability and ultra-high strength of Sandvik Nanoflex® allow for new design possibilities and improved performance of high-end sports applications.

Sandvik Nanoflex® is available as thin-wall tubing, wire and strip, manufactured with high precision tolerances.

2 Sandvik Nanoflex®

Sandvik Nanoflex® is a precipitation hardenable corrosion resistant steel with a basic chemical composition of 12%Cr-9%Ni-4%Mo-2%Cu.

Conventional wisdom in materials engineering says that extreme strength and toughness are contradictory. Sandvik Nanoflex® defies this rule by combining the attributes in a stainless steel. The good toughness ensures a safe use of critical or slender devices without risk of brittle fracture.

2.1 Mechanical Properties

The ultra-high strength of Sandvik Nanoflex® is the result of precipitation at heat treatment of extremely hard, nanometer-sized particles in a semi-hard and ductile metal matrix.

In the heat treated "aged" condition the material exhibits a tensile strength well in excess of 2000MPa, whilst retaining good ductility, which provides for an excellent strength-to-weight ratio that excels even compared to aluminum and titanium. The final ultra high strength level from cold worked condition is obtained with a very low amount of distortion, implying that calibration of shape is generally not required.

2.2 Formability

Sandvik Nanoflex® is manufactured in a various conditions, i.e., in the soft (solution annealed), the semi-hard (cold worked) or the hard (aged) condition. Each condition is tailor-made to match specific final product requirements. It has a remarkably good formability in the annealed condition, which enables advanced deep drawing and bending operations. Strain hardening at forming is used to reach the final hardness. Even in mildly cold worked condition, the increased ductility compared to other materials enables further forming operations, such as bending, flaring or flattening.

2.3 Joining

Sandvik Nanoflex® is easily weldable using common methods such as TIG, laser or micro-plasma welding, producing perfect, crack-free welds.

Mechanical joining methods such as riveting or adhesive joining are also suitable joining methods. To get a strong bonding in adhesion, the surface requires mechanical or chemical activation prior to assembly.

2.4 Surface Properties

Sandvik Nanoflex® offers better corrosion resistance in aqueous solutions than ordinary stainless steels of type AISI 304, and much superior resistance vs. hardenable chromium stainless steels of type AISI 410.

Thanks to its high hardness and corrosion resistance, even thin-wall sections of Nanoflex can effectively be glass/sand beaded or grounded to obtain a resistant satin look without layers of paint; highly suitable for active use in outdoor environments.

Where extra smooth surfaces and appearance are required, Sandvik Nanoflex® is suitable also for polishing by electrochemical or mechanical means. It stainless nature eliminates the need for hard-chromizing, thus offering a valuable environmental benefit. Sandvik Nanoflex® also lends itself to PVD coating, allowing, e.g., a low-friction or wear-resistant surface layer to be applied.

3 Light-Weight Materials – Comparison of Mechanical Properties

Low-alloy Cr-Mo steels, such as AISI 4130, in the as-delivered condition is reasonably formable. After hardening and tempering it will reach typically 1000-1200 MPa tensile strength. These steels are brittle at low temperatures if hardened to higher strength levels, but are still used in various high-performance applications, primarily where the wear resistance and rigidity - thanks to the high E-modulus of steel - are requested. The high density generally excludes Cr-Mo-steels from the toughest light-weight applications, and they cannot normally compete technically with aluminum, titanium and carbon fiber.

Aluminum alloys, such as the 6000 and 7000 series, are not particularly strong, but are still regarded as strong lightweight alloys due to their low density. Major drawbacks of aluminum are poor fatigue properties and low hardness, making equipment hazardous to use as its performance after long active use tends to fade.

Titanium is a useful material for strong and lightweight applications. Its major draw-back is high-cost, both in purchasing and in manufacturing. Furthermore, as titanium has a relatively low hardness, it is not suitable for applications subjected to heavy wear.

When designing rigid, yet lightweight sports items, it is important that the geometrical limitations and the material properties are combined in an optimized way. By comparing the specific E-modulus (E-modulus/density) and the specific Rm (tensile strength/density) for different materials, their suitability for use v.s. different design criteria, such as minimum weight for a given stiffness or strength, can easily be assessed. From Table 1, it appears that Sandvik Nanoflex® will provide high strength for a given load with a high degree of rigidity.

Table 1 Material Data

	Aluminium 7075 T6	Titanium Gr 9	Cro-Mo Steel	C455	Stainless Steel 316L	Sandvik Nanoflex®
E-modulus [GPa]	70	108	200	203	201	205
Density [kg/m^3]	2700	4540	7800	7840	7800	7750
Poinssions ratio [-]	0.3	0.3	0.3	0.3	0.3	0.3
Yield strength [MPa]	528	850	1000	1350	210	1800
Tensile strenght [MPa]	581	950	1280	1500	690	2000
E-modul / Densitet	0.026	0.024	0.026	0.026	0.026	0.026
Density / Yield Strength	0.196	0.187	0.128	0.172	0.027	0.232
Density / Tensile Strength	0.215	0.209	0.164	0.191	0.088	0.258
Yield Strenght / E-modulus	7.54	7.87	5.00	6.65	1.04	8.78
Tensile Strenght / E-modulus	8.30	8.80	6.40	7.39	3.43	9.76

4 Designing with Different Materials

In order to demonstrate the design advantages of Sandvik Nanoflex® for use in sports equipment, a comparison with Cr-Mo steels (type AISI 4130), high strength aluminum (type 7075 T6) and titanium (Grade 9) is made for a basic design case, i.e., a 3-point bending test, with a load of 900N applied on the middle of a tube of a specific outer diameter of 20 [mm].

The results are shown in Table 2. It appears that when designing for a given loading capacity by alternating the wall thickness and material, the considerably higher E-modulus vs. density of Sandvik Nanoflex® makes it the lowest-weight solution, reduced by 17-57% compared to the other materials. To compensate for the 4% lower rigidity compared to aluminum, a minor increase in outer diameter can be made; still making Sandvik Nanoflex® the lightest solution.

Table 2 Stiffness and Weight Comparison, Load on Tube of 900N

	OD [mm]:	WT [mm]:	Stiffness E*I [kNmm2]		Weight [g]	
Aluminium 7075	20	2.05	344	4%	312	17%
Titan Gr5	20	1.35	360	8%	356	33%
AISI 4130	20	0.90	506	53%	424	59%
Sandvik Nanoflex®	20	0.56	331	0%	267	0%

5 Applications

Lightweight, stiffness, strength, toughness – these are all features demanded by today's sports and outdoor consumers for even more active usage.

Below, some examples are shown of use of Sandvik Nanoflex® in the application and fabrication of a number of sports equipments.

5.1 Example 1 - Bicycle Parts

A low weight and a comfortable ride are two primary desired features of advanced bicycle saddles. The flexibility of the rails is of great importance as the saddles are subjected to high fatigue loads. Furthermore, the rail shall contribute to an energy absorbing spring effect, thus providing the comfortable ride. Compared to current materials such as low-alloy steel and titanium, Sandvik Nanoflex® provides a weight reduction for the rail of 28% and 7%. An example of a saddle made by the Italian company Selle San Marco fitted with Sandvik Nanoflex® rail is shown in Figure 1.

Fig. 1 Bicycle Saddle with Sandiv Nanoflex® rail

Other examples of bicycle applications highly suitable for Sandvik Nanoflex® are, for example, frame tubing, seat posts, handle bars and break/gear shift wires.

5.2 Example 2 - Climbing Gear

The Italian manufacturer of mountaineering equipment, CAMP, has radically improved the performance of crampons and ice-axes by utilizing the ultra-high strength of Sandvik Nanoflex®, see Figure 2. Low-alloy steels, such as 39NiCroMo3, are quite formable in the delivery condition, but cannot be hardened to more than 40 à 42 HRc without loosing too much ductility at low temperature.

The considerably higher strength and ductility of Sandvik Nanoflex® allows application at above 50 HRc hardness, which will reduce the wall thickness of the crampon tips from 2.75 to 1.8 mm. This provides a weight reduction of as much as

Fig. 1 Camp Nanotech Vector Crampon with Sandvik Nanoflex®

18% in total, corresponding to almost 200 grams gain per pair of crampon. Thanks to the thinner and stronger tips, the grip and ease of penetration into the ice is also improved, as is the wear resistance. Furthermore, the corrosion resistance of Sandvik Nanoflex® eliminates rust formation and thus the drawback of requiring wiping off the crampons after use.

5.3 Example 3 – Thin Wall Tubing and Shafts

Designing lightweight, yet rigid products with demands for carry high loads, the properties of Sandvik Nanoflex® comes into good use, often enabling a considerable weight reduction. Ideal applications are for example, hockey sticks, tent poles and back packs frames and trekking poles. The high hardness, corrosion resistant properties and high strength ensures both weight reduction and an increased safety margin, even after considerable usage in rough environments.

6 Summary

Sandvik Nanoflex® exhibits an ultra-high tensile strength exceeding 2000MPa in thin-wall tubing, strip and wire. This remarkable strength is attributed to a semi-hard matrix together with the precipitation of nanoparticles at heat treatment. A good ductility and toughness is retained in the heat-treated condition.

Combined with a high E-modulus and a good corrosion resistance, these properties are ideal for application in lightweight, robust high performance sports equipment.

Acknowledgement

This paper is published with the permission of Sandvik Materials Technology. The assistance from Selle San Marco and CAMP is gratefully acknowledged.

List of Literature

1. Anna Hultin Stigenberg & J-O Nilsson: Corrosion Resistant Steel for Use in Medical and Dental Equipment, Sandvik Publication S-40-11-ENG, March 1992.
2. Holmqvist M, Nilsson J-O, Stigenberg A H.: Isothermal Formation of Martensite in a 12Cr-Ni-4Mo Stainless Steel, Scripta Metallurgica *et* Materialia, Vol. 33, No. 9, pp 1367-1373, 1995.
3. US Patent # 5,512,237
4 Sandvik Steel Corrosion Handbook Stainless Steel, 1999.
www.smt.sandvik.com/nanoflex

Manufacturing of Bobsled Runners

Michael F. Zaeh[1] and Paul Gebhard[2]

[1] TU Munich, Institute for Machine Tools and Industrial Management,
Michael.Zaeh@iwb.tum.de
[2] TU Munich, Institute for Machine Tools and Industrial Management,
Paul.Gebhard@iwb.tum.de

Abstract: InnoBay is a project sponsored by the Bayerische Forschungsstiftung (a none-profit sponsor for research and development). Goal is the invention of a new bobsled. Involved in the project team are faculties of mechanical engineering and sport science of the Technische Universitaet Muenchen, industrial partners and the Bavarian Bobsled Association. The partners analyse different topics like the interaction between athletes and sports equipment, the aero dynamical behaviour of crew and bobsled as well as problems concerning production. The Institute of Machine Tools and Industrial Management (*iwb*) is responsible for the manufacturing of the runners. They are important parts of the bobsled, because they significantly determine the runtime. Due to their slender and complex geometry they are difficult to manufacture. During the project various criteria like the choice and order of the production processes or necessary clamping systems will be evaluated regarding their influence on the runtime. In combination with the results of other subprojects, dealing with material selection and geometry optimisation, new bobsled runners will be designed, manufactured and tested. This paper highlights the planned activities and presents first results.

1 Bobsled as a complex mechanical system

The difference between victory and defeat in bobsled races is a matter of several thousandths of a second. The biggest influence on runtime apart from the driver and the pushers has the bobsled and it is therefore of vital importance for the final ranking of the team.

On first sight, a bobsled appears to be a relatively simple mechanical racing device. On closer examination it becomes evident, that this is not the case and that a bobsled is a complex mechanical system that influences the resulting runtime by the interaction of many physical principles. These principles include the aerodynamic resistance of the bobsled, the vibration behavior of the whole structure, as well as issues concerning the friction of the runners on the ice of the race track. Because of the multidisciplinary nature of such mechanical systems, the project "InnoBay" was created. It is sponsored by the Bayerische Forschungsstiftung (a none-profit sponsor for research and development) and has the goal of inventing innovative methods and technologies to solve challenges in different fields of engineering, sports and material science. Because of its complexity, the development of a new bobsled is used to validate the research results of that project. Involved in the project team are faculties of mechanical engineering and sport science of the Technische Universtitaet

Muenchen, a chair of the Friedrich Alexander Universitaet Erlangen-Nuernberg as well as industrial partners and the Bavarian Bobsled Association.

2 Challenges regarding the manufacturing of bobsled runners

One of the significantly decisive parts of the bobsled concerning runtime are the runners. They determine the friction on the race track and therefore influence the speed of the bobsled. Optimal friction behavior is achieved by advantageous combination of material, surface finish and geometry of the runners. However, due to their slender and complex geometry, runners are very difficult to manufacture and their production is accompanied by the generation of residual stresses and in consequence dimensional deviations. These deviations have a great impact on the performance of the bobsled. For this reason, the Institute of Machine Tools and Industrial Management (*iwb*) together with the Chair of medical Engineering is working on the development and testing of concepts to minimize dimensional deviations generated during the manufacturing process of a bobsled runner.

3 Influence of the manufacturing process on dimensional deviation

Dimensional deviations can occur in all steps of a manufacturing process. Figure 1 shows the procedure to manufacture bobsled runners and presents the sources of stresses and dimensional deviations that can occur during the different steps (Hainzlmaier 2005).

As apparent in Fig. 1 residual stresses already arise during processes that precede the actual manufacturing process. They originate from molding processes like extruding, casting, rolling or, as shown in the presented application, forging (Springer 1991).

Another source for possible dimensional deviations is the structure of the applied machine tools. Their elastic deformation and vibration behavior during the production process significantly influence the accuracy of the produced parts. These topics have already been object for intensive research (Minges 1993, Feinauer 1998).

Next to the machine tools the stock removal process itself can influence the product's quality. It creates plastic deformation in a thin surface layer of a part which results in residual stress. Despite the low thickness of this surface layer, these stresses are effective throughout the whole part (Kiethe 1973) and can cause deviations in the desired geometry.

Yet another aspect significantly influencing the dimensional accuracy is the clamping of the parts in the machine. This is especially relevant for long and thin walled parts like bobsled runners. Because of their geometry these parts have relatively low section modules and therefore tend to bend under the forces that are applied on the part by for example the stock removal process (Kaufeld 1987). Because of that, appropriate clamping is important to ensure the necessary support. The disadvantages of a robust clamping strategy are the dimensional deviations that can occur. They either originate from deviations in the clamped parts or the clamping mechanism itself. Especially forged and cast parts cause problems during clamping.

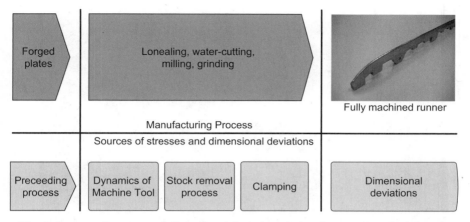

Fully machined runner

Manufacturing Process

Sources of stresses and dimensional deviations

Fig. 1: Sources of stresses and dimensional deviations during the manufacturing of runners

Due to their inexact geometrical shape, these parts are elastically deformed by their fixed support. Despite machined in a theoretically exact clamping mechanism, the parts have considerable dimensional deviations because of elastic rebound after removing the part. The same effect can be observed if an exact part is deformed by an inexact clamping mechanism.

4 Objectives of the *iwb*

The aim of the *iwb* is to analyze the influence of the preceding process, the stock removal process itself and the clamping mechanism on the amount of dimensional deviation of long and thin walled parts. For this purpose an experimental setup will be created that allows to investigate the different effects separately. Strategies and concepts to minimize these deviations will be developed and applied on a bobsled runner (Fig. 2) to demonstrate their utilizability.

5 Approach to minimize dimensional deviations

The first step in the approach to minimize dimensional deviations is the choice of suitable material, tools and sample geometry. Concerning costs and effort the shape of the samples will not be that of a whole bobsled runner, but a simpler slender part showing at least its design features.

For that sample geometry different low stress clamping mechanisms will be designed and constructed. For example, innovative techniques like vacuum or casting clamping systems will be analyzed regarding their applicability. To evaluate these mechanisms milling experiments will be conducted and the machined parts will be measured. Based on these experiments, a mechanism will be chosen that minimizes the effect of the clamping. Thereby, it is possible to evaluate the effects of stock

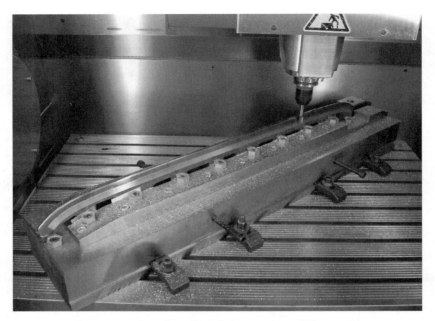

Fig. 2: Manufacturing of bobsled runner in a milling machine

removal process and preceding process without mixing the results with the effects of clamping.

The stock removal process will be evaluated by measuring the residual stress that is generated by different milling strategies. These tests cover different milling parameters as well as different tool paths. Measurement of residual stress will also be performed to find out how preceding processes influence the accuracy of the parts.

Based on these results a general strategy will be developed to minimize deformation of long, slender parts that arise in the manufacturing process. For demonstration purposes, the strategy will be applied on runners for a four-man bobsled.

References

Feinauer, A. (1998) *Dynamische Maschineneinflüsse auf die Werkstückqualitaet beim Hochgeschwindigkeitsfraesen.* Diss. Medien Verlag Köhler, Tübingen.

Hainzlmaier, C. (2005) *A new tribologically optimized bobsleigh runner.* Diss. Technische Universitaet München, Munich.

Kaufeld, M. (1987) *Hochgeschwindigkeitsfraesen und Fertigungsgenauigkeit dünnwandiger Werkstücke aus Leichtmetallguß.* Diss. Hanser, München Wien.

Kiethe H. (1973) *Oberflaechengestalt und Eigenspannungsausbildung beim Walzfraesen von Flachproben aus Ck45.* Diss. Universitaet Karlsruhe (TH), Karlsruhe.

Minges, R. (1993) *Verbesserung der Genauigkeit beim Fünfachsigen Fraesen von Freiformflaechen.* Diss. Universitaet Karlsruhe (TH), Karlsruhe.

Springer, H.-J. (1991) *Abbau von Eigenspannungen in Gesenkschmiedestücken durch spezielle Waermebehandlungen.* Diss. VDI Verlag, Düsseldorf.

Carbon Fiber Reinforced Plastics – Trendsetting Material for High Performance Racing Bike Chassis

Michael Kaiser, Norbert Himmel

Institut fuer Verbundwerkstoffe GmbH, michael.kaiser@ivw.uni-kl.de

Abstract. The sports and leisure area is an extensive field of application for fiber reinforced plastics (FRP). The remarkable material and manufacturing properties like high stiffness and strength at low density, corrosion and chemical resistance as well as the possibility to realize complex geometrical shapes exceed the possibilities of conventional structural materials, like metals. While a few decades ago steel was substituted by aluminum as chassis material for racing bikes primarily due to the considerably lower density, nowadays, FRP, and therein especially carbon fiber reinforced plastics (CFRP), appear to be the trend setting material nowadays. The application of FRP in load bearing structures, such as bike chassis, is more complex and demands a very close interaction between all disciplines involved in the development process such as material science, design, manufacturing and testing. This paper intends to highlight the potential of CFRP for ultra-light chassis applications in an adequate structural design by considering the Canyon Carbon Ultimate F10 racing bike chassis as one of today's lightest and stiffest racing bike chassis worldwide. After a brief introduction into FRP and CFRP a comparison of the anisotropic CFRP with conventional isotropic metals is shown. An overview of the requirements for road and racing bike chassis is given and results of finite element design analyses are presented to show the influence of the lay-up configuration on the frame stiffness. Finally, a comparison with the leading aluminum and CFRP racing bike chassis shows an outstanding lightweight quality of the developed frame which proves the ultra-light weight potential of CFRP in the racing bike chassis sector.

1 Introduction

Beside of the aviation and automotive industry the sports and leisure area is an extensive field of application for fiber reinforced plastics (FRP). The remarkable material and manufacturing properties like high stiffness and strength at low density, the high fatigue life, the good corrosion and chemical resistance as well as the possibility to easily realize complex geometrical shapes exceed the possibilities of conventional structural materials, like metals. While a few decades ago steel was substituted by aluminum as chassis (frame and fork) material for racing bikes primarily due to the considerably lower density, nowadays FRP and especially carbon fiber reinforced plastics (CFRP) appear to be the trend setting material due to their extraordinary high stiffness. The annual growth rate is immense and amounts to more than 200,000 CFRP frames and 1 million forks in the year 2005. Beside of the racing bike chassis, a multitude of other components are made of CFRP, e. g., seat posts, handle bars, rims and cranks.

2 Carbon Fiber Reinforced Plastics

FRP consist of fibers embedded in a matrix whereas each component takes certain functions. The fibers carry the load due to their high strength and stiffness. The matrix protects the fibers, provides load transfer and stabilization against buckling under compression loads parallel to the fiber direction. High-performance applications mostly require the use of thermosetting matrices (e. g. epoxy) reinforced by continuous fibers. The majority of racing bike chassis is manufactured from prepregs. Prepregs are produced from continuous fibers with unidirectional or woven fiber architectures which are impregnated with a matrix resin resulting in a drapable and tacky semi-finished product. One sheet of a prepreg represents a single layer, and a laminate is produced by stacking several layers usually with different fiber orientations onto each other.

In general, the mechanical behavior of a laminate is anisotropic and, therefore, differs considerably from conventional isotropic engineering materials. The mechanical properties of a laminate mainly depend on the mechanical properties of the fiber and the polymer matrix, the fiber volume content and the stacking sequence (lay-up). Table 1 includes a comparison of the tensile modulus E and E_x as well as the shear modulus G and G_{yx} of metals and CFRP with different lay-up configurations (L1-L4). The last two columns quantify the light-weight quality of the material in terms of the modulus to density ratio. The best lightweight quality for purely axial loading (e. g. for a tension rod) can be achieved with a $[0°]$-laminate or for pure torsion loading (e. g. a cylindrical torsion tube) with a $[+45°/-45°]_s$-laminate. A multi-axially loaded laminate requires a quasi-isotropic lay-up configuration, e. g a $[+45°/90°/-45°/0°]_s$ configuration resulting in a decrease of the modulus-to-stiffness properties compared to metals. The light-weight benefit using CFRP is higher the more anisotropic the laminate is.

Table 1: Comparison of mechanical properties of metals and CFRP with different lay-up configurations (x-direction equivalent to $0°$ direction)

Material	Specification	Density ρ	Tensile modulus E, E_x	Shear modulus G, G_{yx}	$(E, E_x) / \rho$	$(G, G_{yx}) / \rho$
		[g/cm³]	[GPa]	[GPa]	[GPa/(g/cm³)]	[GPa/(g/cm³)]
Steel	25CrMo4	7.8	205	79	26	10
Titan	TiAl6V4	4.5	110	42	24	9
Aluminum	AlMg1SiCu	2.8	69	27	25	9
CFRP-L1*	$[0°]$	1.5	154	8	101	5
CFRP-L2*	$[+45°/-45°]_s$	1.5	11	39	7	25
CFRP-L3*	$[+45°/0°/-45°/0°]_s$	1.5	84	22	55	14
CFRP-L4*	$[+45°/90°/-45°/0°]_s$	1.5	56	21	37	14
	* intermediate modulus carbon fiber/epoxy resin, 60 % fibers by volume					

3 Design and Manufacturing

The miscellaneous requirements of racing bikes include the durability, driving behavior, weight, design, ergonomics, aerodynamics and comfort as well as regulations from biking associations like the Union Cycliste International (UCI) which demand the diamond frame geometry. The mechanical loadings include pedaling, braking and seat loadings. In extreme situations the chassis has to withstand jumps, vibrations and impacts from serious road unevenness or frontal crashes. The driving stability can be improved by increasing the lateral stiffness of both frame and fork. Important performance characteristics to quantify the driving stability are the torsional stiffness around the longitudinal frame axis (so called HT stiffness C_{HT}) as well as the side stiffness C_{ST} of the fork. Furthermore, the BB stiffness C_{BB} rates the energy loss caused by the elastic lateral frame deformation under pedal loading.

For the development of the Canyon Carbon Ultimate F10 racing bike chassis (Canyon F10) parameter studies were carried using the finite element method to optimize the frame geometry, the tube cross sections as well as the lay-up configurations to provide maximum stiffness and strength at minimum weight (Kaiser, Arnold, Himmel 2005). Particular design features are the innovative unsymmetrical seat tube of the frame which increases the bottom bracket stiffness compared to standard cylindrical tubes and the steer tube of the fork with a trumpet-like shape to align the part geometry accurately to the force flow which results in an increase of the flexural stiffness and a considerable reduction of the material effort.

Figure 1 shows a result of a parameter study which investigated the effect of the stacking sequence of the down tube, the seat tube and the seat stay on C_{BB}. As the loading of the tubes by axial forces, bending and torsional moments requires the use of 0° and ±45° layers, the starting configuration of the laminate of each tube was 50 % 0° and 50 % ±45° layers.

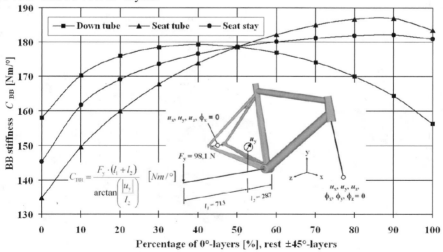

Fig. 1. Influence of down tube, seat tube and seat stay lay-up configuration on the BB stiffness C_{BB}. Starting configuration of all tubes 50 % 0° and 50% ±45° layers

C_{BB} was determined relative to the percentage of the 0° layers. With 40 % 0° layers in the down tube the maximum C_{BB} can be achieved whereas the optimal configuration of the seat tube and the seat stay are 90 % 0° layers. Selecting the optimal configuration for all frame tubes a maximum C_{BB} of 192 Nm/° can be achieved theoretically. Considering the poorest lay-up configuration for each tube a minimum C_{BB} of 86 Nm/° was calculated which is less than 50 % of the maximum stiffness. As C_{BB} is only one requirement of a racing bike chassis other structural requirements demand different lay-up configurations and, at the end, structural compromises. Adjusting the tube wall thicknesses by the density ratio of CFRP to Al, applying the material properties of Aluminum (Al) from Tab. 1 and using the same outer frame shapes as well as an equivalent frame weight results in a C_{BB} of 127 Nm/° of a comparable Al chaiss.

CFRP racing bike chassis structures are mainly produced from prepreg laminates which are laid up in an open mold. After closing and heating the mold a bladder which is arranged inside the laminate is blown up by air, hence pressing the laminate against the inner mold surface. After curing the resin the part is removed from the mold. This technique is rather attractive because of the achievable outer surface quality of the part and the possibility to change the wall thickness without modifying the mold.

The Canyon F10 frame is built in differential design, where pre-manufactured tubes are fixed in a bonding jig, wrapped with prepreg material in the joining areas and cured in a second bonding step. The advantage of the differential design compared to integral design is that a relatively small number of molds are necessary to produce the tube set. The tubes can be produced in excess length and cut to length as required by the specific frame size. Furthermore, the precise positioning of the pre-pregs is easier in straight structures. Finally, the intermediate production step allows an additional quality control. Disadvantages of this procedure are the increased production time and the poor surface quality in the tube-to-tube joining area. Figure 2 shows the frame tube set of the Canyon F10 frame and the final racing bike.

Fig. 2. Canyon Carbon Ultimate F10 frame tube set and racing bike chassis

4 Comparison of the leading racing bike chassis

Figure 3 shows a comparison of the leading racing bike chassis with CFRP and Aluminum frames. All racing bike chassis were equipped with full CFRP forks or forks with CFRP blades and an Al steer tube. The frame weight refers to frame size 57, the fork weight is measured with an equivalent steer tube length to ensure comparability. The weight measurement was always carried out with the same attachment parts. The upper diagrams show the C_{BB} and C_{HT} stiffnesses as a function of the frame and fork weight. The average chassis weight is 1,584 g for CFRP and 1,824 g for the Al chassis at a C_{BB} average of approximately 101 Nm/° and 98 Nm/°, respectively. In contrast to this the average C_{HT} of the Al frames is higher compared to the CFRP frames (83 Nm/° against 74 Nm/°). Generally, a larger scatter of the CFRP chassis data can be observed. In this comparison, the Canyon F10 chassis achieves a C_{BB} of 129 Nm/° and a C_{HT} of 100 Nm/° at a total weight of 1,329 g.

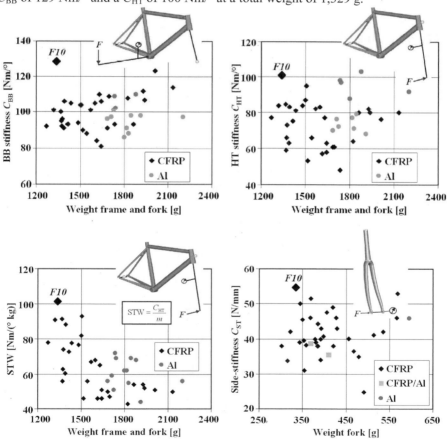

Fig. 3. Comparison of Canyon Carbon Ultimate F10 with leading Aluminum and CFRP racing bike chassis (Musch 2004; Zedler 2003; Zedler 2004; Zedler 2005)

The quantification of the lightweight quality using the stiffness-to-weight coefficient (STW) specified by the C_{HT} to frame weight ratio is more descriptive. The averaged STW value of 64 Nm/(° kg) of the CFRP frames is only a little bit higher than that of Al frames (60 Nm/(° kg)). Although the scatter of the CFRP frame data is rather high again, it can be seen that the best frames in terms of an STW of higher than 75 Nm/(° kg) are all made of CFRP. In this category the Canyon F10 frame achieves the highest STW of over 100 Nm/(° kg). The fork comparison shows the superiority of CFRP as only one of the tested forks was made of Al. Here, in the high-end racing bike sector, the weight saving benefit of CFRP replaced Aluminum almost completely. Fork weights of around 300 g are possible with a high lateral stiffness compared to 566 g weight of the shown pure Al fork.

Finally, extensive test rig experiments and riding tests on the Canyon F10 successfully proved that ultra-light weight CFRP design also can be consistent with durability requirements.

5 Conclusions

CFRP offer the potential to produce excellent light-weight structures due to their specific material properties. Considerable improvements in stiffness and strength are possible compared to metallic materials if the anisotropic behavior can be utilized and the design and the manufacturing technology are guided by the requirements associated with this material. Due to the loading conditions in the tubes which require highly anisotropic laminates racing bike chassis are appropriate structures for the application of CFRP. The capability of CFRP was impressively demonstrated in the Canyon F10 which is one of the lightest and stiffest racing bike chassis worldwide today. With an STW value of more than 100 Nm/(° kg) which is almost 40 % higher compared to the best Al frame a new world-class record could be achieved. Especially in the high-end racing bike chassis sector, it can be expected that Al will continuously be substituted by CFRP.

Acknowledgement

The project is financially supported by the Stiftung Rheinland-Pfalz für Innovation, Germany, and Canyon Bicycles GmbH, Germany. Special thanks are also due to Topkey, Taiwan, for the very good and successful cooperation.

References

Kaiser, M., Arnold, R. und Himmel, N.(2005) *Entwicklung eines Hochleistungs-CFK-Rennrad-Chassis*. 11. Nationales Symposium der SAMPE Deutschland e. V., Darmstadt.
Musch, T. (2004) *Carbon-Rahmen*. TOUR, 01/2004, pp. 14 – 29.
Zedler, D. (2003) *Carbongabel-Test*, TOUR, 12/2003, pp. 30 – 35.
Zedler, D. (2004) *Alu-Rahmen*, TOUR, 02/2004, pp. 16 – 25.
Zedler, D. (2005) *Die neue Leichtigkeit*, TOUR, 02/2005, pp. 14 – 26.

Development of Equipment to Compare Novel Ice Skate Blade Materials

Nick Hamilton[1] and Terry Senior[1]

[1] Sports Engineering, Centre for Sports and Exercise Science. Sheffield Hallam University, N.Hamilton@shu.ac.uk

Abstract. Developments in the manufacture of ice hockey skate blades have allowed novel materials to be used in their production. Qualitative tests by players showed the benefits of the new blades but to quantify the performance gain a system capable of measuring small differences in their coefficients of friction was required. Experimental equipment was developed that enabled measurements of real skate blade samples to be tested in game play conditions. A set of sample results are given to show the application of the equipment.

1 Introduction

Ice skates have come a long way since man first strapped the leg bones of large animals to the soles of his feet. Used for transport, entertainment and competition the first metal blades were introduced in the 14th Century using iron runners. Blades are now manufactured from carbon and stainless steels and hollow ground to produce the twin edge skate seen today.

An innovative application of an existing manufacturing technique has been identified and applied to the manufacture of ice hockey blades. Chiefly laser cut from flat steel stock they can now be manufactured from alternative and novel materials with the potential for performance enhancement. Ice hockey players can reach speeds of up to 14 ms^{-1} in a game with continual changes of direction which necessitates both minimal sliding resistance to maximise acceleration and a constantly sharp edge to maximise deceleration and turning.

The resistive forces experienced by an ice hockey player of mass, m, consist of aerodynamic losses, F_{aero}, and frictional forces where the blades contact the ice, F_{ice}. The ratio of the frictional force, F_{ice}, and the players resultant normal reaction force, N, is equal to the coefficient of friction (CoF), μ. Where $\mu = F_{ice} / N$. The CoF is kept low due to the mechanisms of sliding involved in skating on ice and the presence of a thin film of water acting as a lubricant between blade and ice (Bowden and Hughes 1939; Bowden 1953).

There has been much debate about the mechanism of release of the water film, but it is generally accepted that the frictional heating of the ice surface by the moving blade melts some of the ice. This generates the film and has been experimentally demonstrated (Bowden et al 1939; Bowden 1953). An alternative mechanism was initially suggested where the melting is due to applied pressure but has generally been dismissed at temperatures below 0°C.

The aim of this study was to develop a test apparatus to measure the CoF between ice and skate blades.

2 Literature Review

Qualitative testing of the new blade materials by ice hockey players revealed a significant and noticeable performance enhancement. In a blind test, players reported a perceptible increase in the attainable speed, acceleration and cornering ability when using the new blades. Comments were also made about the longevity of the enhancements and the reduction in the need for resharpening, both very beneficial in competition.

To quantify these improvements an experimental apparatus was required to investigate and measure the CoF for blades of different materials. Prior literature reveals several measurement methods.

2.1 Turntable

A temperature controlled horizontal copper turntable in a vacuum chamber had a thin film of ice frozen on to the upper surface (Bowden and Hughes 1939). Rotating at a regulated speed, driven by an induction motor, a sled of the sample material was mounted on an arm and placed in contact with the upper ice surface. The sled was set up to enable controlled movement in radial, circumferential and vertical directions. Circumferential motion was restrained by a flat spring, the deflection of which gave a measure of the frictional force, F_{ice}. The normal force, N, was supplied by means of weights applied to the sled arm and a radial motion was applied to the sled to ensure a virgin contact.

The results of the study demonstrated no discernable difference between experiments performed in normal and evacuated states showing no influence from contaminants in the air. The experiments were performed at relativity low speeds, 4 ms^{-1}, and at low loads, up to 10N, although a large variety of materials were tested by this method. Only flat and cylindrical samples were tested, rather than an ice hockey blade with a twin edge, but differences in the CoF between materials were found.

2.2 Pendulum

A cylinder of ice was mounted horizontally and revolved at a constant speed along its longitudinal axis within a temperature controlled environment (Evans, Nye and Cheeseman 1976). A pendulum arrangement mounted over the cylinder held two cylindrical samples and an applied load, the position of which was used to adjust the sensitivity of the instrument. Once the cylinder surface was smoothed a 90° V-groove was cut to act as a guide for the samples to run in. In motion the frictional force acting on the samples caused a small rotation of the pendulum, the magnitude of which could be measured via an optical lever and related to CoF. Two methods of damping unwanted oscillations were applied to the pendulum; rotational oscillations were damped via a dashpot and the samples themselves were held in rubber mountings.

Experiments were also made to simulate actual hollow ground ice skate geometry. A curved blade was manufactured and matched to an ice cylinder size that reproduced the interaction of the rocker of a real skate on flat ice.

The experiments were performed at much higher speeds, $10ms^{-1}$, very similar to real playing conditions, at loads of 45N. This technique is preferable as it allows simulated ice skate geometry and a large variety of materials; Copper, Perspex and different grades of steel, to be tested at higher speeds.

2.3 Instrumented Skate

An instrumented skate system was developed and used to measure instantaneous normal and ice frictional forces during actual speed skating conditions (Koning, Groot et al 1992). A measurement system, consisting of three bespoke strain gauge elements, was mounted between the shoe and blade of a speed skate. The arrangement was designed to minimise cross talk between the large normal force and the relatively small frictional forces. The output from all strain gauges was amplified, filtered and logged to a portable data acquisition system for later CoF analysis.

As tests are performed in race like conditions this is the optimum way to measure the CoF. However, the development time and cost of such a system are prohibitive in this case.

2.4 Sled

Two speed skate blades were mounted parallel and held rigid relative to one another in a frame also carrying a camera system and a light projector (Kobayahsi 1973). The system was loaded to 500N and propelled by a catapult system along the ice. The catapult used a mass dropped from a known height to provide the same initial speed to the sled on each run. A reflective strip was frozen below the surface of the ice at 1m intervals which was recorded by the camera as the sled passed over it. This recorded the total distance travelled and, through analysis of the time intervals, recorded changes in velocity and therefore deceleration. Aerodynamic losses were neglected.

These experiments were carried out at slow speeds, with an initial velocity of $1.4ms^{-1}$, but advantageously used real ice conditions and actual blade geometry. The method of recording the displacement required specialist preparation of the ice surface but allowed CoF measurements to a precision of 0.0001.

2.5 Velocity and Temperature

The CoF for ice skates has been shown to be dependant on both velocity and ice temperature (Bowden et al 1939; Bowden 1953; Kobayashi 1973; Evans et al 1975; Koning et al 1992). Where CoF was measured with actual ice blades a decrease was found with decreasing temperature down to -8°C (Koning et al 1992). A similar relationship was found for velocity where CoF increased linearly with velocity. Therefore measurements had to be undertaken in a temperature controlled environment over a range of velocities.

2.6 Coefficient of Friction

A wide range of CoFs have been reported for ice skates. For speed skating, where blades are approximately 1mm thick and perfectly flat, CoF ranges from 0.004 – 0.006 (Koning et al 1992) and 0.003 – 0.010 (Kobayashi 1973) were stated. For a wider blade, 5mm thick, with a radius of curvature, a range from 0.010 to 0.056 was found. These variances are due to the range of velocities and temperatures tested.

3 Experimental Design

To enable the testing of skate blades in realistic ice conditions, a modified version of the sled design was developed.

3.1 Sled Design

Unlike speed skates, ice hockey blades are not perfectly flat, therefore it was necessary to mount 4 blades in the sled for stability. Blades of each test material were manufactured to the same standard profile and hollow ground to the same radius. The blades were held parallel to each other and level by the main aluminium plate and fixed in position by two outer strips, see Fig. 1. Reference pins were integrated to ensure blades could be set in the same position each time and mountings were incorporated for the masses and timing equipment.

Fig. 1. Four blade Sled design

3.2 Measurement and Timing

The experiments were conducted at a commercial ice rink, 60m long, which meant that the ice surface could not be altered in the manner previously described. Therefore a less invasive timing system was required. It was suggested that a series of light gates could be constructed which would time the path of the sled accurately enough to measure the CoF up to playing velocities, approximately 10 ms^{-1}.

Due to the bright lighting conditions within the rink a light gate system based on 630-680nm laser diodes was constructed with a fast acting phototransistor used as the basis of the sensing element. The sensitivity of the phototransistor was matched to the frequency of the laser and a wide half angle selected to minimise alignment issues. The rest of the circuit was constructed to produce a 5V TTL pulse each time the laser was interrupted and sampled by a National Instruments USB-6009 running specifically written software on a PC laptop. The sensing element was tuned to be able to time the breaking and restoring of the beam to ±10µs.

A frame was constructed to accept a laser diode and sensing circuit allowing the laser to be aligned accurately to the phototransistor. Multiple light gates were manufactured and positioned along the path of the sled at known intervals on the ice, typically 11m apart. The output from each light gate was linked together and connected to the sampling equipment.

To measure velocity at locations along the path of the sled an accurately machined plate 1m long was mounted on the sled. As the sled moved along its path the plate broke the beams of the light gates, this time interval was used to calculate the velocity of the sled at each location and the change in velocity over the path of the sled. If the CoF:

$$\mu = \frac{F_{ice}}{N} = \frac{ma}{mg} = \frac{a}{g} \tag{1}$$

where, a, is the deceleration of the sled mass, m, measured by the change in the velocity, Δv, over time Δt. The velocity of the sled was plotted against time and the gradient of the resulting line gave the deceleration and therefore the CoF. To minimise random error an average CoF value was measured over a series of experiments using the same initial velocity.

3.3 Propulsion

To give the sled an initial velocity a propulsion system was developed using a pneumatic ram. Rigidly braced against the side of the rink a high pressure air cylinder fed a regulated air supply to a pneumatic ram which acted centrally on the sled. A guide rail arrangement attached to the main frame was used to ensure the sled did not drift under application of the force. A range of initial velocities was used from 1 to 10 ms^{-1} with the temperature of the ice recorded with an infrared thermometer each time to ensure continuity.

4 Sample Results

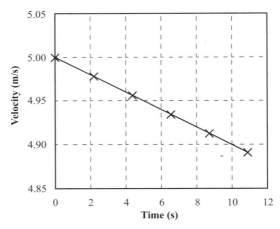

Fig. 2. Sample velocity results for skate blade material B at -4°C

With the equipment developed it was possible to make measurements of blades made from several different materials and compare their CoF. Figure 2 shows an example of these results for a blade, velocity and temperature combination resulting in a deceleration of 9.8×10^{-3} ms^{-2}. Using Eq.1 this equates to a CoF of 0.001.

5 Conclusions

A system has been developed that enables the measurement of the CoF of ice skate blades of different materials and the improvement in their performance.

References

Bowden, F.P. and Hughes, T.P. (1939) The mechanism of sliding on ice and snow. Proceedings of the Royal Society of London, A. 179, 280-298.

Bowden, F.P. (1953) Friction on snow and ice. Proceedings of the Royal Society, A. 217, 462-478.

Evans, D.C.B., Nye, J.F. and Cheeseman, K.J. (1976) The kinetic friction of ice. Proceedings of the Royal Society of London, A. 347, 493-512.

Kobayashi, T. (1973) Studies of the properties of ice in speed-skating rinks. ASHRAE Journal. 73, 51-56.

Koning, de J. J., Groot, de G. and Jan Van Ingen Schenau, G. (1992) Ice friction during speed skating. Journal of Biomechanics. 25(6) 565-571.

4 Safety

Synopsis of Current Developments: Safety

Jasper E. Shealy, PhD

Department of Industrial and Systems Engineering, Rochester Institute of Technology, Rochester, New York, USA, jeseie@rit.edu

Research and Development Issues in the field of Safety in Sports Engineering

Safety in Sports Engineering is a broad and diverse topic. The papers in this section reflect that diversity. Historically safety in sport has been a stated goal, but often was secondary to performance. As we become more sophisticated in our ability to analyze and anticipate problems, we also become more proficient in the application of analytical tools such as qualitative analysis, simulation, application of new materials to old problems, statistical analysis, experimental field and lab studies, as well as epidemiological data to determine the degree to which a product might work.

From an engineering perspective, research and development issues often begin with an effort to try and characterize and define the nature of the problem so that further work will have a clear focus and set of objectives. The first paper in this section falls into that category. Franklin, et al[1] have set out to use a qualitative assessment model to determine potential problems associated with hydrostatic weighing equipment. Such an assessment is very useful in anticipating possible problems with either existing or proposed equipment or processes. In their analysis they found many problems with existing hardware and the methodology associated with hydrostatic weighing. They ultimately came up with a set of recommendations to redesign the entire hydrostatic weighing system.

Simulation is another technique that is useful in design and refinement of designs before commitment to a final form. Jin, et al[2] have used finite element modelling as a means of simulation to determine the optimum mix of mechanical properties for the shell and liner of a helmet to achieve optimum shock absorption. They modelled the shell of the helmet as an elastic-plastic material, taking into account brittle fracture; they modelled the liner as a non-linear visco-elastic material. This allowed them to determine the optimum combination of materials such that the shell will fail without the liner bottoming, thus improving the shock absorption ability of the helmet.

A mix of field data acquisition, analysis, simulation and statistical analysis was used by Schwiewagner, et al[3] to determine the effect of various heights of protective glass at ice hockey arenas in order to protect spectators from getting hit by the puck. They determined by field observation what the typical puck trajectories are for twelve different take-off angles. They used 29 different locations on the hockey ice to determine the frequency of the path that the puck would take, given various hockey trajectories. From this they determined what percentage of the paths would be stopped by different heights of protective glass around the ice playing field.

New materials with different properties from traditional materials may call for the reconsideration and redesign of safety product. Greenwald, et al[4] have examined the potential for engineered textiles to offer differential stiffness, and other properties in ways that traditional textiles cannot. They looked at the potential for the prevention of wrist injuries by a design of a wrist protector that would provide directional stiffness to prevent terminal wrist extension during a dynamic wrist impact without restricting mobility in a non-fall situation. As they point out, the current generation of wrist protective devices are uncomfortable and bulky to the point that they are rarely used. If their innovative design with engineering materials is effective it could be extremely useful in snowboarding where distal radius wrist fractures are the predominant injury.

Epidemiological evidence is often useful in evaluating the ultimate effectiveness of safety devices. Shealy, et al[5] examined the epidemiological evidence that the use of a helmet in recreational winter sports will result in fewer fatalities. They have looked at fatality data for skiers and snowboarders in the US from 2000/01 season through the 2004/05 season. They noted that about 40% of the fatally injured were observed to be using a helmet at the time of death whereas the rate of helmet utilization within the general population is lower. They have computed the incidence of fatal injury in the US over the period of time that helmet utilization has gone from practically zero to 33.2% in the 2004/05 season. That rate has not changed. What has changed is the noted primary cause of death. For non-helmeted fatals, some sort of head injury is noted in 72% of the cases versus only 45% for helmeted fatals. The conclusion is that the helmet may be altering the distribution of the primary cause of death without altering the ultimate outcome. A suggested reason for this is that the magnitude of the circumstances that typically lead to death may simply overwhelm the degree of protection that a helmet may offer.

---

[1] Franklin, K., Williams, S., and Gordon, R., (2006) Safe Use and Redesign of Hydrostatic Weighing Equipment, ISEA Conference July 2006.

[2] Miyazaki, Y., Ujihashi, S., Jin, T., Akiyama, S., and Woong, K. C., Effects of the Mechanical Properties of the Shell and Liner on the Shock Absorption of Helmets, ISEA Conference July 2006.

[3] Schwiewagner, C., Bohm, H., and Senner, V., Simulation of Puck Flight to Improve Safety in Ice Hockey Arenas, ISEA Conference July 2006.

[4] Greenwald, R., Chu, J. J., and Jessiman, A. W., Dynamic Wrist Protector for Sports Applications, ISEA Conference July 2006.

[5] Shealy, J. Johnson, R. J., and Ettlinger, C. F., Do Helmets Reduce Fatalities or Merely Alter the Pattern of Death?, ISEA Conference July 2006.

Safe Use and Redesign of Hydrostatic Weighing Equipment

Kathryn Franklin, SimonWilliams, Rae Gordon

University of Glamorgan, klfrankl@glam.ac.uk

Abstract. The aim of this study was to examine critically the safe usage of equipment used for research tests on athletes and to propose improvements. The equipment examined in this study is a bought-in hydrostatic weighing unit which is used in a research environment. Hydrodensitometry is a technique which is widely used to assess body composition. It uses Archimedes Principle to calculate the body density of a subject and this gives a measure of body fat levels. Levels of body fat and muscle are used to assess body composition to optimize muscle mass and as a performance indicator within professional sport environments. Body fat levels are used as training guidance in many sports and optimal limits are set for different sporting applications. The safety of subjects while undergoing testing is paramount. The responsibilities of testers to their subjects, as well as to themselves and their co-workers, has grown in importance from both an ethical and safety stance. With a general increasing consciousness of safety, the user needs to be aware of their legal duties but also all testers have a moral obligation to their subjects to keep them safe from harm. This coupled with the increasing numbers of personal injury claims means that the researcher is required to demonstrate a reasonable level of care for the subject's health and safety. Awareness of health and safety is an absolute requirement. The hydrostatic weighing equipment was examined in line with relevant European based legislation and safety requirements and these are discussed along with good practice for international testers. A full equipment and task risk assessment was carried out and hazards were identified. The main hazards found were slips and falls, exposure to chemicals, drowning and electrocution. The outcome from this study was, in the short term, safe working procedures were needed with safety controls. These included a winch to remove subjects in an emergency, movement of an electronic display, chemical usage assessment and guidelines. In the long term, redesign of the equipment was necessary as the hazards and risks were found to be significant. The redesigned equipment changed the orientation of the subject from vertical to horizontal position. The elements that were addressed were the access method into the hydrostatic system became less hazardous and adding a mechanism which removes subjects from the water should they encounter difficulties during the testing. The redesign should also improve the accuracy and repeatability of the readings.

1 Introduction

This paper examines the health and safety issues raised by the hydrostatic weighing of individuals using a standard piece of laboratory equipment, provided by a UK based company and commonly used. Densitometry is the methodology used for estimating body composition based upon body density and hence muscle and fat levels. Several methods can be used to estimate body density but one of the most common methods is underwater or hydrostatic weighing. This technique employs

Archimedes Principle (Massey 1986) which states that the up-thrust or buoyancy of a body immersed in a fluid is equal to the weight of the displaced fluid. The hydrostatic weighing technique involves the complete immersion of a subject in water.

Fig. 1. Hydrostatic weighing equipment

1.1 Equipment

The hydrostatic system, shown in Fig. 1, is designed for use in exercise physiology laboratories and allows the determination of body density using the underwater measurement method.

This system is comprised of a large thick walled, cylindrical, high density polyethylene tank, with a gantry to support the fabric seat attached to the gantry by a chain seat. The tank is 1.5m high and has a diameter at the entrance point of 1.2m and at its widest point a diameter of 1.3m. The measurement of buoyant weight is made using a load cell with a digital display unit. Water enters the tank via a 50mm diameter pipe and there is a 12mm drain pipe at the base of the tank. The water temperature is kept at approximately 37°C by an electrically powered heater and is chemically treated with calcium hypochlorite. The tank is accessed via internal and external vertical ladders with narrow treads on each rung. The subject once in the tank is required to sit and balance on a material sling, after exhaling and placing their head under the water, their buoyant weight is recorded on a load cell.

2 Method

2.1 Health and Safety Legislation

In order to assess the hazards involved in hydrostatic weighing the equipment and current procedure was examined. Based on these examinations a risk assessment (HSE 1996; HSE 1998) was carried out as required by The Management of Health and Safety at Work Regulations 1999(HMSO 1999) and based on European Directive89/391/EEC. The risk assessment highlighted hazards encountered and recommended actions to minimise or eliminate these hazards. There are many pieces of UK and European Legislation in place that may impose legal requirements on the equipment and procedure of hydrostatic weighing. The Health and Safety at Work Act 1974 - Section 2(2)(a) requires the provision and maintenance of safe "systems of work that are so far reasonably practicable, safe and without risks to health" (HSMO 1974). This is required for both employees and visitors.

A safe system of work is a formal procedure which results from a systematic examination of a task in order to identify all the hazards and assess the risks and which identifies safe methods to be followed to ensure that hazards are eliminated or the remaining risks minimised.

Later legislation, The Management of Health and Safety at Work Regulations 1999 states that 'Every employer shall make a suitable and sufficient assessment of -

- the risks to the health and safety of his employees to which they are exposed whilst they are at work; and
- the risks to the health and safety of persons not in his employment arising out of or in connection with the conduct by him of his undertaking'.

A risk assessment is a systematic examination of a piece of equipment or task which could cause harm to people so that reasoned conclusions can be made as to whether any or enough precautions have been taken in order to prevent that harm and whether further actions are required. They are not complex activities but an awareness of hazards and health and safety legislation is required in order to make the risk assessment valid and worthwhile. It is essential that the individual that carries out a risk assessment is competent. A specific piece of legislation that applies to the hydrostatic measurement system is The Confined Space Regulations 1997 (HMSO 1997) which is based upon European directive 92/57/EEC. These UK regulations define a confined space as 'any place, including any chamber, tank, vat, silo, pit, trench, pipe, sewer, flue, well or other similar space in which, by virtue of its enclosed nature, there arises a reasonably foreseeable specified risk.' These Regulations include drowning as a specific risk. The tank by the nature of its access and content of water becomes a confined space with the foreseeable specified risk being the drowning of the individual. The tank, having been defined as a confined space, is then subject to further requirements such as a written safe system of work and emergency procedures. The risk assessment and the recommendations may be subject to change for different individuals as the general health and mobility of the subjects could be different.

3 Results

3.1 Risk Assessment

Following a qualitative risk assessment of the hydrostatic weighing of a subject which is shown in Table 1 the areas were highlighted as serious issues for concern for this and possibly other systems.

Hazard	Harm	Existing Actions Precautions	Recommendations
Access	Slip or Fall	Small treads, covered in rubber matting, have been added to the rungs.	A platform is required at the top of the ladders to enable ease of movement between internal and external ladders. A

			sturdier handrail is also required.
Water	Drowning	There is no mechanism for retrieval of a person from the tanks should they encounter difficulties	A tripod, winch and harness mechanism is needed to remove subjects in difficulty. Emergency and resuscitation equipment and procedures are required.
	Chemicals	Currently the water is tested for pH value between 7.4 and 7.6. Additives to the water calcium hypochlorite.	Material safety data sheet for calcium hypochlorite is required and a COSHH assessment is necessary. This information should be displayed near the tank. Amount required to dose tank to be specified.
	Slips	The floor surrounding the tank is tiled and prone to water spillages	The floor must be covered with non slip flooring so that slips on smooth, wet floors with wet feet cannot occur
Electrical supply	Electrocution	The electronic display for the weighing system is positioned near the tank and is supplied by 230V.	This display needs to be remote from the tank incase it becomes detached and falls in to the water or is taken in by a gantry collapse. A lower voltage system or even a battery supply could be used.
Information	Lack of knowledge of process by subject	The subject is currently talked through the process and signs an ethical consent form	It is important to ascertain whether the subject has medical problems such as epilepsy or suffers from conditions such as claustrophobia which may induce panic attacks. A pre-weighing questionnaire and a written safe system of work are required. This must be read and understood by the operator and subject.
Support Gantry	Gantry Collapse	The gantry from which the subject is weighed is composed of 1" thin walled conduit. This is also used as a handrail.	A more substantial gantry is needed particularly if heavy subjects are being weighed. This would be provided with a tripod system. The existing gantry is insufficient to be used as a handrail.

People at risk are operators and subjects

Table 1. Qualitative Risk Assessment for Hydrostatic Weighing

3.2 Access

Falls from height onto the concrete floor outside the tank or into the tank while accessing or egressing is likely and the subject could fall upto 1.5m. The treads on the ladders on the outside and inside of the tank are narrow and for the subject to move from the outside to inside ladder and vice visa is difficult. This is due to the arrangement of the top rungs and currently requires a high level of mobility. A person with bare, wet feet may easily lose their balance, slip or fall. Health and Safety Executive statistics state that between 1997 and 2001 29 people were killed and over 14,000 injured by falls from heights of 2m or less in industrial accidents (HSE 2003).

3.3 Emergency Procedures

The lack of any emergency procedures leads to a situation where a subject could not be removed being from the tank if they were taken ill or lost consciousness. If a subject become unwell and slipped from the sling while in the tank there is no current method for effecting their removal. If they are unconscious then to attempt to remove a wet, semi-naked person from such an awkward position would be very difficult. Their head would be under water and drowning becomes a distinct possibility. Going (1996) outlined the requirements for equipment to carry out underwater weighing of subjects and recommended that the tank was at least a 3mx3mx3m.

There is a requirement for a safe method for extricating a subject from the tank should they suffer difficulties. If a redesign of the system is not feasible then this is most easily achieved by a tripod, winch and harness mechanism. Sufficient headroom above the tank is required for this arrangement. This system will not assist with the risk of fall but will effect the removal of an unconscious person of any size out of danger and give enough time for the alarm to be raised. Resuscitation instructions and equipment are also required.

3.4 Chemicals

The chemical used to disinfect the water in the tank is calcium hypochlorite. It is commonly used to disinfect swimming pools. It is toxic if ingested, inhaled or comes into contact with skin in its powder form hence it must be handled carefully and the correct amount must be added to the tank. The Control of Substances Hazardous to Health Regulations (COSHH) 2002 (HMSO 2002) require assessments to be carried out on any chemical used in the workplace and instructions on how to handle the substance safely must be provided. The COSHH assessment must be referenced in the risk assessment. The amount to be used in the tank should be specified.

3.5 Safe System of Work

A step by step outline of hydrostatic weighing is required highlighting any hazards in the process. The safe system should be communicated effectively to the subject and operator involved in the hydrostatic weighing so that the hazards are recognized and precautions met. The subject should be fully briefed as to the procedure and the medical history of the subject should be investigated. Fear of confined spaces, heights or vertigo should be discussed. Each individual should be assessed as to their ability to manoeuvre themselves into the tank. The minimum number of operators required to carry out a hydrostatic weighing test should be detailed.

4 Redesign of Hydrostatic Weighing System

The hydrostatic weighing system was redesigned to reduce the risks posed by the current system and can be seen in Fig.2 (Jones 2003).The recumbent design removes the need for the subject to be at height, allows emergency evacuation of the tank if the subject becomes distressed.

Fig 2. Redesigned Hydrostatic Weighing System

Accuracy and repeatability of buoyant weight readings should be improved.

5 Conclusions

Risk assessments by a competent person are required for many activities and hydrostatic weighing would require a full risk assessment so that legislation is complied with and personnel are safe. The hydrostatic weighing system and the act of carrying out the weighing process are both very simple. They do however expose the subject under going the process to some severe hazards. These need to be recognised and addressed to ensure the safety and health of any individuals being weighed. A safe system of work is required as well as an emergency procedure. These should be written and communicated to all the people involved. An emergency mechanism to remove an individual from the tank is necessary. An initial assessment of the physical and mental health of individual is required in order to ensure that the subject is fit enough to undertake the assessment. Redesign of the tank is required and a recumbent weighing system would reduce some of the risks.

References

Going Scott B (1996) Human Body Composition, Chapter 1 - Densitometry, Edited by Roche, Heymsfield and Lohman, ISBN 0-87322-638-0

HMSO, Her Majesty's Stationery Office (1997) The Confined Space Regulations 1997, Statutory Instrument 1997 No. 13 ISBN 0 11 064643 6

HMSO, Her Majesty's Stationery Office (2002) The Control Of Substances Hazardous to Health Regulations 2002, Statutory Rule 2002, No. 2

HMSO, Her Majesty's Stationery Office (1974) The Health and Safety at Work Act 1974

HMSO, Her Majesty's Stationery Office (1999)The Management of Health and Safety at Work Regulations 1999, Statutory 1999 No. 3242 ISBN 0 11 085625 2

HSE, Health and Safety Executive (1996) A guide to Risk Assessment Requirements INDG218L, HSE Books

HSE, Health and Safety Executive (1998) Five steps to risk assessment INDG163 (rev1), HSE Books ISBN 07176 1565 0

HSE, Health and Safety Executive Statistics, Internet address www.hse.gov.uk/statistics/index.htm, accessed September 2005

Jones, L., B.Sc.Undergraduate Engineering Thesis, University of Glamorgan 2003

Massey,B.S., Mechanics of Fluids, page 52,Van Nostrand Reinhold(UK), ISBN:0-442-305

Effects of the Mechanical Properties of the Shell and Liner on the Shock Absorption of Helmets

Yusuke Miyazaki[1], Sadayuki Ujihashi[1], Tomohiko Jin[2], Shinichirou Akiyama[2] and Ko CheolWoong[3]

[1] Department of Information Science and Engineering, Tokyo Institute of Technology, Tokyo, Japan, miyazaki@hei.mei.titech.ac.jp
[2] Toyota Motor Co., Aichi, Japan
[3] Iowa Univeristy , Iowa, USA

Abstract. A combination of the characteristics of the shell and liner of a helmet, that makes it possible to improve the shock absorption capacity was investigated by constructing the finite element model of the helmet modeling both the bottoming of the liner and the fracture of the shell. The material model of the shell was defined as an elasto-plastic material, taking into account brittle fracture. The liner was modeled as a non-linear visco-elastic material, expressing both energy absorption by buckling of the cell foam, and the extreme increase of the stiffness caused by bottoming. The results of the parametric study, where the foaming ratio of the liner and the shell-thickness were varied, indicated that there is an optimum combination where the shell part fails without the liner bottoming, improving the shock absorption ability of a helmet.

1 Introduction

When a helmet is subjected to an impact load, the inner liner, made of EPS foam, can absorb a part of the energy of the impact by crushing. In addition, the outer shell can fail, absorbing some of the impact energy.

However, even if fracture of the shell occurs, when the liner has deformed at the bottoming point the stiffness of the liner increases greatly, which causes a higher risk of head injury.

Consequently, it is important to control the energy distribution upon impact, both in order to avoid bottoming of the liner, and to have the shell absorb more of the impact before fracturing. This will improve the overall shock absorption ability of the helmet. However, it seems that this phenomenon has not been taken into account in the design of the helmet or clarified in previous studies (Yettram, Godfrey, Chinn, 1994; Cheng, Chang, Chang, Huang, Wang, 2000; Ko, Ujihashi, Inou, Takakuda, Ono, Mitsuishi, Nash, 2000; JIS T 8133 2000).

Therefore, the purpose of this research is to investigate the combination of characteristics of the shell and liner that makes it possible to improve the shock absorption of a helmet, by using an FE model that models both the bottoming of the liner and the fracture of the shell.

2 Construction of a Helmet Finite Element Model

The FE models used in this study consist of three components; helmet, head-form and surface. PAM-CRASH v.2000 (ESI) was used as a solver.

The shape of the shell part of the helmet was measured by using a 3D scanner (Vivid900, Konica Minolta). Then, the FE model of the helmet, consisting of 3mm shell thickness (using shell elements) and 30mm liner thickness (using solid elements), was constructed as shown in Fig.1. Both the head-form model (Fig. 2) and the hemisphere anvil model (Fig.3) used in JIS T 8133:2000 were also created from the measured three-dimensional coordinates.

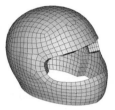

Fig.1. Helmet model   **Fig.2.** Head-form model   **Fig.3.** Anvil model

The stress-strain relationships of the shell, including its fracture point, were obtained by conducting the bending tests on specimens cut from the helmet. As the results of the test show, the stress–strain curve of the specimen is linear-elastic before brittle fracture occurs. Therefore, the material model of the shell used for the finite element analysis was defined as an elasto-plastic material, taking into account energy absorption by brittle fracture. The Young's modulus is 7.98Gpa, and the fracture strength is 260MPa, as shown in Fig.4.

Impact tests were carried out to obtain the material properties for the liner, made from EPS foam, as shown in Fig.5. The four test pieces whose foaming ratios varied from 20 to 30 were 60-mm square and 30mm thick. The impactor, whose mass is 4.5kg, fell onto the specimen from 1.25m so that the impact speed is 5m/s. The impact load was obtained by multiplying the acceleration response which was obtained from an accelerometer attached to the impactor with the impactor's mass. Displacement responses of the impactor were obtained by using an optical displacement transducer. The stress-strain relationship of the liner obtained from the tests shows the characteristics of both energy absorption by buckling of the cell foam, and the extreme increase of the stiffness caused by bottoming, as shown in Fig.6. Therefore, the material model for the liner was defined as non-linear visco-elastic (Fig.7). After an iterative process, the mechanical behavior of the liners was identified, as shown in Fig.6. These results show that bottoming of the liner was modeled accurately. In addition, the relationship between the foaming ratio and the visco-elastic properties was obtained, as shown in Table1.

Both FE models for the head-form and anvil were modeled as a rigid body. Contact conditions were defined between the liner and head-form, and the shell and anvil.

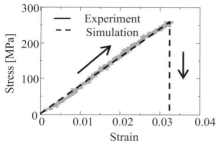

Fig.4. Stress-strain curve of shell

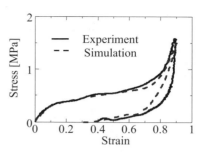

Fig.6. Stress-strain curve of liner

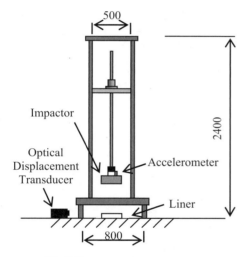

Fig.5. Impact test apparatus

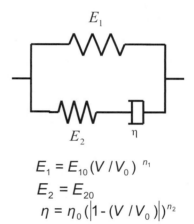

$$E_1 = E_{10}(V/V_0)^{n_1}$$
$$E_2 = E_{20}$$
$$\eta = \eta_0(|1 - (V/V_0)|)^{n_2}$$

Fig.7. Non-linear 3 elements visco-elastic model (V: Volume)

Table1 Relationship between foaming ratio and visco-elastic parameters

Foaming Ratio	E_{10} [MPa]	n_1	η_0 [kPa·s]	E_{20} [MPa]	n_2
f	$\dfrac{30}{f} - 0.6$	0	20	$2f$	0.2

The results of the JIS test simulation were compared with the experimental results to validate the FE models. The liner acceleration response and Head Injury Criterion (HIC) value of the rigid head-form model were compared with those tested under the same conditions, where a helmet is dropped from 2.5m height, and both temporal and frontal parts of the helmet impact an anvil (Fig.8). As shown in Fig.9, although the peak accelerations and HIC values from the simulations were almost consistent with those from the tests, these responses showed little difference. This might be due to the lack of the comfort foam, and its contact condition.

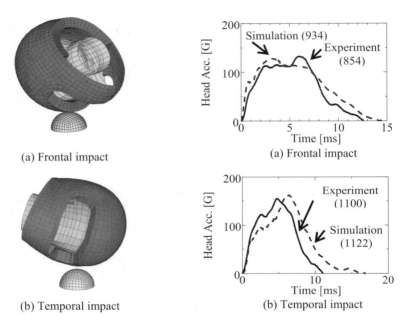

(a) Frontal impact   (a) Frontal impact

(b) Temporal impact   (b) Temporal impact

Fig.8. Impact postures of JIS test simulation for validation of FE helmet model

Fig.9. Head acceleration responses and HIC values

3 Parametric Study

Simulations with varied combinations of liner foaming ratio and shell thickness were conducted in order to investigate the effects of the mechanical properties of the shell and liner on the shock absorption ability of helmets. Visco-elastic parameters for five foaming ratios of the liner were calculated by using the equations shown in Table1. Shell thickness varied in 0.5mm increments, from 2.0-4.0mm.

HIC values for each combination of shell thickness and foaming ratio are shown in Table2. For the frontal impact case, the best combination in these combinations was a 2.5 mm shell thickness and liner foaming ratio of 25. In the case of side impact, the combination of 3.5mm shell thickness and 35 foaming ratio is thought to be the best solution in these combinations, as shown in Table 2(b). Figure10 shows the acceleration responses for these conditions as compared to the normal helmet (shell thickness: 3.0mm, foaming ratio: 20). Peak acceleration in the best combination is lower than that in the normal helmet for both frontal and temporal impact.

Although a thinner shell can absorb more impact energy due to its fracture, the bottoming of the liner occurs because behavioral function describing shell's impact force distribution no longer applies. Therefore, peak acceleration of the head should be higher, and consequently the ability of the helmet to absorb the shock of impact is lost. In the case of a thicker shell, the same impact force distribution function still applies because shell fracture does not occur. However, although liner behaves well, absorbing energy and deforming as expected, the overall ability of the helmet to absorb impact energy before shell failure degrades. It is therefore important that a helmet design should take into account the behavior of both the shell and liner, so that the shell may fail without causing the liner to bottom.

Table 2 HIC values for each combination of shell thickness and foaming ratio of liner

(a) Frontal impact

Frontal Impact		Shell Thickness [mm]				
		2.0	2.5	3.0	3.5	4.0
Foaming Ratio	15	912	819	1009	1563	1655
	20	866*	707	1000	1238	1354
	25	883*	674	969	1009	(1106)
	30	886*	775*	906	904	(918)
	35	889*	952*	920*	(900)	(875)

*: Maximum Strain of Liner>0.8, ( ): No fracture of Shell occurred, ☐ : Minimum HIC Value

(b) Temporal impact

Temporal Impact		Shell Thickness [mm]				
		2.0	2.5	3.0	3.5	4.0
Foaming Ratio	15	1492*	1295	1273	1274	1475
	20	1482*	1263	1171	1164	1324
	25	1684*	1272*	1167	1106	1243
	30	2133*	1306*	1112	1060	1140
	35	2674*	1481*	1182*	1039	1094

*: Maximum Strain of Liner>0.8, ☐ : Minimum HIC Value

150   Yusuke Miyazaki et al.

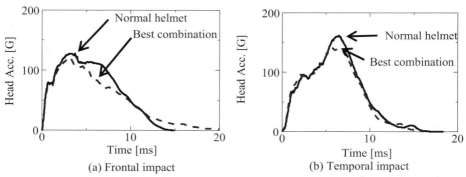

Fig.10. Comparison of the acceleration responses between best shell and liner combinations and the normal helmet design

4 Conclusion

A combination of the characteristics of the shell and the liner of a helmet, which makes it possible to improve the shock absorption capability, was investigated by constructing an FE model and modeling both bottoming of the liner and fracture of the shell.

The FE model of the helmet was constructed from points measured by a 3D scanner. The material model of the shell was defined as an elasto-plastic material, taking into account brittle fracture. The liner was modeled as a non-linear visco-elastic material, expressing both energy absorption caused by buckling of the cell foam and the extreme increase of stiffness caused by bottoming. The results of the JIS test simulation were compared with the experimental results to validate the finite element models.

As the parametric study varied the combinations of the foaming ratio of the liner and shell thickness, the results indicate that there is best combination of these components whereby the shell fails but bottoming of the liner is avoided. This combination improves the overall shock absorption ability of a helmet.

References

Yettram,A.L., Godfrey,N.P.M., Chinn,B.P. (1994) Materials for Motorcycle Crash Helmets – A Finite Element Parametric Study, Plastics Rubbers & Composites Processing & Applications, **22**-4, pp. 215-221.

Cheng ,C., Chang ,L., Chang ,G., Huang ,S., Wang ,C. (2000) Head Injury in Facial Impact –A Finite Element Analysis of Helmet Chin Bar Performance. Journal of Biomechanical Engineering. 122,640-646

Ko, C.W., Ujihashi, S., Inou, N., Takakuda, K., Ono, K., Mitsuishi, H., Nash, D. (2000) Dynamic Responses of Helmets for Sports in Falling Impact onto Playing Surface. The Engineering of Sport, 399-406

JIS T 8133 (2000) Protective Helmets for Vehicle Users. Japanese Industrial Standards

Simulation of Puck Flight to Improve Safety in Ice Hockey Arenas

Schwiewagner C., Böhm H., Senner V.

Technische Universitaet Muenchen, schwiewagner@sp.tum.de

Abstract. Ice hockey boards are equipped with protective glass to prevent the spectators from being hit by the puck. Although the height of the board with protective glass is 2.05 m, this is not high enough to prevent the spectators from puck collisions - severe accidents have occurred in the past. This study investigates, which increase of height of the safety glass is necessary to significantly reduce the risk of severe puck injuries for the spectators.

For this reason the flight of the puck is simulated using a rigid body puck model. Flight curves from 29 different initial positions on the ice rink were calculated, using 12 different initial take off angles and the maximum initial speed and spin that has been determined experimentally with three top level players.

The simulations show that an increase of the security glass by 0.8 m will lower the relative frequency by 37 % of those shots with a potential to hit a spectator. The maximum velocity of such dangerous shots was decreased from 22.6 to 15.9 m/s. However, this reduced velocity and number of dangerous shots do not protect the spectators completely from injuries. The simulation model suggests that a barrier of 6,37m protective glass leads to a 100 % reduction of all dangerous shots.

1 Introduction

Ice hockey is a dangerous game – you can read that on the instruction manuals for hockey equipment. This warning is not only valid for the players, but for the spectators of ice hockey games, too. Some serious face injuries happened to viewers of this game. For this reason ice hockey boards are equipped with safety glass but the existing height of 2.05 m seems not high enough to prevent the spectators from being hit by the puck. The purpose of this study is therefore, to investigate which increase of the height of the protective glass is necessary to significantly reduce the risk of severe puck injuries for the spectators.

2 Method

Puck flight is described by Newton-Euler differential equation for a three dimensional rigid body (Haug 1989). The differential equations of motions are integrated using MatLab (The Mathworks, Inc. Natic, USA). The external aerodynamic forces on the puck are the sum of the drag, lift and sideward forces F_D, F_L, and F_S respectively. The external aerodynamic moment consists of two parts, the pitch moment (T_M) and the spin down moment (T_N). The pitch moment is dependent on the factor q "equation 5", which is the dynamic pressure, multiplied by the planform area of the disc, as were lift and drag. The spin down moment is estimated from the decay of the spin rate measured at real puck shots described subsequently.

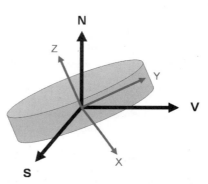

Fig. 1: The puck local body fixed coordinate system (X, Y, Z) has its origin in the puck's center of mass. The Z axis of the local coordinate system is the main spinning axis, it is pointing upward at the beginning of each shot. V is the direction of flight, N is in the plane of Z and V, perpendicular to V. S is pointing sidewards and is perpendicular to N and V.

Due to spin of the puck the air speed in the direction of the spin is greater than at the side where the spin is opposite to the air stream. Consequently by Bernoulli's theorem the pressure at the one side is lower than at the other. This effect causes the puck to move transversely to the left or the right depending on the direction of the spin. Assuming there is no wind and the puck is spinning around its local Z axis the side force (F_S) is caused by the Magnus force "equation 4", which is proportional to the spin and the stream velocity v, relative to the spinning rim (de Mestre 1990). This dependency is accounted for in a cos dependency of the sideward force on the angle of attack.

$$q = 0.5 A \rho v^2 \qquad (1)$$

$$F_D = q C_D(\alpha) \vec{e}_v \qquad (2)$$

$$F_L = q C_L(\alpha) \vec{e}_n \qquad (3)$$

$$F_S = 0.2 \rho v 2 \pi r^2 \omega_z \cos(\alpha) \vec{e}_n \times \vec{e}_v \qquad (4)$$

$$T_M = q C_M(\alpha) \vec{e}_S \qquad (5)$$

$$T_N = 1.8 * 10^{-5} \omega_z \vec{e}_z \qquad (6)$$

In general the drag and lift coefficient C_D and C_L respectively are dependent on three quantities: the Reynolds number, the Spin parameter and, in case of an asymmetrical body such as a Frisbee or a puck, the Angle of attack (α). In a wind tunnel experiment "Fig. 1", drag and lift Coefficients were determined for a puck and it is shown that C_D and C_L are independent of the Reynolds number in the range of 1.24*10e5<Re<2.85*10e5, corresponding to a tunnel speed range of 13 m/sec to 30 m/sec. Potts and Crowther (2000) showed that the spin rate has only a small effect on the aerodynamic load of a rotating disc wing, therefore the aerodynamic coefficients used in this study are assumed to be independent of the spin rate of the puck. The aerodynamic coefficients in equations 2-6 were measured for different angles of attack ranging from 0 degree to 90 degrees. The aerodynamic measurement setup is shown in figure 2, the measured aerodynamic coefficients in figure 3.

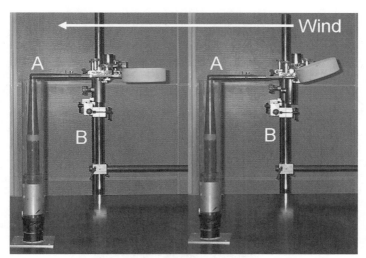

Fig. 2: Wind tunnel experiment with an enlarged 2:1 puck model to receive suffi-cient forces on the measurement scale. The setup shown on the left and right side of the picture shows an angle of attack of 0° and 40° respectively. The wind speed was up to 30 m/sec. The aerodynamic forces were measured with a force plate attached to pole "A". The additional aerodynamic resistance of pole "A" itself was determined attaching the puck to pole "B".

To obtain more information on existing puck velocities and spin at different take off angles, puck shots of top level players were additionally measured using a Vicon MX-460 camera system with 6 cameras operating at 400 Hz "Fig. 2". The focal volume of the camera is of 3.0 m length, of 1.5 m width and of 2 m height so that the first flight phase from the take off of the puck is recorded. The range of the takeoff angle of the velocity vector with respect to the ground was 3° to 63° for all 107 shots measured. With the initial conditions of 107 shots (spin, angle of attack and total velocity) the flight width is calculated with the simulation model. From 0° to 64° (in steps of 4°) respectively one shot, leading to the longest flight distance, was selected. This leads to 12 most dangerous shots within the range of take off angles. Flight curves from 29 different initial positions on the ice rink were calculated, using the 12 initial conditions distributed over the complete range of possible takeoff angles as described before.

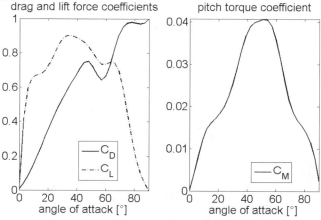

Fig. 3: Measured aerodynamic coefficients.

Fig. 4: Player shooting the puck on an equivalent surface to ice made of "Nierolen" (Nierolen GmbH, Lenggries, Germany). Three reflecting markers of 6 mm diameter are attached to the puck surface to determine flight trajectories.

3 Results

The simulations show that an increase of the protective glass by 0.8 m will lower the relative frequency by 37 % of those shots with a potential to hit a spectator. The maximum velocity of such dangerous shots was decreased from 22.6 to 15.9 m/s.
A remaining risk for puck over flight using the measured data can be eliminated at a safety glass height about 6.37 m (+board height).

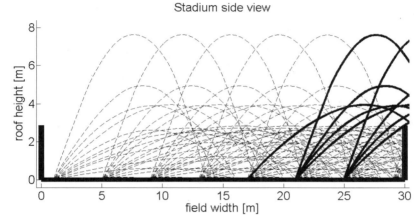

Fig. 5: Simulated puck trajectories within the ice hockey arena. The dangerous shots are plotted with bold lines; the safe shots within the boundaries are plotted with dashed lines. Shown are 88 simulations starting from eight different locations on the ice field under 12 different angles.

4 Discussions

The computer simulation showed that an increase of 0.80m in safety glass height can't protect the spectators of getting hit by the puck. Puck overflight is possible and estimated cranial impact load is much higher than stated in Güllich's (1988) threshold.

The model suggests that preventing the spectators from being hit, does not end with an increase of 0.80 m safety glass. Raising the glass-wall to a height of above 6.00 m is difficult to construct. The present situation with high nets on the side-boards doesn't seem satisfying, because of interference with television broadcast. Therefore other solutions must be found. We recommend investigating whether the distance between the spectators and the side-board as well as the height of the first seat row could be changed to reduce the number of dangerous shots. This can be done with the simulation model presented in this study.

References

De Mestre N. (1990). The Mathematics of Projectiles in Sport, Cambridge University press, NY, USA, Chapter 7.5

Gülich, H.A.(1988). Biomechanische Belastungsgrenzen. Aktualisierte Literaturstudie zur Belastbarkeit des Menschen beim Aufprall. *Bundesanstalt für Straßenwesen Bereich Unfallforschung*. Bergisch Gladbach, pp 35-53

Haug, E. J. (1989). Computer aided kinematics and dynamics of mechanical systems. Allyn and Bacon Series in Engineering Vol.1 Basic Methods, Allyn and Bacon, Boston, MA. Chap. 9.3

Potts J.R. & Crowther W.J. (2000). The Flow Over a Rotating Disc-wing, RAeS Aerodynamics Research Conference Proc., London, UK.

Dynamic Wrist Joint Protector for Sports Applications

Richard M. Greenwald, Jeffrey J. Chu, Alexander W. Jessiman
Simbex, Lebanon, New Hampshire, USA, rgreenwald@simbex.com

Abstract. A novel brace for preventing wrist fractures in snowboarding was developed and validated. Commercially available wrist protection devices do not consistently prevent wrist fractures during high-energy impacts, limit mobility, and are used infrequently. A wrist protector consisting of engineered textiles combined with energy absorbing materials to prevent terminal wrist extension during a dynamic wrist impact without restricting mobility in non-fall situations was constructed. The fibers of the textile were integrated with molded components that allowed them to be firmly affixed to the forearm and hand.

Impact testing was performed with a linear rail drop tower to test the efficacy of the design concept for preventing wrist hyperextension. The prototype wrist guard and a commercially available wrist guard were mounted on a prosthetic arm hinged at the wrist and impacted on a snow-like surface at 2.4 m/s. Impact force, peak pressure on the forearm, and wrist flexion angle and angular velocity were recorded.

The prototype guard prevented terminal extension of the wrist for all drops. Loads transferred to the forearm were relatively low and were distributed across the contact area between the attachment and the surrogate forearm. The commercially available wrist guard was unable to prevent terminal extension at the loads and velocities used in testing. The prototype wrist guard was validated for preventing terminal wrist extension under dynamic loading conditions, and further development and testing of this engineered textile design is warranted.

Introduction

Significance. Wrist fractures due to impact following a fall occur frequently in sports, particularly in-line skating and snowboarding. Wrist fractures may lead to lost productivity, early arthritis in the wrist joint, and loss of functional independence. The most common wrist fracture injury mechanism is rapid hyperextension of the wrist during a fall onto an outstretched hand. Commercially available wrist protection devices have not been shown to consistently prevent wrist fractures during high energy impacts, and are used infrequently by athletes, and in particular youths, because they are often bulky and limit normal range of motion. These wrist guards are typically constructed with rigid aluminum or plastic plates or splints crossing the wrist joint on either the dorsal or volar surface of the hand as a protection mechanism against abrasions and/or carpal fractures, including the wrist. They typically constrain the normal range of wrist motion.

Epidemiological and biomechanical studies support the conclusion that the use of wrist protection for in-line skating and snowboarding significantly reduces the incidence and severity of wrist injuries and in some cases prevents wrist fractures (Scheiber et al, 1996, Staebler et al, 1999, Idzikowski et al, 2000, Ronning et al., 2001, Machold et al, 2002, O'Neill, 2003, Bladin et al, 2004). The problem is that very few people actually wear the protective devices because they limit normal range of motion and are generally bulky and unappealing. Additionally, wrist guards have limited effectiveness at high impact energies because they do not prevent terminal wrist extension (Greenwald et al., 1998).

Dynamic impact testing showed no significant difference between maximum wrist extension angle during an impact for guarded and unguarded cadaveric forearms (Greenwald, unpublished data, 1997). Fracture mechanisms were similar for guarded vs. unguarded wrists, implying that while the brace absorbed some energy during the impact, the wrist was still driven to terminal extension, where the wrist itself was significantly more stiff than the guard. Further bending of the wrist towards extension leading to fracture was not resisted by the guard at all. Impulse and force data from these tests revealed that current wrist guard designs may have some prophylactic effect at lower impact energies as they provide additional resistance to falling motion during initial loading. However, they appeared to have little effect in reducing loading rates at higher loads (Greenwald et al, 1998).

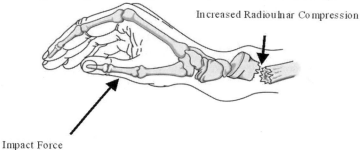

Increased Radioulnar Compression

Impact Force

Fig. 1 Typical distal radius fracture from fall on an outstretched hand. This is just one of many fracture types commonly seen.

Clinical Relevance. Distal radius fractures are commonly seen by orthopedists, and account for a vast majority of upper extremity fractures. Approximately 40% of hand fractures are either radius and ulna fractures or carpal fractures (Barnaby, 1992). The primary mechanism of injury for distal radius fractures is excessive compression resulting from extreme extension (dorsiflexion), which is typical of falls onto an outstretched hand (See Fig. 1). Complex fractures of the distal radius can lead to significant long-term disability if not treated adequately. While most low-energy fractures are easily treated, complex fractures typically associated with high-energy impacts often require sophisticated surgical techniques and significant rehabilitation.

We hypothesize that wrist fractures in in-line skating and snowboarding can be significantly reduced through the use of a device that prevents terminal wrist extension during a dynamic wrist impact, yet which permits relatively normal wrist range of motion during use.   Previous impact research on protection offered by commercially available wrist guards using a human cadaver model demonstrated limited ability of the guards to prevent terminal wrist extension during both quasi-static and dynamic loading (Greenwald et al, 1998). These guards also reduced passive range of motion by more than 50%.

Proposed Solution. We propose to overcome limitations of existing wrist guards by creating an engineered textile wrist protector (ETWP), which 1) allows relatively unconstrained range of wrist motion until near terminal extension, 2) prevents terminal wrist extension, 3) allows for incorporation of a thin, flexible skid pad, 4) absorbs a limited amount of the impact energy, and 5) transfers the remainder of the impact energy to bending of the mid-shaft radius and ulna at levels and rates below their fracture limit and without increasing the likelihood of radial head fractures.

Braided fibers demonstrate load-displacement characteristics similar to our specifications under loading conditions. Braided fibers apply little resistance in tension (Stiffness 1) until the fibers 'lock up' and the structure becomes substantially stiffer (Stiffness 2). This two-phase stiffness is what allows the ETWP to be unconstrained until a defined locking point, designed to be near terminal extension. Fiber parameters that can be adjusted to tune the behavior to the desired load-displacement characteristics include material, number of fibers, braid angle, fiber thickness, and braid diameter. Fig. 2 shows an example of how braid angle can be varied to alter the location of the lock-up point of various carbon fiber braids.

The goal of this study was to compare wrist kinematics for a prototype ETWP and a commercially available wrist protector under simulated impact conditions mimicking a fall onto an outstretched hand.

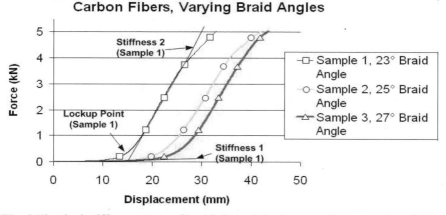

Fig. 2 The dual stiffness nature of braids is exploited to tune the properties of the Engineered Textile Wrist Protector (ETWP). Shown is an example of how the lockup point of the textile varies with braid angle.

Methods

A prototype ETWP (Fig. 3a) was developed using braided fabric materials. A bicycle glove was used for the finger attachment, and a flexible carbon fiber impregnated shell was used as the forearm attachment. The braided fiber components were fixed to the two attachments with epoxy and formed a crossing pattern on the palmar surface of the hand. The braided pieces used were not developed specifically for this application, but were sufficient to demonstrate proof-of-concept. Our goal was to create a prototype that could be attached to a surrogate arm for functional testing. To demonstrate technical efficacy of the prototype, it needed to prevent terminal wrist extension. While the terminal extension angle for humans wrists is typically given as 59°, the terminal extension angle of the hinged surrogate arm was considered 90°, due to the lack of muscles and connective tissue.

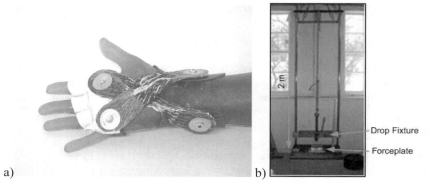

a)   b)

Fig. 3 a) ETWP prototype, shown attached to surrogate arm, and b) Linear drop rail system with an AMTI 6-DOF forceplate mounted in base.

A surrogate arm modeled from a human arm was constructed from 80 Shore A cast polyurethane with a single hinge joint in the flexion/extension axis. The surrogate arm was attached to a linear rail system (Fig. 3b). Two devices (ETWP and commercially available guard used by the primary author in previous wrist guard research) were impacted (n=3 trials, each) at 2.4 m/s (~30 cm drop height). To simulate characteristics of snow, a laminate of latex and neoprene sponge (Mecham, 1997) was attached to the surface of a forceplate (AMTI MC6-4000, Watertown, MA, USA). Impact force was measured with a LabView data acquisition program collecting at 10kHz. Pressure distribution between the forearm and the forearm shell was measured with pressure sensitive film (Fujifilm Prescale, Fujifilm).

Results

The prototype ETWP device prevented the wrist from rotating beyond 74±3° of extension compared with 99±5° of extension for the commercially available wrist guard (Figure 4). Impacts with the ETWP guard generated a pressure distribution pattern that was very evenly distributed with an average pressure less than 1 MPa. Displacement of the forearm attachment was minimal.

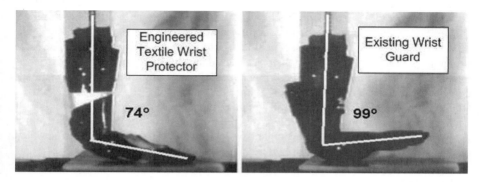

Fig. 4 Comparison of terminal extension angle between ETWP braided fiber prototype (left) and commercially available wrist protector (right).

Discussion and Conclusion

A prototype wrist protective device was developed to prevent terminal extension of a surrogate wrist in laboratory drop tests simulating a forward fall onto an outstretched arm. The material and geometry of braided fibers influence their material and mechanical properties, and are tuned to provide desired load-displacement characteristics under impact loads. Our design goals include placement of the braided fibers on the palmar surface of the hand to allow relatively unconstrained wrist rotation in all directions except for wrist extension beyond 50° in the sagittal plane. Between 30-50° of wrist extension, the braided fibers stiffen rapidly, and provide resistance to further rotation. Energy is transmitted through the braid and to the attachment points on the forearm. These degrees of freedom are desirable attribute during activities such as snowboarding or in-line skating. Convention wrist protectors typically limit normal wrist flexion/extension and other wrist rotations significantly during normal use.

Braided fiber constructs are thin, lightweight, low profile, relatively low cost, highly tunable, and easy to manufacture and integrate - making them a suitable candidate solution for a wrist orthosis. The mechanical characteristics of braided fibers afford high flexibility, modulus of elasticity, and yield strength, characteristics required by the device. The forearm attachment provided an anchor point for the braided fiber construct. This first test of the ETWP device demonstrates efficacy of the method – this was achieved without tuning the load-displacement properties of the braids. The braided fiber prototype "locked up" as desired and prevented palmar contact with the impact surface that would lead to increased forearm bending loads.

We hypothesize that user compliance will increase significantly with a wrist protective device that allows relatively unconstrained wrist motion except when activated during a fall event. User acceptance of any wrist protection device is predicated on comfort, freedom of motion during normal use, ease of application, removal, and style. We plan to integrate wrist protection into gloves typically worn during snowboarding and in-line hockey. Standalone wrist guard designs will utilize the same mechanism in a low profile, comfortable package.

Limitations of the current work include differences between the surrogate arm (hinge joint with only flexion/extension) and human arms with musculature/tissue dynamics. Furthermore, the braided fibers used were not optimized for specific load displacement curves that more closely mimic real-world falls for preventing extension after 50° as would be desired for human application.

We conclude that the prototype Engineered Textile Wrist Protector shows promise for providing active sports participants with an increased level of protection against wrist fracture due to wrist hyperextension.

Acknowledgments

This development project was supported by NIH SBIR R44AR049959 through the National Institute of Arthritis and Musculoskeletal and Skin Diseases.

References

Barnaby W, *Fractures and dislocations of the wrist.* Emergency Medicine Clinics of North America 1992; 10, 133-151.

Bladin C, McCrory P, Pogorzelski A. *Snowboarding Injuries Current Trends and Future Directions*, Sports Med 2004;34 (2) 133-138

Greenwald RM, Janes PS, Swanson SC, MacDonald TR. *Dynamic Impact Response of Human Cadaver Forearms Using a Wrist Brace*, American Journal of Sports Medicine. 1998; 26(6): 1-6, 1998.

Idzikowski J, Janes P, Abbot P, *Upper extremity snowboarding injuries: ten-year results from the Colorado Snowboard Injury Survey.* The American Journal of Sports Medicine 2000; 28(5), 825-832.

Machold W, Kwasny O, Eisenhardt P, Kolonja A, Bauer E, Lehr S, Mayr W, Fuchs M, *Reduction of Severe Wrist Injuries in Snowboarding by an Optimized Wrist Protection Device: A Prospective Randomized Trial.* The Journal of Trauma: Injury, Infection, and Critical Care. 517-520.

Mecham MD, *Incidence and severity of head impact during freestyle aerial ski jumping..* Master's thesis, University of Utah, Salt Lake City, UT, 1997

O'Neill DF, *Wrist Injuries in Guarded Versus Unguarded First Time Snowboarders.* Clinical Orthopaedics and Related Research, 2003, 409, 91-95

Rønning R, Rønning I, Gerner T, Engebretsen L, *The Efficacy of Wrist Protectors in Preventing Snowboarding Injuries.* The American Journal of Sports Medicine 2001; 29(6), 581-585.

Schieber R, Branche-Dorsey C, Ryan G, Rutherford G, Stevens J, O'Neill, J. *Risk factors for injuries from in-line skating and the effectiveness of safety gear.* The New England Journal of Medicine , 1996; 335, 1630-1635.

Staebler M, Moore D, Akelman E, Weiss A, Fadale P, Crisco J, 1999. *The effect of wrist guards on bone strain in the distal forearm.* The American Journal of Sports Medicine 27, 500-506.

Do Helmets Reduce Fatalities or Merely Alter the Patterns of Death?

Jasper E. Shealy[1], Robert J. Johnson[2] and Carl F. Ettlinger[3]

[1] Rochester Institute of Technology, USA, jeseie@rit.edu
[2] University of Vermont, USA
[3] Vermont Safety Research, USA

Abstract. The use of helmets has been proposed as a means of reducing the incidence of fatality in skiing and snowboarding. This paper presents results that suggest that while helmets may be effective at preventing minor injuries, they have not been shown to reduce the overall incidence of fatality in skiing and snowboarding even though as many as 40% of the population at risk are currently using helmets. The results indicate that the use of a helmet will indeed influence the primary cause of death, but perhaps not the ultimate outcome.

1 Introduction

Death in alpine sports (skiing and snowboarding) remains a relatively rare event. In the US, from the time period 1978 to 2004, the incidence rates have been established as 0.75 deaths per million visits (DPMV) in skiing and 0.53 DPMV in snowboarding (Shealy, Johnson and Ettlinger 2006). The inclusion/exclusion criteria for these deaths are: in-bounds; trauma related while skiing or snowboarding; non-lift related; to a customer; and not employment related. Over this period of time the incidence rates have not shown any significant trending up, or down. The most common primary injury has been some sort of a head injury, at approximately 60% of all deaths (Shealy, Ettlinger and Johnson 2000). While some sort of head injury is usually the first listed cause of death, most of the fatalities also involve multiple, or secondary trauma sites; single causes of death are not common. Helmets have been recommended as a means of reducing the incidence of death since head injuries are the most commonly noted cause of death (U. S. Consumer Product Safety Commission 1999).

Helmets have been noted in clinical and case-control studies to reduce the incidence of any head injury by 35 to 50% (Hegel, Pless, Goulet, Platt, and Robitaille 2005; and Macnab, Smith, Gagnon, and Macnab 2002). In a prospective epidemiological case-control study at Sugarbush VT, a similar reduction was noted when all head injuries are considered, however the salutary effect was limited to the less serious head injuries; such as scalp lacerations and mild concussions; no significant effect noted for the more serious head injuries such as concussions more severe than mild, closed head injury, skull fracture and death due to head injury (Shealy et al 2006).

Helmet utilization within the population of alpine sports participants in the U.S. has shown a steady increase since the year 2000 of about 4 to 5% per year per year. For the 2004/2005 season the overall utilization rate within the U. S. was estimated to be 33.2% (RRC Associates 2005). In terms of helmet user age, the utilization rate in the 04/05 season was highest for those under nine years of age at 66%, next highest for those over 65, at 46%, Helmet utilization was higher for males than females (35.2% versus 30.4%), and for the most skilled participants (45%).

2 Objectives

The objective of this study is to examine the distribution of stated primary cause of death as a function of helmet utilization for the last five winter seasons in U.S. to determine what effect the use of a helmet has on the pattern of primary cause of death.

3 Methods and Materials

Since 1985, the authors have tracked deaths to skiers and snowboarders in the U. S. by a variety of means. The US Consumer Product Safety Commission purchases abstracted death certificate information from most of the more than 6,000 counties in the U.S. There is often a lag of many months to as much as a few years in that process. If skiing or snowboarding is not listed as a product related to the death, then that death may go unreported. It is frequently difficult to determine if the death took place within the bounds of a resort or not. The National Ski Areas Association (NSAA) releases annual summary statistics on the number of deaths to skiers and snowboarders, but does not provide detailed information on the individual incidents. NSAA also provides annual data on the number of skier and snowboarder visits.

With the advent of the internet, it has become possible to electronically survey media reporting of events such as skiing and snowboarding fatalities nationwide on a more or less real time basis. The web based incident specific information is the basis for the fatality data in this report.

4 Results

According the NSAA, from the 2000/01 seasons through the 20004/05 season there were a total of 215 deaths within the U.S. skiing and snowboarding population that met the study inclusion/exclusion criteria. From the internet, incident specific data were obtained on 140 of those reported deaths. Not all 140 cases had complete data. Only 76 of the 140 cases had specific information on helmet usage. 28 of those 76 fatally injured (36.8%) were said to be using a helmet at the time of death, and 48 (63.2%) were said to not be using a helmet at the time of death.

The incident specific information has not been independently verified by such means as a professional autopsy or coroner's report.

The mean age of the known helmet users and non-helmet users was not significantly different (30.8 versus 26.0 years of age respectively), nor was there a significant difference in the gender distribution (22.4% female overall).

For 22 of the 28 known helmet users, the primary cause of death, or primary body part injury was given. Ten of those 22 helmet users (45.5%) had head or some sort of head injury listed first.

For 36 of the 48 known non-helmet users, the primary cause of death, or primary body part injury was given. Twenty-six of those 36 non-helmet users (72.2%) had head or some sort of head injury listed first.

The difference in stated primary cause of death for the two groups (known helmet users and known non-helmet users) is significant by means of a Chi-Square 2x2 contingency test (C-S = 5.845, p-value = 0.0156; 95% confidence interval 4.535>2.300>1.167). Non-helmet users were 2.3 times more likely to be reported to have died as a result of some sort of a head injury than those who died while wearing a helmet.

5 Discussion

Previous research (Shealy, et al 2006) has shown that even though the prevalence of helmet utilization is rising by 4 to 5% per year in the U.S., there has been no statistically significant observable effect on the incidence of fatality. Research (Hagel, et al 2005, and Macnab, et al 2002) has shown that helmets are an effective means of addressing minor head injuries such as scalp lacerations, and may even be effective in reducing the prevalence of mild concussion (Shealy, et al 2006). In order to understand why helmets may not be as effective at addressing the more serious head injuries that result in death one needs to understand the circumstances of the typical fatal scenario and the physical limitations of practical helmets.

Typical maximum speeds (Shealy, Ettlinger and Johnson 2005) for all skiers and snowboarders under normal circumstances average 43.5 km/h (27 mph). Skilled young adult male skiers and snowboarders go even faster than the rest of the population. Skilled young adult male skiers and snowboarders are the most commonly fatally injured. Those speeds occur on wide, smooth well-groomed intermediate skill level trails, which is also where most fatalities occur. A review of the narrative of most fatality reports shows that the typical fatality occurs to an experienced male between late teens and late 30s in age, while traveling at a relatively high speed on the margins of what are called 'Blue Square" trails in the US (Shealy, et al 2000).

The American Society for Testing and Materials has a standard for snow sports recreational helmets (ASTM F2040). The test criteria require the helmet to limit the measured g forces to an accelerometer embedded in the head form to not more than 300g following a 2.0m drop onto a steel surface. The impact speed following a 2.0m drop is 6.2 m/s, or 22.3 km/h (13.9 mph).

A generally accepted reasonably safe impact g-limit to the brain is on the order of 180g (Mertz, Irwin and Prasad 2003). Research indicates that minor brain damage may begin to occur at levels as low as 150g, and by 275g significant injury of some sort to the brain is highly probable (Patrick, Mertz, and Kroell 1993).

An insight as to why this study found a difference in patterns of death as a function of helmet utilization can be found in the following study. A field simulation (Scher, Richards and Carhart 2005) was done of a snowboarder going 30 km/h (18.5 mph), catching an edge and falling head first onto; soft snow, icy snow or a fixed object (28 cm upright wooden post). This simulation was done to assess the effect of wearing a helmet or not under the three different impact conditions. The helmet in question met the requirements of ASTM F2040. The g-loads to the headform were measured and the associated Head Injury Criterion (HIC) values were computed. They found that if the impact is onto a soft snow surface, both the measured g-loads (under 100g) and the computed HIC values (less than 220) are well within acceptable limits regardless of whether a helmet is used or not. When the impact was onto icy snow, the helmet reduced the average measured g-load from 329 to 162, and the HIC value from 2235 to 965. When the impact was against the fixed object, the helmet reduced the values from 696g to 333g and the HIC from 12185 to 3299. They concluded that under the circumstances of impact with soft snow, the use or non-use of the helmet had no significant effect; the impact was not likely to result in any significant injury. In the matter of the impact with a fixed object, while the use of a helmet was associated with a significant reduction in both the g-load and the HIC, the likely outcome remained that of a fatal injury. With an impact on icy snow, the use of a helmet could be the difference between a significant head injury (possibly life threatening) or a minor head injury.

Thus we can see that if the accident scenario involves a skier or snowboarder traveling at average, or above average speeds, and they experience a direct impact with a fixed object such as a tree trunk, the outcome is likely to be same, i.e., death, regardless of the use of a helmet or not, but the severity of the fatal head injury is likely to be less.

6 Conclusions

Most fatalities appear to occur under circumstances that are likely to exceed the protective capacity of current helmets designed for recreational snow sports. While helmets can reduce the impact to the head, it is entirely possible to overwhelm that degree of protection. It appears that the kinetic energy in most death scenarios is so massive as to overwhelm the degree of protection that a helmet can offer. Data from sources outside the scope of this study indicate no decline in fatality incidence even though helmet utilization within the high risk group of skilled/experienced young adult participants is on the order of 40% or more. It is clear that the pattern of death is different as a function of helmet utilization, but there is no evidence of any effect on the incidence fatality rate.

References

Macnab, A. J., Smith, T., Gagnon, F. A., and Macnab, M., (2002) Effect of helmet wear on the incidence of head/face and cervical spine injuries in young skiers and snowboarders, *Inj. Prev.* 8;324-327.

Hegel, B., Pless, I. B., Goulet, C., Platt, R. W., and Robitaille, Y. (2005) Effectiveness of helmets in skiers and snowboarders: case-control and case crossover study, *BMJ*, 330:281

Mertz, H. J., Irwin, A. L., and Prasad, P. (2003) Biomechanical and scaling bases for frontal and side impact injury assessment reference values. Stapp Car Crash J. 47:155-158.

Patrick, L. M., Mertz, H., and Kroell, C. K., (1993) Cadaver knee, chest and head impact loads, *Biomechanics of Impact Injury Tolerances of the Head-Neck Complex, PT-43,* Stanley H. Backaitis, Ed., Society of Automobile Engineers, Inc., pp. 255-263.

Scher, I., Richards, D., and Carhart, M. (2005) Head contact after catching an edge: An examination of snowboarding helmets. Presented at ISSS 15, Akai, Japan April 2004.

Shealy, J. E., Ettlinger, C. F., and Johnson, R. J. (2000) Rates and modalities of death in the U. S.: Snowboarding and skiing differences- 1991/92 through 1998/99. *Skiing Trauma and Safety: Thirteenth International Symposium, ASTM STP 1397*, American Society for Testing and Materials, West Conshohocken PA, pp. 132-138.

Shealy, J. E., Ettlinger, C. F., and Johnson, R. J. (2005) How fast do winter sports participants travel on alpine slopes? *Skiing Trauma and Safety: Fifhteenth International Symposium, ASTM STP 1464*, American Society for Testing and Materials, West Conshohocken PA, pp. 59-66.

Skiing Helmets: An Evaluation of the Potential to Reduce Head Injury, (January 1999) U. S. Consumer Product Safety Commission, Washington, D. C., 20207, 17 pages.

5 Computer Applications in Sports

Synopsis of Current Developments: Computer Application in Sports

Arnold Baca

University of Vienna

Research activities are strongly affected by actual developments in computer science, such as in the areas of hardware (increasing processor speed and storage capacity, advances in communication technology), software (powerful tools), information management (data bases, data mining) and media (internet, eLearning, multimedia).

Currently, the following main areas of research are distinguished:

- Data acquisition, processing and analysis
- Modelling and simulation
- Data bases and expert systems
- Multimedia and presentation
- Learning, coaching and teaching technologies

Most current research activities can be assigned to one of the following topics:

- Motion analysis
- Game and competition analysis
- Training and performance analysis
- Pattern recognition
- Complex systems
- Pervasive / Ubiquitous Computing
- Instruction, training, Virtual Reality

All contributions accepted for this topic area may also be assigned to the research areas and, moreover, to the research activities listed above.

In addition to a topic overview, eight papers illustrate recent developments. Three contributions fall into the category *Pervasive / Ubiquitous Computing* (research areas: *Data acquisition & processing* and *Presentation)*, two into the categories *Training and performance analysis / Game and competition analysis (Modelling)* and three into the category *Instruction & Training (Data bases, learning and coaching technologies)*.

Pervasive / ubiquitous computing

The integration of sensor-, information- and communication technologies provides new means for developing systems to acquire data during sports performance, in training and competition. Various small ("smart") sensors and devices are incorporated into the sports equipment or attached to the athlete. Mobile computers store and present the data recorded, other systems use telemetric methods to transmit the data acquired to receiving stations, which then process and adequately present them. Portable devices, which are not bound to laboratory conditions, are particularly useful. Numerous developments in elite, mass and health sport are expected in the near future.

Three papers included in this section address applications of these technologies. New options offered for computer-supported training are discussed by Wiemeyer. A prototyping example from mountain biking is presented. Stevens, Wulf, Rohde and Zimmermann introduce a framework for developing concepts for ubiquitous fitness support. Relevant contexts for an intelligent adaptation of devices offering support for ubiquitous fitness are derived. A practical example of the development of an innovative computer-supported fitness device is presented in the contribution by Henneke, Hoisl, Schönberger and Moritz.

Analysis of training & performance / game & competition

Various models are developed and utilised in sports. Their application ranges from analyses and simulations of human motion to investigations of the players' behaviour in game sports and includes studies of the interaction of training and performance. Soft computing methods provide interesting and promising alternatives in this area.

Two papers related to this field are included in this section. Ganter, Witte and E-delmann-Nusser compare two computerized methods for modelling the relationship between training load and performance, Sierksma studies the relation between the performance of an individual s player and the performance of the whole team.

Instruction and Training

In order to assist athletes in training or for instructional purposes sophisticated hard- and software systems are developed.

Three examples illustrate these activities. Miyaji, Ito and Shimizu present an internet based data base for sports movements. They use streaming technology and metadata allowing the user to retrieve selected scenes from movie files. A computer based training aid for cricket bowling is introduced by Justham, West and Cork. Vernadakis, Zetou, Avgerinos, Giannousi and Kioumourtzoglou investigate the effect of multimedia instruction when learning certain skills in volleyball.

Development and Integration of a Novel Cricket Bowling System

Laura Justham, Andrew West, Alex Cork

Loughborough University, L.Justham@Lboro.ac.uk

Abstract. Cricket bowling is a complex skill which cannot easily be re-created using machines. Bowling machines are used by players of all standards to provide an unlimited number of deliveries, launched towards the batsman in a repeatable manner for training purposes. However they have a number of limitations which results in them not being used as extensively or as productively as they could be if they were more realistic. The purpose of this paper is to explore the requirements of players and coaches when considering the ideal bowling machine and the use of the systems engineering approach of Quality Function Deployment (QFD) to ensure a novel bowing system fully caters for these needs. A state of the art bowling system is being designed at Loughborough University using this methodology; a prototype machine has been built and tested at the England and Wales Cricket Board's National Cricket Centre.

1 Introduction

Cricket bowling is a dynamic and complex skill which involves both the physical delivery of a cricket ball and the tactical considerations used to outwit the batsman and force them to make errors during a match. There are three main types of bowling; fast, medium and spin. The fast bowler relies upon the ball traveling at speeds between 75 mph and 100 mph such that the time between the ball leaving the bowler's hand and reaching the bat is less than half a second. The medium paced bowler, who generally bowls at speeds between 60 mph and 75 mph, uses the protruding external seam to affect the boundary layer and air flow around the ball during flight, which causes the ball to drift sideways. There is an overlap between medium and fast paced bowlers, leading to descriptions such as fast-medium being used to describe the technique. Spin deliveries are the slowest with speeds of 40 mph to 60 mph being observed. The ball is launched with a fast rifle style spin rather than the generic backspin which is imparted onto the quicker bowled deliveries. The spin delivery may take up to 1 second to reach the batsman, so the ball tends to be pitched closer, which increases the uncertainty in the delivery as the ball has an unpredictable motion off the pitch. Batsmen attempt to preempt the characteristics of the ball delivery from the bowler's run-up. They tend to use previous knowledge of the bowler and notable changes observed during the run up and delivery stride to predict variations in the upcoming delivery and adjust their shot accordingly (Abernethy 1981; Abernethy 1984; Penrose 1995).

Engineering and technology is playing an ever increasing role in sport with elite level athletes using computer based performance analysis tools, training aids and development techniques. In cricket this includes tools such as the Hawk-Eye ball tracking system, video analysis software such as SiliconCOACH, high speed video motion capture techniques such as the Vicon Peak system, ball by ball video and data capture for match analysis using programs such as Feedback Cricket or Crickstat and training aids such as bowling machines to name but a few.

2 Bowling Machines

Player requirements: ▐▐◼◼◼◼◼►

•Range of ball release speeds

•Types of deliveries possible

•Variability in the machine set-up and use

•Ease of use and portability

•Machine cost

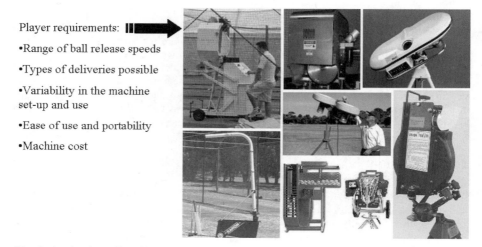

Fig. 1. A selection of bowling machines which are currently available on the market. (pictures taken from Manufacturer and media websites (from top left clockwise): www.bbc.co.uk, www.probatter.com, www.bola.co.uk, www.kdsports.com.au, www.mlb.com, www.masterpitch.com, www.kanon.co.za, (center) www.jugs.com).

Bowling machines are a valuable source of an unlimited number of deliveries which allows the batsman to carry out training drills and basic match preparation. They are used extensively when the batsman wishes to develop new skills or practice a specific type of shot because he can concentrate fully on the areas of most importance and ensure he trains purposefully for his own needs. They are able to create a limited range of deliveries but their design and mechanical reproduction of a delivery causes the batsman's response to be altered due to a lack of realistic visual cues (Gibson 1989) and therefore they are used sparingly during the playing season as they are not a realistic representation of an actual human bowler.

The original bowling machines were catapult style ball launching devices which impart a large backspin component onto the ball but inhibit any other type of spin. It was not until the early 1970's that a different machine design was developed, which used two counter rotating wheels to launch the ball (Berry 1987). The two rotating wheels and catapult arm designs are not the only bowling and pitching machines available; one, three and four rotating wheels, concave rotating discs, compressed air cannon style ball launching devices and simple 'toss practice' machines are also manufactured globally. Similarities between baseball and cricket means that with only slight adaptation pitching machines may be used for cricket training, which opens up a much larger market of available equipment. Figure 1 shows a selection of both baseball and cricket machines which are available on the market. Each type of machine has different functionality including different ranges of ball release speed, deliveries which may be created and variability in the machine settings such as height or angle of ball release. In part they all provide the functionality required by the end users and with the use of techniques such as Quality Function Deployment (QFD) the optimal design for a novel bowling machine may be developed.

3 Quality Function Deployment

QFD is a structured approach to develop and represent product design strategies. Matrices are used to communicate each step of the QFD process and since it is based upon a systems engineering approach, active involvement from a number of inter-related disciplines is required in order to successfully complete the process.

The first step in the QFD process is to develop a product plan which helps to define the overall strategy, consider target markets and prospective customers, consider the product line which will be available, investigate existing competitor products and to plan, schedule and organize the project. The second, and arguably the most significant step is to capture the 'voice of the customer' (VOC). The VOC is diverse because individuals have different opinions and something which is very important for one customer may be felt to be surplus to requirements by another. Therefore it is important to define the target group and ensure that their requirements are used. In this research the target groups of customers are elite cricket players and coaches. They were interviewed and their general requirements were broken into more specific areas before being organized into the key customer needs and grouped into logical clusters.

For this paper the next step in the QFD process, the product planning phase, has been carried out for the design and development of a novel cricket bowling machine. Many of the currently available machines have been developed to simulate bowling rather than to accurately recreate a technically correct delivery, identical to a real bowler. The novel system presented in this paper has been developed using a systematic approach to accurately recreate any mainstream cricket bowling delivery using the VOC requirements of elite level players and coaches based at the England and Wales Cricket Board National Cricket Centre (ECB-NCC) and a technical evaluation of existing machines used at the NCC for training purposes.

4 Development of the Novel Bowling System Using QFD

	A	B	C	D	E	Max.
Delivery capabilities	45.5	42.5	73	75.5	85.5	100
Accurate ball feeding	11.5	12.5	11.5	14	18	20
Visualization of delivery	8	8.5	8.5	8.5	37.5	40
Miscellaneous	33.5	25.5	36.5	37	66	75
Total Score	**98.5**	**89**	**129.5**	**135**	**207**	**235**

Table 1: The results obtained from the competitor evaluation carried out to measure how existing machines and the design for the novel system respond to customer requirements. The final column details the highest score possible in each category

Delivery Capabilities	A	B	C	D	E
Ability to bowl any common delivery	4	5	2	2	1
Technically correct deliveries	4	5	3	2	1
No deterioration or increased variation with time	5	1	1	1	1
Pace deliveries from 25mph to over 100mph	3	5	4	2	1
Spin in all realistic orientations	4	4	2	2	1
Leg and off spin deliveries	4	4	1	1	1
Controllable spin speed	4	4	3	2	1
Controllable spin orientation	4	4	3	2	1
Realistic spin deliveries (flight turn and dip)	4	4	1	3	1
Variable angle and height of ball release	4	5	2	3	1
Variable line and length of delivery	4	5	1	1	1
Controllable delivery parameters	4	5	2	2	1
Controllable line and length	4	5	3	1	1
Variation simulating that of human bowling	5	4	2	3	1

Table 2: A weighting table showing how well existing machines and the novel bowling machine designs respond to the VOC for the specific objective of recreating an accurate bowling delivery (where 1 is best and 5 is worst).

The requirements identified from the VOC were prioritized depending on their relative importance using a 1 to 5 rating. The major needs and requirements of players and coaches were established and grouped logically according to the area of training in which they fell. Over 45 requirements were identified which were based around three major objectives; the accurate and realistic recreation of bowling deliveries, the use of a real cricket ball which is correctly oriented to allow the batsman to 'pick the seam' during the ball's flight through the air and the use of visual cues to allow the batsman to pick up advance information about the delivery.

The customer requirements were then used to compare and measure the prototype of the novel bowling system against existing competitor products which are already in use at the NCC and other cricket clubs. To ensure an accurate understanding of the competitor machines, controlled testing of the machine capabilities was carried out as well as a web-based survey and discussions directly with the manufacturers. Three competitor machines (Machines A, B, and C), the first generation prototype (Machine D) and the second generation design (Machine E) were considered for the competitor evaluation and the final scores obtained for each of the objectives above are displayed in Table 1.

Table 1 shows that the first generation prototype novel bowling machine (machine D), which does not yet include all the capabilities that have been designed for the new system, is already marginally better than the best current competitor machine. It can be seen that the areas in which current machines suffer most are those of the visualization, which enables the batsman to judge the delivery and obtain accurate visual cues before the ball release as they would from a real bowler, and the miscellaneous section which included requirements such as "variable delivery sequences", "Easy delivery input to allow targeted training" and "Easy to set-up and use". Table 2 shows a more specific breakdown of selected requirements for the recreation of bowling deliveries objective in the form of a weighting table, where 1 is classified as the best and 5 is considered the worst machine of the five.

Product requirements and technical characteristics were then established to respond to the customer requirements. Interactions between the customer and product requirements were determined to highlight trade-offs needed to obtain the optimal product design. A number of improvement goals were then developed which focused on an autonomous system that could be programmed to carry out complete deliveries including a full visualization to allow the batsman to predict the delivery from real advance cues rather than unrealistic triggers currently associated with bowling machines.

5 Discussion

A complete house of quality QFD matrix was developed to help analyze the requirements and finalize the design and planning stages of a novel bowling system. After interviews with players and coaches at the ECB-NCC an extensive set of customer needs and technical characteristics were drawn up from the functionality and requirements of an ideal bowling machine. These were identified with respect to the advantages, disadvantages and capabilities of existing bowling machine technology.

The initial machine (Machine D in the QFD matrix) was built and tested alongside the other existing machines at the ECB-NCC but does not yet incorporate the full range of capabilities required for the final system (Machine E in the QFD matrix). It includes the essential elements to correctly orient the ball and reproduce technically correct deliveries at a manually controllable angle (pitch and yaw) and height of release. It does not include the full visualization system or fully automated control, as shown diagrammatically in Fig. 2. QFD has enabled the machine to be designed and built to accurately meet the critical customer requirements and technical characteristics to greatly improve existing cricket bowling machine technology. Maintenance of the QFD matrix has led to further development and a refined product design to incorporate a fully synchronized visualization system with the novel machine. QFD has simplified the design process and enabled an optimal system to be developed directly.

Machine D

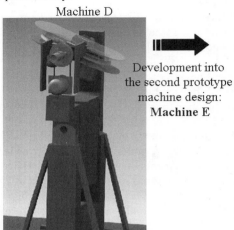

Development into the second prototype machine design:

Machine E

•Automated variation of ball release position

•Simulation of a human bowler's variation

•Synchronised visualisation of the bowler's delivery to provide visual cues

•Portable machine

•Easy to set-up and use through a graphical user interface

Fig. 2. A CAD model of the first prototype design which has been developed using QFD, and the major development areas required for the second prototype machine.

Acknowledgements

The authors would like to acknowledge the financial support of the Engineering and Physical Sciences Research Council of Great Britain (EPSRC) and the IMCRC at Loughborough University. They would also like to thank the players and coaches at the ECB-NCC and the technical staff at Loughborough University for their ongoing support and assistance.

References

Abernethy, B. (1981). "Mechanisms of skill in cricket batting." Australian Journal of Science and Medicine in Sport **13**(1): 3-10.

Abernethy, B., Russell, D.G. (1984). "Advance cue utilization by skilled cricket batsmen." Australian Journal of Science and Medicine in Sport **16**(2): 2-10.

Berry, S. (1987). "Bowling Machines." Cricketer International **68**(9): 15.

Gibson, A. P., Adams, R.D. (1989). "Batting stroke timing with a bowler and a bowling machine: a case study." Australian Journal of Science and Medicine in Sport **21**(2): 3-6.

Penrose, J. M. T., Roach, N.K. (1995). "Decision making and advanced cue utilization by cricket batsmen." Journal of Human Movement Studies 29(5): 199-218.

Creating the SMART system—A Database for Sports Movement

Chikara Miyaji, Koji Ito, and Jun Shimizu

Japan Institute of Sports Sciences, Tokyo, Japan, Chikara.MIYAJI@jiss.naash.go.jp

Abstract. The authors built a database for sports movement, called the SMART system-- Sports Movement Archiving and Requesting Technology system. The main characteristics of the system are: (1) Features sports oriented streaming, (2) Using metadata for searching the scene, (3) Versatile authentication system, and (4) Open system architecture. This article describes the overall system construction—streaming, searching strategies, data structure, and authentication system.

1 Introduction

Improving one's skill is one of the main goals of training and teaching sports technique, and videos are the most commonly used tools. Today, using videos on computers have become popular because of the benefits of faster access and easier handling. But a lot of problems still exist—a need for large disk space to hold many videos, difficulty in searching for a particular scene, and rare versatile analysis software for the movement.

The SMART system solves these problems:

- **Solves disk space problem**: This system uses distributed streaming servers for video resources.
- **Sports user-friendly browser**: The system's viewer provides step and slow motion on streaming.
- **Search scene by query**: This system uses tagged metadata to search scenes.
- **Provides annotation system**: The users can add annotations: drawings, text comments, and voice to the scenes.
- **Provides authentication system**: This system provides flexible authentication on distributed streaming servers.
- **Open system architecture**: this system provides viewer software and components freely available for further enhancement.

The SMART system is a server-client system; it contains a user server, search servers, annotation servers, a movie information server, streaming servers, download servers and client browsers. Basic processes are summarized : (1) the browser sends a query to the search server, (2) then the search server returns the list of scene information which match the query, (3) the user selects a scene to browse, (4) browser accesses the streaming server to present the scene, and show it with annotations, (5) this scene information can be saved as an object and reused later.

2 Using Streaming

Streaming is a technology that sends the movies as sequences of packets over the network. It benefits our movie database in several ways.

- No files are necessary to view movies on the local computer
- Easy-to-protect copyrighted content
- Content can be shared on a network

Recent improvements in compression technology have significantly decreased bandwidth requirements for high-quality streaming media. We found that good quality sports movies can be sent at 1Mbps bit rate with Windows Media 9, possible on the ADSL network environment commonly used in Japan. This gives great possibilities for our system. Although the format of the streaming of our system is not restricted to a certain type, we choose Windows Media for the first target of the streaming to implement our system.

Current use of streaming is one-way data transmission from a server to a client, but this is not suitable for browsing sports movement. Users want more interaction with the movie clips, such as pause, fast forward, step-by-step and slow motion replay. Unfortunately, Windows Media SDK (System Development Kit) cannot handle step-by-step and slow motion functionals for programming streaming. Instead of using Windows Media SDK, we choose the DirectShow library to program functionals of streaming and succeeded to realize the step-by-step and slow motion interactions. Compared to other browsing applications for streaming, our system provides to be the best suitable browser for sports movement.

3. Searching the Scene

To find a specific scene in a movie file, we defined metadata—sports event information. Although a complex combination of queries for these metadata can find a scene, users do not want to construct such queries. Instead of asking such complex queries, the SMART system provides a simple query guide that asks users only necessary arguments to define a query.

3.1 Sports Event Information

Sports event information is a set of time intervals where an event occurs and the content is defined as a single or multiple pair of names and values to specify a sport event. These names and values are special keywords that are defined by and familiar to a sports community, and these events are attached to all movies by specialists.

`event = {mid,ts,te,name,value}`

`mid`:	movie ID to specify a movie
`ts`:	start time of the event
`te`:	end time of the event
`name`:	name of the event
`value`:	value of the name to specify the event

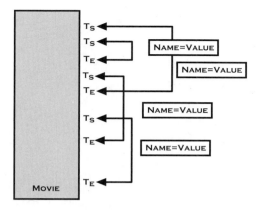

Fig. 1 Sports event information attached to a movie.

3.2 Interval Arithmetic

All events are stored in a relational database and a simple SQL command can find the list of $\{$`mid,ts,te`$\}$ where the interval matches with `name = NAME` and `value = VALUE`. This can be written as an expression `match(NAME,EVENT)` and is defined as below.

> `match(NAME,VALUE) = {{mid, ts, te}, {mid, ts, te},…}`---- (1)

Using this expression, the scene where $\mathbf{NAME}_1$ is $\mathbf{VALUE}_1$, and $\mathbf{NAME}_2$ is $\mathbf{VALUE}_2$ written as follow;

> `match(NAME₁,VALUE₁) AND match(NAME₂,VALUE₂)` -------------- (2)

This is calculated from the interval arithmetic of the AND conjunction for each interval list of a movie.

3.3 Pseudo Query

The interval arithmetic can be applied not only to AND but also to OR. We have defined several simple expressions we call pseudo queries to express complex queries.

- **match(name, value)** returns where name=name and value='value' with ambiguous match.
- **exact(name, value)** returns where name=name and value='value' with exact match.
- **sequence(exp1, exp2)** returns if an interval in exp1 is just before the interval in exp2.
- **anytext(text)** returns where value=text with forward/backword match.

> `pseudoQuery ::= exp;`

```
exp ::= func
   | ( exp )
   | exp AND exp
   | exp OR exp;
func ::= match(name,'value')
   | exact(name,'value')
   | sequence(exp1,exp2)
   | anytext('text');
```

For example, the query "A scene of Ichiro hitting a single to a pitcher whose name contains 'Randy'" is described as a pseudo query "match(batter,'ichiro') AND exact(result,'single') AND match(pitcher,'Randy')".

This pseudo query is sent to the search server where each expression is converted to a SQL command and the returned values are calculated as an interval arithmetic of AND/OR. Finally a list of intervals is returned that satisfy the query condition.

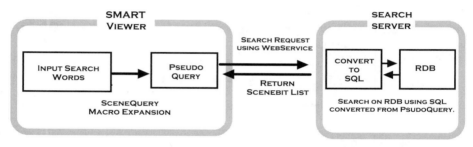

Fig. 2. Pseudo query is sent to the search server and displays the result.

The SMART viewer uses graphical inputs to define a pseudo query which is described as sceneQuery format. These graphical inputs consist of pulldown menus and simple text input that ask to fill necessary information to complete the pseudo query. Fig. 3 shows an output of such graphical input interface.

Fig. 3 Graphical input of the SMART viewer.

4 Data Format

The SMART system uses several definitions of information to hold and exchange the information via the Internet or the files. Below is the list of such information:
- **Scenebit**: Describes information of a movie scene.
- **Annotation**: Collection of annotation information for a movie.
- **Event**: Collection of event information for a movie.
- **SceneQuery**: Definitions that display queries on the SMART viewer.
- **QueryResource**: Definitions where SceneQueries are located.

We use XML (eXtended Markup Language) to express the above information because of its flexibility and readability. The main format of the system is called scenebit, which expresses information of a movie clip with additional appended information. Scenebit format also includes definitions of annotations and events.

5 System Construction

The SMART system consists of distributed servers and client software--the SMART viewer. Fig. 4 shows the information streams between servers and a client of the system.

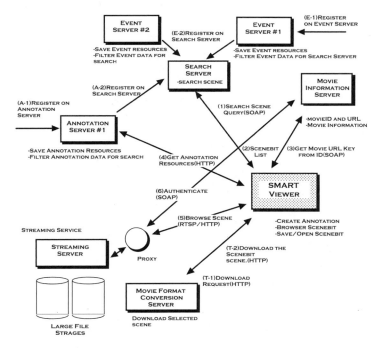

Fig. 4 Overall construction of the SMART system

184   Chikara Miyaji et al.

Several servers perform different tasks in the system:
- **Movie information server**: contains movie id and source URL.
- **Search server**: contains event and annotation information for searching scenebit.
- **Annotation server**: contains annotation information and resources.
- **Event server**: contains event information.
- **Proxy**: a java proxy to perform authentication of a streaming file.
- **Tibenecs server**: converts compressed movie scenes to an AVI files for download.
- **Streaming server**: contains movie file and services streaming.

Proxy server is placed before the streaming server to authenticate the user's access. Each streaming server authenticates access by the proxy, and this brings flexibility to construct distributed streaming servers without registering all users on every computer.

6 SMART Viewer

The SMART viewer is a client software that asks the search server queries and browses the returned scenebits. Below, a text and an arrow are displayed on the snapshot of a movie, which were added by the annotation capability of the software.

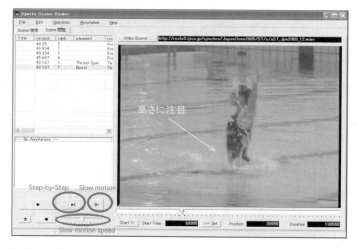

Fig. 5 SMART viewer—a client software to search and browse the scene

7 Conclusion

SMART viewer and other components are distributed by JISS and will be freely accessible to the sports community on the site of JISS.

Computer Application in Sports

Arnold Baca

University of Vienna, arnold.baca@univie.ac.at

Abstract. The paper gives an overview of the history of the application of computers in sports and of current research activites in this field. Consequences for university education in sports related disciplines are discussed.

1 Introduction

During the last decades, computer science has become an important interdisciplinary partner for sport. This is due to the fact that the use of data and media, the design of models, the analysis of systems etc. increasingly require the support of suitable tools and concepts which are developed and available in computer science. Research activities in this field are strongly affected by current developments in computer science. In particular, progress in hardware (processor speed, storage capacity, communication technology), software (tools), information management concepts and methods (data bases, data mining) and media (internet, eLearning, multimedia) are of great importance.

2 Computers in Sport – Historical Survey

First computer applications in sports are reported in the mid of the '60s (Lees 1985). Concepts and experimental methods applied in other scientific disciplines were adopted. Statistical analyses and numerical calculations of biomechanical investigations were performed with the mainframe computers available at that time. The computer programs developed by Plagenhoef (1969) published in the very first edition of the Journal of Biomechanics may serve as example. In the biomechanical literature numerous examples can be found propagating the use of computers since then (cf. Lees 1985). Measuring devices acquiring biomechanical data (reaction forces, EMG data, etc.) using analog-digital converters and opto-electronic systems for human motion analysis (cf. Furneé 1989) have been developed for decades. Moreover, modern computer technology was of crucial importance for solving numerical problems in biomechanical modelling and simulation (cf. Hatze 1980 and 1983).

Another early use of computers in sports originates from notational analysis. Rudimentary forms of notating or coding events in games can be found for centuries (Hughes and Franks 1997). Downey (1973) presented a complex notational system for lawn tennis matches. The objective method of game analysis in squash proposed by Sanderson and Way (1977) was basis for a computerised system developed by Hughes (1985).

Pioneering work in this area was also done by Miethling and Perl (1981). Their concepts for modelling sport games process oriented date back to the early '80s.

Computers have also long been in use for information and documentation purposes. A historical survey of the International Association of Sports Information (IASI 1994) reports 1967 as the year, where an automatic processing of sports documentation was first demonstrated at an IBM computer in Graz, Austria.

Applications of computer technology in Physical Education are documented since the mid of the '80s (Sharp and Paliczka 1984; Donnelly 1987; Skinsley 1987; Sharp 1988). On the one hand, it was sought to integrate the computer into lessons (Skinsley, 1986), on the other hand learning software ("computer aided instruction") was developed. In combination with modern multimedia technology such software has nowadays obtained a quality, which was not foreseen at that time.

More details of this historical development are given in Baca (2006).

3 Research Activities

According to the development of computer science, contemporary sports related research can take advantages in recording, analysing, and handling data. This improvement does not just mean "more" and "faster". The improved quality of data handling also and in particular allows for overtaking or developing new concepts and methods – as e.g. soft computing (Perl 2006).

Currently, the following main areas of research can be distinguished:

- Data acquisition, processing and analysis
- Modelling and simulation
- Data bases
- Multimedia and presentation
- IT networks

There are many research activities that cut across these areas. Most of them can be assigned to one of the following topics:

- Motion analysis
- Game and competition analysis
- Training and performance analysis
- Pattern recognition
- Complex systems
- Unconventional modelling / soft computing
- Pervasive Computing
- Instruction, training

- Virtual Reality (for a recent review see Katz, Parker, Tyreman, Kopp and Chang 2006)

An attempt has been undertaken to assign each oral presentation given at the 1st (1997) to 5th (2005) International Symposium on *Computer Science in Sport* to one of the areas of research listed above. With few exceptions, a best matching field of research could be identified and resulted in the following distribution: Data acquisition, processing and analysis; IT – 29%, Modelling and simulation – 35%, Data bases (and expert systems) – 8%, Multimedia (and educational tools) – 29%.

Two selected working fields of current research activities will be briefly introduced in the sequel, thus highlighting the fruitful interdisciplinary cooperation in *computer science in sport*.

3. 1 Pattern Recognition / Soft computing

Patterns in sport can be found as tactical patterns in a game, as motor patterns from movements, as training or performance patterns in training analysis and so on. Such patterns reduce the complex real information onto the most relevant parts – e.g. trajectories of a time-dependent process – and so help for easier analyses (Perl 2006).

In this way movements can be classified and represented by trajectories, as is shown in Fig. 1, where the trajectories of a rower exercising with two different speeds (20 strokes per minute – SR 20, 30 strokes per minute – SR 30) on a rowing ergometer can be compared under the aspect of intra-individual variability. Kohonen feature maps have been used to map state vectors of biomechanical parameter values (Perl and Baca 2003). It can easily be seen, that all the trajectories are rather similar to each other, but the rowing at higher speed is more stable than that at lower speed.

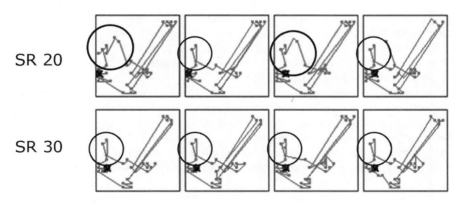

Fig. 1. Trajectories of ergometer rowing: Each node represents the 2-dimensional mapping of the corresponding high-dimensional vector of biomechanical attribute values. Each trajectory represents one cycle of the rowing motion at different times. Connected nodes of the trajectories represent successive states in time (Perl and Baca 2003)

3.2 Pervasive Computing

The integration of modern sensor-, information- and communication technologies provides new means for developing systems to acquire data in training and competition. Various sensors and devices are incorporated into the sports equipment or attached to the athlete. Mobile computers acquire and present the data recorded, other systems use telemetric methods to transmit the data acquired to receiving stations, which then process and adequately present them. Portable devices, which are not bound to laboratory conditions are particularly useful.

Moreover, such technological solutions are of relevance, which enable real-time data acquisition and/or integration of specific sensors, such as devices for position measurement.

As an example, the HP iPaq 5550 Pocket PC with PC Card Expansion Pack can be used for mobile data acquisition and wireless data transfer of biomechanical and/or physiological data.

Mobile GPS navigation systems allow to estimate positions, paths and tracks of motions and can be connected to a PDA or laptop using wireless Bluetooth technology. Because of inherent restrictions of such systems and methodological limits the interpretability of the data acqired is, however, restricted. For instance, even if the path of the sensor/receiver could be measured with highest precison, the distance covered by the sensor/receiver cannot be assigned to the running distance covered by the athlete, to whom the device is attached.

High accuracy can be achieved by navigation systems with locally positioned – for example along the edge of the court – receiving units (Fischer, Pracherstorfer, Stelzer and Söser 2003). Due to the availability of such technologies potentials not known so far arise for the automated analysis of the tactical behaviour of teams.

Extensions and/or variants of components, which are currently used or developed in mobile telemedicine ("*Mobile Health*") might be applied in health and mass sport in the near future. The *Mobile Health Toolkit* propagated by IBM can be seen as a typical representative of such system modules. This toolkit acts as the hub between Bluetooth supported terminal equipment (heart freuqency monitors, blood pressure gauges, etc.) and the service infrastructure (GSM/GPRS). Since merely standard components and technologies are used, a high acceptance and widespread utilization can be expected. For *sports engineering* and *computer science in sport* the challenge arises to design and implement adequate terminal equipment for systems based on such technologies.

Generally, the application of standardized components – in particular interfaces – appears to be of specific importance for a longterm, sustainable use of systems applied in sports related science.

4 Education

The development of *computer science in sport* and the increasing importance of knowledge in information and communication technologies and computer skills for career perspectives of students of sports studies show consequences in curricula. Lectures not only on basic tools of informatics but more and more also on complex tools, concepts and advanced methods are integrated into education programs (Wiemeyer and Baca 2001).

Moreover, several universities have started to offer specific course programs related to *computer science in sport*. Two approaches can be distinguished. Course programs are either set up supplementary to sports related studies (e.g. Technical University Darmstadt, Germany) or supplementary to studies of computer science as is currently put into practice at the author's university (University of Vienna).

Throughout the last years increasing efforts in developing multimedia based courses and materials can be observed in sports related disciplines (Katz 2003; Sorrentino, 2001; Igel and Daugs 2005; Wiksten, Spanjer and LaMaster 2002). Multimedia eLearning will facilitate distance learning and the international exchange of study modules.

It is expected that (national) master programs will be increasingly offered in the very near future. The master program (MSc Performance Analysis) established at the University of Wales (MSc Performance Analysis 2006) may serve as an example.

In addition, the evolution of international master programs (e. g. European Master) is expected.

5 Conclusion

There have been impressive developments in the field of computer application in sports and sport science throughout the last decades. Computer science in sport has established as a scientific discipline, fields of research have evolved, regular congresses are organized, and an international journal is published (IJCSS, http://www.iacss.org/ijcss/iacss_ijcss.html [14. 2. 2006]). Because of these advances consequences in education can be observed, which in their turn will show catalytic effects on the scientific progress.

References

Baca, A. (2006) Computer science in sport: An overview of history, present fields and future applications (part I). Int. J. Comp. Sci. Sports, Spec. Ed. 2, 25-35.

Donnelly, J. E. (Ed.). (1987) *Using microcomputers in physical education and the sport sciences.* Human Kinetics Publishers, Champaign.

Downey, J. C. (1973) *The singles game.* E. P. Publications, London.

Fischer, A., Pracherstorfer, G., Stelzer, A. and Söser, A. (2003). Local position measurement system for fast and accurate 3-D monitoring. In: S. R. Doctor, Y. Bar-Cohen, A. E. Aktan (Eds.), *Proceedings of SPIE, vol. 5048*, pp. 128-135.

Furneé, E. (1989) *Computer motion analysis systems: The first two decades*. Diss., Techn. Univ. Te Delft, Faculty of Applied Physics.

Hatze, H. (1980) Neuromusculoskeletal control systems modeling - a critical survey of recent developments. IEEE Trans. Autom. Control AC-25, 375-385.

Hatze, H. (1983) Computerised optimisation of sports motions: an overview of possiblities, methods and recent developments. J. Sports Sci 1, 3-12.

Hughes, M. D. (1985) A comparison of patterns of play in squash. In: I. D. Brown, R. Goldsmith, K. Coombes and M. A. Sinclair (Eds.), *International Ergonomics '85*, Taylor & Francis, London, pp. 139-141.

Hughes, M. D. and Franks, I. A. (1997) *Notational analysis of Sport*. E. and F. N. Spon, London.

IASI (1994) *Yesterday, today, tomorrow: Better sport documentation through international cooperation*. Available at http://www.iasi.org/publications/pdf/YesterdayTodayTomorrow.pdf [4. 1. 2006]

Igel, C. and Daugs, R. (2005) *Handbuch eLearning (Handbook of E-Learning)*. Hofmann, Schorndorf.

Katz, L. (2003) Multimedia and the internet for sport sciences: applications and innovations. Int. J. Comp. Sci. Sport 2, 1, 4-18.

Katz, L., Parker, J., Tyreman, H., Kopp , G. and Chang, E. (2006) Virtual Reality in Sport and Wellness: Promise and Reality. Int. J. Comp. Sci. Sport, Spec. Ed. 2, 5-16.

Lees, A. (1985) Computers in sport. Appl. Erg. 16 , 1, 3-10.

Miethling, W. and Perl, J. (1981) *Computerunterstützte Sportspielanalyse (Computer-assisted game analysis)*. Czwalina, Ahrensburg bei Hamburg.

MSc Performance Analysis (2006) Available at http://www.uwic.ac.uk/new/courses/sport/MSc_Performance_Analysis.asp [4. 1. 2006]

Perl, J. (2005) Computer science in sport: An overview of history, present fields and future applications (part II). Int. J. Comp. Sci. Sport, Spec. Ed. 2, 36-47.

Perl, J. and Baca, A. (2003). Application of neural networks to analyze performance in sport. In: E. Müller, H. Schwameder, G. Zallinger, F. Fastenbauer (Eds.), *Proc. 8th Annual Congress of the European College of Sport Science*, ECSS, Salzburg, p. 342.

Plagenhoef, S. C. (1968) Computer programs for obtaining kinetic data on human movement. J. Biomech. 1, 221-234.

Sanderson, F. H. and Way, K. I. M. (1977) The development of an objective method of game analysis in squash rackets. Br. J. Sports Med. 11, 188.

Sharp, B. (1988) Computing in physical education II. Scottish journal of physical education 16, 1, 13-16.

Sharp, B. & Paliczka, V. (1984) Computing in physical education. Scottish journal of physical education 12, 2, 10-18.

Skinsley, M. (1986) Computing and physical education. Br. J. Phys. Educ. 17, 3, 96.

Skinsley, M. (1987) *Physical education and the computer*. The Ling Publishing House, London.

Sorrentino, R. M. (2001) Designing computer enhanced instruction for sport. In: M. Hughes and I. M. Franks (Eds.), *Proc. Computer Science and Sport III and Performance Analysis of Sport V*. Centre for Performance Analysis, Cardiff, pp. 57-62.

Wiemeyer, J. and Baca, A. (2001) Education in computer science. In: J. Mester, G. King, H. Strüder, E. Solakidis and A. Osterburg (Eds.), *Proc. of the 6th Annual Congress of the European College of Sport Science*, ECSS 2001, Cologne, p. 69.

Wiksten, D. L., Spanjer, J. and LaMaster, K. (2002) Effective use of multimedia technology in athletic training education. J. Athl. Train. 37 (suppl.), 213- 219.

Ubiquitous Fitness Support Starts in Everyday's Context

Gunnar Stevens, Volker Wulf, Markus Rohde and Andreas Zimmermann
University of Siegen, Germany, gunnar.stevens@uni-siegen.de

Abstract. The paper describes the results of ethnographic studies of fitness and sports activities, and introduces a framework for developing concepts and solutions in this area.

1 Introduction

Today mobile devices deliver information that is adapted to the current location, network bandwidth or screen size of the user's device and commercial solutions push the mobile market to providing content and services on a variety of devices and visual displays. Future communication technology must take a step out towards new "smart" information services that use sensor fusion and multi-modal user interfaces for contextualized user interactions.

The concept of context-aware adaptable systems combines context-awareness with end-user development, in order to create highly adaptable and adaptive applications. Such applications of context-aware adaptable systems range from the fitness and wellness domain to technical support for device maintenance.

This contribution describes the results of ethnographic studies of fitness activities, and illustrates a framework for developing concepts and solutions in this area. Concepts for ubiquitous fitness support will be developed and validated by means of prototypical implementations. We follow the principles of a human-centered development process (ISO13407, 1999) and conform to an evolutionary and cyclic process model. We have started with qualitative analysis of the application domain and conducted a three month ethnographical study in fitness centers. We also conducted seven semi-structured interviews with trainees and three with fitness trainers. Finishing the first project phase, we identified several characteristics and will discuss their consequences for a ubiquitous fitness support.

2 Physical activities

Sports in a traditional understanding has been defined as "organized play that is accompanied by physical exertion, guided by a formal structure, organized within the context of formal and explicit rules of behaviour and procedures, and observed by spectators" (Anshel 1991, p.143). Still widely spread, this formalizing definition coerces sports into a specific scheme and strangely strangles the scope for innovation with respect to social and individual use value. However, there are also more context-sensitive approaches, defining sports as a "specific expression of human movement behaviour" (Haag 1996, p.8) that becomes "sports" only by "a situation-specific reception and an attribution of meaning" (Heinemann 1998, p.34). Eventually it is the purpose an individual assigns to a movement which he considers being sportive (which in many cases encompass "physical exertion"), that defines sports. Doing sports and playing games has many similarities, especially the voluntary character of the activities motivated by a perception of fun.

Following Webster, fitness is the quality or a state of being fit. Today the term 'fitness' normally refers to performing some physical exercises. People mostly do these exercises at a specific place like fitness centers or a room dedicated to fitness practices at home.

However, we believe that a ubiquitous fitness support will fall short, if it takes only physical exercises into account. In order to understand the fitness domain in a better way, it is useful to compare the fitness domain with the domain of sports, since both domains handle the issue of physical activities. Based in our previous experience in supporting cooperative sports (Moritz and Steffen 2003, Moritz 2004, Wulf et al. 2004) and taking into account the analysis of our empirical work in the fitness domain, we will argue that fitness and sports are two different styles doing physical activities.

In post-industrial societies like Europe or North America, physical activities are no longer part of the daily work practice. Moreover, the changing rhythm of work influences the opportunities and preferences for conducting sports. Due to lacking physical activities, people do not stay fit in an unobtrusive and natural way. As a result doing fitness becomes a distinct activity and people getting aware of this issue: Fitness becomes a life style and an attitude of human actors.

Observing and interviewing people who subscribe to fitness centers, we have identified five different motives in our empirical data:

1. Weight reduction (in particular at so called "problem-zones")
2. Extra fitness training to complement other sport activities
3. Muscle invigoration after injuries
4. Muscle growth (bodybuilding)
5. General fitness (staying healthy)

The motives and intentions of the people may be different, but it is remarkable that for most people these exercises are not a value by its own – and in particular not fun– but a way to reach the goals.

Although there do exist several other motivations to visit a fitness studio, such as meeting friends, a technical solution should take into account the question of how a trainee is supported in reaching her personal fitness goals. This does not mean that we should just look to the actual training activities. The opposite is the case. As the specific need for fitness activities is created by the pattern of the everyday's life in modern societies, we also consider this context.

3 Relevant contexts for ubiquitous fitness support

If we want to provide people with devices, that offer support for ubiquitous fitness, we need to analyze possible application areas and points in time. From this analysis we will be able to derive different contexts, in which the devices for fitness support might be applied. During our studies and observations in the fitness domain, we identified three relevant contexts for ubiquitous fitness support:
- During the actual training
- Staying at the fitness studio or similar places
- Anywhere else in everyday's life

Figure 1 makes these three contexts and their relation to each other transparent.

Fig. 1. Three Relevant Contexts for the Fitness Domain

These three contexts are the source for an intelligent adaptation of the device. The following paragraphs describe the contexts investigated and give examples for activities that occur in these contexts and that are taken from the interviews and observations we conducted.

3.1 During the Actual Training

"During the actual training" describes the context in which the trainee practices at the fitness device. Examples for tasks that occur in this context are:
- The trainee conducts fitness activities within the studio and wants to get information about his activity, e.g. his heart rate, in an ongoing way.

•   The trainee wants to get information about the settings of the fitness device (e.g. time, power level, tempo, calorie consumption)
•   The trainee wishes to adapt the settings of the fitness device to the personal needs and goals. Thus, it is possibly to perform exercises more efficiently.

3.2 Staying at the Fitness Studio

"Staying in the fitness studio" describes the context, in which the trainee resides in a fitness environment, but is not actually conducting training activities. Examples of activities which occur in this context are:
•   The trainee notices an improvement of training efficiency. So he asks the trainer to adapt the workout-schedule. They discuss about the new goals to reach though exercises. After this clarification the trainer re-works a new plan
•   The body shaping course is over. Now the trainee wants to refresh himself at the fitness bar. He orders a fitness drink and talks to the other trainees sitting at the bar about her training activities or other private issues.

3.3 Anywhere Else in Everyday's Life

Here fitness related contexts of everyday's life (e.g. home, work and spare time) are spotlighted. Examples of activities that occur in these contexts are:
•   The trainee is a 45 years old project manager. She must react to projects needs very flexibly and often needs to reschedule her training plan. So she wishes a computer support to integrate her fitness goals into her challenging work environment.
•   The trainee lives on the 6th floor and has three children. Everyday the trainee has to go shopping and to carry the shopping goods six floors to her flat, but she is not aware of the fitness related effects of this activity.

4 Support needs in the fitness domain

Orthogonal to the different contexts mentioned above, we have identified three different needs for assistance that should be addressed by ubiquitous fitness support realized by information technology: Motivation, care-taking and advice.

4.1 Motivation

For most people doing work-outs in fitness centers, fitness is not a value in its own. Many people have the strong desire to stay fit, but find several excuses and reasons for not following the training plan in a correct manner. One way to deal with this lack of motivation can be "external" motivation.
The way to create and build up motivation depends on the trainee's preferences and the context the trainee currently is in. For example, when staying at the fitness centre, an integration of computer games with fitness devices may help to develop moti-

vation. This is contrary to the "Everyday's life"-context, in which we will need other forms of motivation. For instance the integration of intelligent reminders for fitness into the personal calendar may encourage motivation.

4.2 Care-Taking

In our interviews we see that in particular novice trainees are afraid that they perform exercises in a wrong way and that they injure their bodies. The partial loss of sensitivity for their bodies that stealthily appears through the typical office work is one reason for that. During the work-out a potential fitness support system should take care of novice trainees and protect them from conducting the exercises in an unhealthy manner.

Some producers of fitness devices experiment with new approaches of force feedback and new materials that allow an immediate adaptation of the fitness device. Such an approach opens new opportunities to take care of a trainee based on specific user models. In addition, a system could create warnings and bio-feedback in a way that the trainee will regain her body sensitivity and improve her own body-control.

4.3 Advice

One reason why people visit fitness studios is the fact that they want a professional advice for how to work-out in an efficient and health-encouraging way. Usually the advices of a trainer materialize in a trainings plan.

The trainings plan has two different levels: an exact plan for the execution on the micro-level that explains in detail which exercise should be executed how often on which fitness device. Moreover, a macro-level plan, gives hints about how often a trainee should visit the fitness studio in order to reach a specific training goal. Usually the macro-level spans over a period of several months.

Training plans play an important role in the fitness domain and should also play a central role in ubiquitous fitness support. They are a good candidate to bridge between the different contexts, e.g. "during the training" and "within everyday's life".

A better computer support creates new opportunities to provide accurate advices and to handle training plans more flexibly in a way that they can take the contingencies of everyday's life into account. Therefore, the monitoring of data about physical activities will be critical. It also offers the chance to compare training as it was planned with physical activities as they were actually conducted and thus a nominal/actual comparison can be reached.

Moreover, persistent histories of the trainings data will be good for motivational reasons. With the help of individually adaptable reporting functionality, the trainee will get feedback about the improvement of her constitution in a short-term and in a long-term perspective.

5 Ongoing work

In a next phase, we plan to deduce socio-technical concepts for a ubiquitous fitness support based on our empirical findings. While a technical realization is of critical importance, the concepts must also fit into the social life of the trainees. For instance, they must respect work rhythms and take into account existing infrastructures for fitness activities (like fitness-studios). In addition, the interest of stakeholders, such as educated trainers or producers of fitness equipment, has to be considered. These concepts will be specified by prototypical implementations and evaluated by means of field trials.

References

Anshel, M.A. (Ed.) (1991) *Dictionary of the Sports and Exercise Sciences*; Human Kinetics Books, Illinois.

Haag, H. (Ed.) (1996) *Sportphilosophie*; Verlag Karl Hofmann, Schorndorf.

Heinemann, K. (1998) *Einführung in die Soziologie des Sports*; Verlag Karl Hofmann, Schorndorf.

ISO13407 (1999) *Human-centred design processes for interactive systems*. International Organization for Standardization.

Moritz, E.F. and J. Steffen, (2003) Test For Fun – ein Konzept für einen nutzerorientierten Sportgerätetest; in: Roemer et al (eds) *Sporttechnologie zwischen Theorie und Praxis*; Shaker Verlag, Aachen, pp. 43-63.

Moritz, E.F. (2004) Systematic Innovation in Popular Sports, in: *5th Conference of the International Sports Engineering Association*, September 14-17, Davis CA.

Wulf, V., Moritz, E.F., Henneke, C., Al-Zubaidi, K. and G. Stevens (2004) Computer Supported Collaborative Sports: Creating Social Spaces Filled with Sports Activities, Proc. 3rd Int'l Conf. Educational Computing (ICEC 2004), Springer, LNCS, pp. 80-89.

Application of Different Computerized Methods for Modelling the Training-Performance Relationship

Nico Ganter, Kerstin Witte and Jürgen Edelmann-Nusser

Otto-von-Guericke-University Magdeburg, nico.ganter@gse-w.uni-magdeburg.de

Abstract. In the recent past, computer aided simulative models have been utilised to analyse the interaction of load and performance in training. The aim of the study was the evaluation of two models with antagonistic structure, the fitness-fatigue model and the PerPot model. Both were applied on an empirical dataset, containing training and performance data of ten sport students during an eight week cycling training period. The modelling was conducted using appropriate software tools. Results showed interindividually different qualities in fitting of modelled and empirical performances. Since the fitness-fatigue model resulted in marginal model parameters, interpretation remained questionable. The PerPot model offered reasonable delay values but due to different performance levels, interindividual comparison was difficult. For modelling the training-performance relationship both models revealed specific characteristics with some limitations in the interpretation of the model parameters.

1 Introduction

A main focus in training science lies on the analysis of the athletes' training process. The adaptation process as well as the performance response to physical training are known to be highly individual (Mester and Perl 2000). In the recent past, computer aided simulative models have been utilised to analyse the interaction of load and performance in training. Due to the complex nature of performances in sports, the main focus is set on disciplines that are closely related to physiological adaptations, such as the endurance sports (swimming, running, cycling, etc.).

Systems with an "antagonistic structure" have been established in modelling individual performance characteristics during the training process. The basic assumption of such models is, that the training load (as the input component) has both positive and negative effects on the performance (as the output component) of the "dynamical system" athlete. The quantity as well as the time characteristics of each component are specified by amplitude (potential) and time (delay) parameters.

The aim of the study is the evaluation of two antagonistic concepts, the "fitness-fatigue-model", proposed by Banister (e.g. Banister 1982) and the "PerPot" meta-model introduced by Perl and Mester (Mester and Perl 2000; Perl and Mester 2001; Perl 2001), for modelling the training-performance relationship in cycling. Therefore, an empirical dataset, containing training and performance data of an eight week cycling training period was utilised and modelling was conducted with appropriate software tools. The individual model parameters shall be used to interpret the adaptation behaviour of each subject to the cycling training.

1.1 Fitness-Fatigue Model

The basic assumption of the model is that two factors generated by repeated training (w), called fitness (p) and fatigue (f), affect performance (a). Each of them is mathematically described by a first order differential equation. The difference between the two components represents how well the athlete performs at any time (see Fig. 1 left). The quantity of impulses generated from training is defined by weighting factors ($k1$ and $k2$) and the decline of both components between two training sessions is described by an exponential decay with relative time constants ($t1$ and $t2$). With respect to the initial values ($a0$ – initial performance), individual model parameters are now estimated by fitting the modelled performances to real performance data until a least squares best fit, using Eq. 1 (equation modified after Banister 1982).

$$a(t) = a0 + k1 \bullet [p(t-i)e^{-i/t1} + w(t)] - k2 \bullet [f(t-i)e^{-i/t2} + w(t)] \qquad (1)$$

The values of the weighting factors and time constants allow for interpretation of the individual adaptation behaviour of the athlete, concerning a more fatigue or a more fitness induced performance response to training.

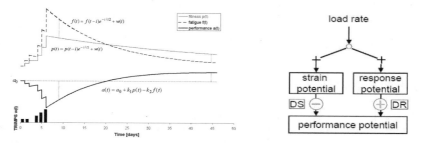

Fig. 1. Graphical representation of the fitness-fatigue model (left, Banister 1982) and the basic structure of the PerPot model (right, Perl 2001)

1.2 PerPot Model

The basic PerPot model implies flows between potentials. Two buffer potentials for strain (strain potential - SP) and response (response potential - RP) are fed by the load rate as the input component of the model (represented by the time series of training load). These potentials influence the performance potential (PP - as the output component of the model) in an antagonistic way: The response potential raises the performance potential and the strain potential reduces it (Perl 2001, see Fig. 1). The dynamical (and non-linear) behaviour of the model is achieved through specific delays of strain flow (DS) and response flow (DR), where flow rates additionally depend on the current states of the potentials. With this basic structure and extensions qualitative phenomena of physiological adaptation (supercompensation, overflow, inverted U) can be quantitatively described. The delay values as well as the initial and maximal values of the potentials are refined during calibration, by best fitting the time series of modelled to real performances. In this way, the model parameters are determined for each athlete with the corresponding time series, representing the individual adaptation behaviour.

2 Methods

2.1 Empirical Data

Training and performance data of ten sport students (9 male, 1 female) were obtained during an eight week cycling training. Subjects provided different prior cycling training experience (recreational to competitive) and performance levels, assessed by an incremental exercise test prior to the training. During training sessions with the personal bicycles on the road, heart rate was continuously recorded. The training programme consisted of periods with varying volumes and intensities, including an overreaching period with high volume and intensity and a taper period with reduced volume. All performance tests as well as the incremental tests were completed on a Cyclus2 ergometer (RBM elektronik-automation, Germany) with a road racing bike mounted. The mean mechanical power output was measured during a 30-seconds all-out cycling exercise as an indicator of the performance three times a week.

Training was quantified using the TRIMP score, where the training impulse is calculated from the duration multiplied by the delta heart rate exercise ratio of the training session (equation: cp. Banister 1982), in order to account for both the training volume and the intensity. The resulting daily scores then represented the time series of training load for each athlete.

The performance values were transformed by dividing the power outputs of the 30-s tests by the maximal value, intraindividually, reached during the training period (according to 100 units).

2.2 Modelling

All data were used to fit the models. The modelling was conducted using respective software tools. For determination of the fitness-fatigue model parameters, a data analysis software (Origin 7.0, OriginLab®) was used. With the aid of an implemented curve fitting algorithm (Levenberg-Marquardt) parameter sets were estimated within limited ranges (cp. Busso, Denis, Bonnefoy, Geyssant, and Lacour 1997) until a least squares best fit of modelled to original performances (Eq. 1). The initial performance ($a0$) was set to the first value of the testing period (after athletes had completed one familiarisation session). The ratio of $k1/k2$ was derived in order to make results interindividually comparable.

The PerPot modelling was done with the aid of the PerPotV10-4 simulation environment (Perl). Basically, the model implements an abstract time axis with equidistant values and requires complete pairs of load and performance data (missing performance values were replaced by 0 and thereby not included in the fitting procedure). To provide equidistant values, the original time series had to be modified. The time unit was chosen as the time between two test sessions, which was two or three days respectively (3 tests per week). The mean daily training load between two test sessions was used as the input time series. During the first step, normalization factors were individually derived in order to scale input and output parameters between 0 and 1. The next step included the calibration of the normalized data, which means that initial and maximal potentials as well as the delays were individually determined by a least squares fitting procedure of modelled and original performances. For interindividual comparison the ratio between the delay values (DS/DR) was computed.

The coefficient of determination ($R2$) was derived for both models between modelled and original performances as an indicator for the goodness of the fit.

3 Results

The subjects indicated different performance developments during the training period, not only depending on their initial performance level but also on the specific training regimen. Besides a trend of increasing performances for almost all subjects, variations in performances also occurred due to variable training loads. The results of the fitting process for subject M7 can be seen in Fig. 2 for the fitness-fatigue model (left) and the PerPot model (right). The estimated model parameters for all subjects are presented in Table 1.

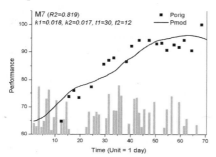 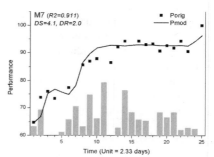

Fig. 2. Results of the model fit (Pmod – modelled, Porig – original performance) with the fitness-fatigue model (left) and the PerPot model (right) for subject M7

According to previous studies, dealing with the fitness-fatigue model (e.g. Busso et al. 1997), time parameters, which represent the individual decays of fitness and fatigue, were estimated in limited ranges ($t1$ between 30 and 60 days, $t2$ between 1 and 20 days). These ranges have shown to be empirically and physiologically reasonable. In the current study at least one of the time parameters is critical for every subject (means equal to the upper or lower limit of the range, see Table 1). Therefore, interpretation of the values is questionable. Theoretically, a fast fatigue decay rate (low $t2$) and a slow fitness decay rate (high $t1$) are positively influencing performance. Concerning the weighting factors only the ratio ($k1/k2$) is of interest, because the quantities of both are only depending on the quantification of training and performance. A ratio >1 is positively affecting performance, since the positive effects of fitness are dominating the negative effects of fatigue. This is the case for subjects M3, M5, M7 and M8. The quality of the performance fit, expressed by R2, is satisfying for two subjects (R2>0.8) but poorer for the remaining ones (0.4<R2<0.7). For subject M2 the performance fit is absolutely unsatisfactory (R2=0, see Table 1).

Table 1. Parameters derived for the fitness-fatigue model (left) and the PerPot model (right), n: performance measures, t1 and t2: time decays of fitness and fatigue, k1 and k2: multipliers of fitness and fatigue, DS and DR: delays of strain and response, R2: coeff. of determination

Subj.	n	t1	t2	k1	k2	k1/k2	R2	DS	DR	DS/DR	R2
				Fitness-Fatigue Model				PerPot Model			
M1	19	32	1	0.018	0.024	0.76	0.842	6.0	3.6	1.7	0.754
M2	16	33	20	0.000	0.002	0.00	0.000	1.5	1.5	1.0	0.305
M3	16	60	20	0.011	0.010	1.11	0.677	4.4	2.5	1.8	0.532
M4	15	60	4	0.002	0.005	0.34	0.548	3.5	2.7	1.3	0.227
M5	18	60	20	0.009	0.008	1.16	0.588	2.5	2.0	1.3	0.743
M6	21	60	20	0.022	0.024	0.92	0.478	2.9	2.5	1.2	0.309
M7	20	30	12	0.018	0.017	1.06	0.819	4.1	2.0	2.1	0.911
M8	18	30	1	0.009	0.000	>>1	0.605	3.5	2.5	1.4	0.819
M9	19	30	5	0.009	0.010	0.92	0.432	3.3	2.8	1.2	0.120
W1	19	60	1	0.006	0.006	0.93	0.520	3.0	2.5	1.2	0.267

The PerPot model parameters provide a good fit of modelled and original performances for subject M7 (R2=0.911, see Fig. 2 right). This is also the case for subjects M1, M5 and M8 (R2>0.7). Poor results are obtainable for subjects M2, M4, M6, M9 and W1 (R2<0.3). The absolute values of delay parameters DS and DR are mainly dependent upon the individual normalization factors of load and performance and, therefore, only the ratio DS/DR is interindividually comparable. With respect to the time scale, the ratio gives an idea on how fast strain is responding compared to recovering abilities of the athlete. A ratio >1 indicates that strain is responding later and thus represents a well trained state of the athlete. This is qualitatively characterized by a transient behaviour of the performance. A ratio equal to 1 (see subject M2) might be the result of lacking dynamical changes in performances during the training period. In this case a high stable performance level of the subject may explain the relatively constant performances, resulting in a poor model fit (as was also detected by the fitness-fatigue model).

4 Discussion

The practical application of the presented models with the empirical data is different. Model equations of the fitness-fatigue model were implemented with the Origin 7.0 data analysis software to obtain model parameters. Empirical data can be handled in the worksheets.

The determination of the model parameters is fully implemented in the calibration component of the PerPot software environment, so only the data preparation is to be handled manually. Using equidistant time series with complete load and performance data is specific to the PerPot model and seems not always easy to realise during daily training practice. However, it might be of necessity when accounting for the training-performance relationship, since the adaptation behaviour of the "dynamical system" athlete needs to be completely reflected by its time series.

Concerning the performance fits, good as well as poor results are obtainable with both models. Since at least one time parameter of the fitness-fatigue model (*t1*, *t2*) is marginal for almost all subjects and one amplitude parameter (*k1*, *k2*) is equal to 0 for two subjects, interpretation of the parameters remains questionable. Therefore, goodness of fit (expressed in the value R2) would be no valid indicator. Probably, the linear characteristic of the model, expressed by its linear model equations is unable to fully account for the adaptation behaviour in all the tested subjects. Due to its non-linear behaviour, one would expect better results with the PerPot model. In fact, the provided delay ratios (*DS/DR*), appear, with one exception, reasonable indicating a well trained state for all athletes. But interindividual comparison of athletes on different performance levels is only feasible when using the same normalization factors, which is only valid on similar performance levels and was not the case for the tested group. Both models are unable to model short time variations in performance (e.g. peaks) due to their delayed dynamics. In order to account for short time dynamics, the use of time variant model parameters becomes necessary. Busso et al. (1997) presented an approach for using time variant parameters with an adapted fitness-fatigue model. The use of time variant delays is already implemented in the PerPot environment, but in either case a deeper analysis of the individual system behaviour of the athlete is required.

In conclusion the application of the fitness-fatigue model for the presented data remains questionable due to marginal model parameters, that do not allow for interpretation of individual adaptation behaviour. The PerPot model parameters allow for interpretation of the adaptation behaviour but are limited in the interindividual comparison due to a non-valid general normalization of the empirical data for athletes on different performance levels.

This project was supported by the German Federal Institute of Sports Science (Bundesinstitut für Sportwissenschaft), Bonn, Germany, (VF 07/05/13/2005).

References

Banister, E. W. (1982). Modeling Elite Athletic Performance. In J. D. MacDougall, H.W. Wenger & H. J. Green (Eds.), *Physiological Testing of Elite Athletes* (pp. 403-425). Champaign, Il.: Human Kinetics Publishers.

Busso, T., Denis, C., Bonnefroy, R., Geyssant, A., and Lacour, J.R. (1997). Modeling of adaptations to physical training by using a recursive least squares algorithm. Journal of Applied Physiology, 82 (5), 1685-1693.

Mester, J. and Perl, J. (2000). Grenzen der Anpassungs- und Leistungsfähigkeit des Menschen aus systemischer Sicht – Zeitreihenanalyse und ein informatisches Metamodell zur Untersuchung physiologischer Adaptationsprozesse. Leistungssport 30 (1), 43-51.

Perl, J. (2001). PerPot: A Metamodel for Simulation of Load Performance Interaction. Electronic Journal of Sport Science, 1, No. 2.

Perl, J. and Mester, J. (2001). Modellgestützte Analyse und Optimierung der Wechselwirkung zwischen Belastung und Leistung. Leistungssport, 31 (2), 54-62.

Computer-Supported Training (CST) – Integrating Information and Communication Technologies (ICT) for Outdoor Training

Josef Wiemeyer

Technische Universität Darmstadt, Germany, wiemeyer@sport.tu-darmstadt.de

Abstract. Recent trends in information and communication technologies (ICT) and sensor electronics offer new and fascinating options for computer-supported training (CST). Particularly miniaturization and integration of several functions within one device (e. g., smartphone or PDA) offer new options for the documentation and analysis of training data in sport. In order to plan, document and analyze outdoor endurance training procedures many parameters can be assessed, e. g., stress and strain indicators like velocity, acceleration, work, power, rate of perceived exertion (RPE), heart rate (HR), and lactate. The purpose of this paper is to describe the procedure of developing and testing a prototype that integrates assessment and further analysis of GPS data, performance data, HR, and RPE. The available products are analyzed and a first step to a solution is presented using the following components: Garmin GSMMAP 60 CS, Polar HR monitor S720i (with special supplements for biking), and HP iPAQ h2210. First test with representative subjects are promising.

1 Introduction

Information technology shows some recent trends that offer new options for outdoor training and coaching in sport (for a more detailed discussion, see Wiemeyer in press):
1. Integration of several information and communication functions
2. Ubiquitous computing
3. Ambient intelligence
4. Miniaturization
5. Complexity

There are several examples how these trends can enhance training and coaching in sport. Particularly for outdoor training there is significant information that can be obtained by using new ICT. This information comprises, e.g., position, time, velocity, performance, stress and strain indicators like heart rate (HR), blood lactate level etc.

Therefore, the purpose of this project was first to develop a prototype to gather the respective data and to test the practical application. A second goal was to develop an internet-based platform to gather, illustrate and analyze the recorded training data.

In this paper we first discuss existing products for navigation and recording of training data. Then we introduce our own prototype and report first experiences with the application of the prototype.

2 Products for navigation and recording of training data

In this section products for GPS navigation and recording of training data are compared.

2.1 GPS navigation

As a first step we compared the features of selected products for GPS navigation (Blumenbach, 2004). These are: Garmin GPSMAP 60-2 (handheld), Magellan Trakman color (handheld), Navman X300 (upper-arm device), Suunto x9 (wrist device), and Garmin Foretrex (wrist device). We obtained the following results (see Wiemeyer in press, for a more detailed discussion):

1. The prices of the products range from 200 to 700 Euro.
2. The devices normally work with sample frequencies of 1 Hz, but with different accuracy values ranging from 1 m to about 15 m.
3. The displays have different sizes and operate with different color depths. Often background illumination is used to improve reading under sunshine and darkness.
4. Most products have an interface (serial or USB) for connecting them to the computer.
5. There is a great variation of size and weight, ranging from watch size to 6.1 x 15.5 x 3.3 cm^3 and 78 to 200 g, respectively.
6. The batteries or accumulators run for 14 to 20 hours, which normally is sufficient for outdoor use.
7. Many products have additional functions qualifying them for special purpuoses, e.g., compass or air pressure registration for hiking.

2.2 Products for recording of training data

In table 2 selected products for navigation and recording of training data are compared. The prices range from 300 to 700 Euro: Garmin Forerunner 301, Ciclosport RDS with CP 41, Timex bodylink, and FRWD O400.
We can discover the following tendencies:

1. With one exception all devices work using GPS.
2. There are great differences concerning size and resolution of the display.
3. With one exception all devices have an interface.
4. With one exception all devices are able to store the recorded parameters.
5. The power supply is sufficient for 14 hours up to 2 years.

As a first conclusion the analysis of products being available for outdoor training reveals that there are several products with different features which allow navigation and recording of training data. It appears interesting to take also subjective parameter into consideration (e.g., rate of perceived exertion, RPE), to integrate all the recording and to process the recorded data in a more animating way.

Table 2. Comparison of products for navigation and recording of training data

Product / Feature	Garmin Forerunner 301	Ciclosport RDS with CP 41	Timex Bodylink	FRWD 0[400]
Device	GPS- Handheld with breast belt	GPS- Handheld with breast belt and GPS antenna	GPS- Handheld with breast belt and GPS antenna	GPS- Handheld with breast belt and GPS antenna
Function	GPS HR recording	Navigation with Doppler-effect (Radar) HR recording	GPS HR recording	GPS HR recording
Antenna	integrated	none	?	?
Display	LCD, coloured, 100 x 64 px Background illumination	LCD, monochrome (watch)	LCD, monochrome (watch)	---
Sample frequency	1 Hz	?	?	0.2 to 1 Hz
Accuracy	Position: < 15m	Approx. ± 7%	± 1%	?
Interface	USB	---	USB	Serial (RS 232)
Memory	100 Route pts. 3,000 Track pts.	watch	Data recorder	Data recorder
Launch time	Warm: 15s Cold: 45s	?	?	?
Size [cm³] Weight	8.4 x 4.3 x 1.8 79g	5.5 x 5.2 x 2.6 ca. 65g (RDS)	"small"	11.5 x 7.3 x 2.6 ?
Shock resistance	6g	yes	?	yes
Power supply	Li-Ion accu 14 h	1.5V AAA 35 – 50 h	Data recorder: 2 years	4 NiMH accus; up to 15 h
Other	Software for data analysis			Barometer, software for data analysis

3 Developing and testing a prototype

The purpose of the project was twofold: first to develop and test a system composed of commercially available components to record and integrate performance parameters and objective and subjective indicators of strain (i.e., HR and RPE), and second to gather, analyse and present the data using an internet-based information system.

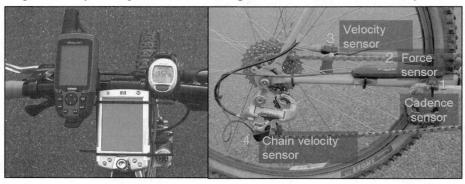

Fig. 1. Recording devices attached to the handlebars of the mountain bike (left side) and biomechanical sensors (right side)

We used the following components and attached them to a mountain bike (MTB; see fig. 1, left side):

1. Garmin GPSMAP 60 CS with a mass of 200 g and a shock resistance of 6 g was used for navigation. The sample frequency is 1 Hz and the accuracy is about 15 m (horizontal) and 3 m (altitude), respectively. The working temperature ranges from -15°C to +70°C.
2. HR and performance parameters were recorded using S720. This watch offers numerous recording functions (ECG-analogue beat-to-beat measuring, HR zones, tour length, velocity, altitude, temperature etc.).
3. Furthermore several biomechanical sensors were used (see fig. 1, right side): cadence sensor, force and velocity sensor, and a chain velocity sensor. According to LMT (2005) the recordable power ranges from 50 to 1000 Watts and the cadence from 0 to 120 cycles per minutes (accuracy: ± 10%).
4. A PDA (HP iPAQ h2210) was used for recording the RPE values and further features of the training course (s. fig. 2). The RPE program was developed using eMbedded Visual Basic 3.0.

Using this arrangement the system underwent a two-stage field test. The purpose was to evaluate the acceptance, performance, and usability of the whole system in general, to discuss the recorded parameters and additional options, and to evaluate the usability of the RPE program. One female and three male subjects representing different target groups (age: from 25 to 33 years; level: leisure sport to amateur) were asked to test the system. The subjects had to ride through two different courses. The first course was very easy (length: 10.3 km; altitude difference: 33 m) and was used to accustom to the system. The second course was more demanding (length: 16.2 km; altitude difference: 97 m). The subjects were guided by the GPS system and they had to type their RPE data into the PDA at regular intervals.

After the second test the subjects were interviewed using a semi-structured interview method. They were asked questions concerning the following aspects: usefulness of the recorded parameters, additional options desired, the usability of the components and the whole system, the complexity of the system, the usefulness of special options for special target groups, and the acceptable price of the system.

Fig. 2. PDA-display for recording RPE data (left side) and features of the course (right side)

All subjects liked driving the prototype very much. They appreciated a simple system that is easy to use and is not too expensive.

They also detected several problems:

1. The GPS system failed during rain and in the forest.
2. The RPE software had some deficits, e.g., size of the buttons and storing procedure.
3. Typing during riding the MTB was a problem. Voice recording using a Bluetooth headset might solve this problem.
4. Reading the information on the display of the PDA during sunshine was hardly possible. The use of adhesive transparencies does not really improve the situation.
5. Limited battery capacity of the PDA might cause a problem if the system is used for some hours.
6. The subjects desired an individually adjustable user interface (e.g., options to choose data fields and functions or two different operating modes for beginners and experts). Some of the functions offered to them seemed superfluous (e.g., wind conditions) and some other additional functions were desired (e.g., view points).

Fig. 3 shows a visionary prototype with functions and user interface changed according to the first tests and integrating all the above-mentioned functions.

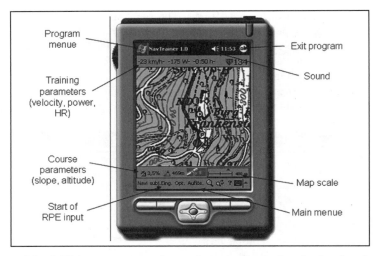

Fig. 3. Visionary prototype integrating recording and navigation functions

4 Conclusions

The developed system proved to be a promising tool to record various subjective and objective data for outdoor training.
However, there are some problems that need to be solved in order to improve the performance and usability of the system.

Acknowledgement

The author likes to thank t-online and particularly Frank Milczewsky for supporting the project. Furthermore a team of teachers and students worked on the project: Christoph Keller, Frank Hänsel, Henning Eifert, Dieter Bremer, Christian Simon, and Dietbert Schöberl.

References

Blumenbach, T. (2004) Zentimetergenaue Messung im Sport – Möglichkeiten und Grenzen von GPS. In: H. Gros et al. (Eds.), *Sporttechnologie zwischen Theorie und Praxis III* (S.61-69). Shaker, Aachen.
LMT (2005) Polar Power Output. Retrieved June 27, 2005 from http://www.conconi.ch/download/polar_power_output.pdf
Wiemeyer, J., Keller, C., Hänsel, O. & Eifert, H. (2006) Integration von Informations-Technologien für das Outdoor-Training im Sport. In: K. Witte, J. Edelmann-Nusser, A. Sabo and E.F. Moritz (Eds.), Sporttechnologie zwischen Theorie und Praxis IV (S.189-198). Shaker, Aachen.

Potentials of Information Technologies for Innovation in Fitness Equipment – A Case Study

Christian Henneke[1], Frank Hoisl[2], Stefan Schönberger[2] and Eckehard F. Moritz[1]

[1] SportKreativWerkstatt GmbH, München, Germany, che@SportKreativWerkstatt.de
[2] Technische Universität München, Germany

Abstract. In this paper the potentials of increasing the attractiveness of fitness workout by integrating information technologies into fitness equipment will be discussed. User needs will be identified, existing solutions will be examined, and the development of an innovative computer-supported fitness device will be presented as an example.

1 Introduction

A lot of people join fitness clubs highly motivated. Despite this initial enthusiasm a huge fraction stops exercising within a few months. This drop-out phenomenon can be traced back to diverse reasons like high membership fees, the necessary time effort and the emotional situation of the members (Rampf 1999). The latter includes the fact that many people quickly get bored by the monotonous fitness training offered in the gyms.

A possibility to make fitness workout more fun and therefore long-term attractive is the utilization of information technologies (Moritz 2003). Computer technologies already are ubiquitous in many areas of our everyday life and the technological evolution indicates that this trend will keep accelerating. Especially computer games, simulations and virtual reality applications achieve a high level of fascination.

Fitness devices can be used as alternative input devices for such simulations, replacing joystick, mouse and keyboard. Physically demanding motions control the simulations and the fitness training gains a playful and diversified character.

First products combining fitness training and computer simulations have been introduced on the market. They can be classified into two main groups. On the one hand there are several solutions adding a simulation to classic fitness devices such as ergometers or stairwalkers. On the other hand there are gadgets which can be connected to a video game console or PC via standardized interfaces. The downside of both approaches is the focus on either the exercising or the gaming aspect.

To bridge this gap between efficient workout and fun and to explore the potentials of Computer Supported Collaborative Sports (Wulf, Moritz, Henneke, Al-Zubaidi and Stevens 2004), a totally new piece of equipment has been developed, combining both aspects in a holistic approach. In the development of this concept, the systematic innovation approach for popular sports developed by Moritz (e.g. 2004a) has been applied; as has been illustrated in (Acosta, Moreno and Moritz 2004; Moritz 2004b). In this paper, some details of the realization of this device will be introduced, and a few first impressions of the utilization discussed.

2 The Concept FlyGuy

2.1 Core Functions and Overall Concept

As required core functions for a computer supported fitness device, fun, efficiency in the training, immersion and collaboration could be identified.

For realizing the fun aspect, a concept has been developed, where the motions a sportsman does during his workout on a training device are being used for controlling a flight simulation. The flight task was chosen because the simulation of an activity that can normally not be performed easily seemed more fascinating than the simulation of an existing sport like cycling. Furthermore, flying is a highly attractive sensation; being able to "fly" is the fulfillment of almost everybody's dream.

As training exercises a butterfly-like motion with the arms and a stepping-exercise with the legs have been chosen. Both motions are performed against a bidirectional resistance. This way large muscle groups of the front and back of the upper body, the arms and the legs can be trained.

By doing the butterfly-like motion with his arms the user controls the pitch and roll of the virtual flight. The stepping-exercise controls the acceleration in the simulation which the user watches on a monitor or head-mounted display. The adaptable resistance is provided by a mechanic and hydraulic system that also gives a force feedback onto the support structure on which the sportsman is lying. Thus he perceives the actual flight status not only optically but also haptically and immerses even more into the virtual environment. Realizing a good correlation of the actual training motions, the motions in the virtual flight environment and the feedback onto the body support structure is crucial for achieving a high level of immersion.

For addressing the collaboration multiple users should be enabled to train in a cooperative or competitive way. This can be realized by connecting several FlyGuys via internet or local network.

Figure 1 shows the overall concept of the training device. The design task has been divided into three main modules representing the major components of the FlyGuy. In the following chapters the basic solutions used in each module for realizing the core functions are being presented.

Fig. 1. Concept FlyGuy

2.2 The Parallelogram Framework

The fitness device is based on a framework that is fixed to a ground plate and is responsible for the guidance and stabilisation while workout exercises are being performed. As the central component it is the interface to all the other modules of the training device.

Besides the stabilization, the framework contains a second main function: In combination with the hydraulic leg module it is responsible for the realization of the feedback of the acceleration in the flight simulation. Representing the acceleration was very challenging as it is not possible to easily accelerate the whole device in an adequate way. Therefore the feedback is implemented by lifting the user upwards and forwards. This change of position is performed by a single-acting hydraulic cylinder connected to the ground and the front side of the framework via joints. The needed pressure is generated by the stepper leg motions of the user. By doing the stepping-motions, oil is being pumped by hydraulic cylinders in the leg module from an oil tank to the hydraulic cylinder which lifts the framework (Fig. 2).

When the user stops moving his legs, the oil flows back into the oil tank, the whole parallelogram framework subsides slowly towards the starting position and the user slows down in the flight simulation. That drain can be regulated for realizing different workout intensities.

Fig. 2. Upward motion of the parallelogram framework

2.3 The Leg Module

As described in the previous chapter, the user does a stepper-like exercise with his legs. The resistance to the motions is being generated by hydraulic cylinders. When stepping faster than a defined minimum frequency, the user is being lifted upwards in the training device and accelerated in the flight simulation and vice versa.

Fig. 3. The leg module

2.3 The Arm Module

With the assembly based on the parallelogram framework the user controls the pitch and roll of the flight simulation. The front part of the body support can be rotated around the longitudinal body axis. This means that the motions of the assembly are completely independent from the framework motions generated by the leg motion. The feedback from the arm motions onto the front part of the body support platform is generated by rotating poles with the arms.

The butterfly exercise can be executed bilaterally or unilaterally. Based on a neutral position the arm bars can be lifted up or pushed down.

With a bilateral rotation of the arm bars the angle of the platform and the pitch in the simulation are being controlled. Starting from a horizontal orientation which represents the neutral position several positions can be obtained. Depending on the motion the pitch in the computer simulation follows the angle of the platform. Figure 4 shows the lower and upper extreme positions of the bilateral motion from the front view.

By doing a unilateral or even counter-lateral motion the roll in the virtual flight is being affected and the user rotates around the body longitudinal axis on the training device. Therefore the breast support is rotating, while the pelvis support is rigid to provide stability and balance (Fig. 5).

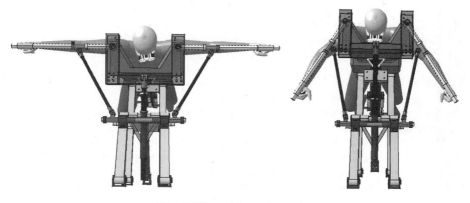

Fig. 4. Bilateral butterfly motion

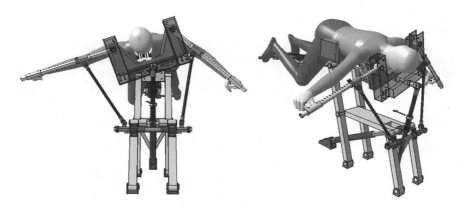

Fig. 5. Unilateral butterfly motion

2.4 Motion Capturing and Signal Processing

For capturing the arm motions two potentiometers are being used. They are mounted in the pivot joints of the arm poles and measure the angle of each pole relative to the breast support structure. The frequency of the stepping-motion is being measured by an incremental encoder which is positioned in a roll guiding a rope that connects the two step-pedals.

The digital signal from the incremental encoder and the analogue signals from the potentiometers are being converted in a microcontroller unit and transmitted to a standard PC interface. A virtual joystick driver processes the data and controls the flight simulation which is based on the open-source software FlightGear.

At the present time the user can only fly in the virtual environment without solving any special tasks. For the future it is planned to implement several flight courses that must be followed by the sportsman. Thus different exercises can be realized where the user must do certain motion sequences to follow the course. The same way, a variety of performance levels can be achieved.

3 Conclusion

The application of information technologies is an interesting approach for enhancing the attractiveness of fitness workout.

Existing products combing sports and computer technologies focus either on the sports or the entertainment aspect. In the presented project a concept has been developed combining both sides: efficient fitness workout and fun. A fitness device is being used as input device for a flight simulation which is being controlled by the training motions.

First results of using the FlyGuy as a training device have shown that this approach indeed offers a totally new scope of sensations for the user as well as a highly increased motivation for those people who like fitness but do not like existing offers of home or gym equipment, and even for those who only like to play and take the fitness benefits as a positive side effect.

As a next step, the ergonomics as well as the industrial design will have to be improved, the results of further evaluations applied to optimize the device, and the whole concept reduced in complexity. The objective is to finally have a product on the market that can attract new target groups as well as enhance the training motivation of already active athletes.

References

Acosta, C., Moreno, E.A., Moritz E.F. (2004) Dojyo: An Innovative Distributed Engineering Project, Contribution to the Intertech Conference 2004, Cincinnati.

Rampf, J. (1999) Drop-out und Bindung im Fitness-Sport: Günstige und ungünstige Bedingungen für Aktivitäten im Fitness-Studio. Sportwissenschaftliche Dissertationen und Habilitationen; Bd. 48, Czwalina, Hamburg.

Moritz, E.F. (2003) The Virtualization of Fitness Training. In: Volvo Sports Design Forum 2003, München.

Moritz, E.F. (2004a) Systematic Innovation in Sports Engineering. In: 5th Conference of the International Sports Engineering Association, September 14-17, 2004, Davis CA

Moritz, E.F. (2004b) Systematische Innovation in der Sporttechnologie; in: Moritz et al. (Eds.): Sporttechnologie zwischen Theorie und Praxis 2; Shaker Verlag, Aachen.

Wulf, V.; Moritz, E.; Henneke, C.; Al-Zubaidi, K.; Stevens, G. (2004): Computer Supported Collaborative Sports - Creating Social Spaces Filled with Sports Activities. In: 3rd International Conference on Entertainment Computing ICEC 2004, Eindhoven.

Computer Support for Coaching and Scouting in Football

Prof.dr. Gerard Sierksma

Department of Econometrics, University of Groningen, The Netherlands, g.sierksma@rug.nl

Abstract. Science and football are two different worlds that nevertheless deal with each other in an increasing manner. Decisions about the effective and efficient use of football's most important human capital, the players, are taken not any more solely on the base of intuition and common sense, but are supported by advanced computer systems. These systems allow the club management to measure and analyze the effectivity of the decisions in a more objective and sound way, resulting in a decrease of wasted money, such as 'bad buys and transfers'. Indeed, football is arrived in the process of professionalization, although - like in all sports - emotion and work of man will never be excluded! In this paper two computer support systems are introduced, namely *Coach&ScoutAssistant* and *EffectivityInAction*. The first system is used to track and trace the development of (youth) players and to calculate the value-added and best position in the team of a scouted player. The second system is used on-line during the match and generates graphs that reflect the effectivity of the competing teams. Both systems are answers to the intriguing question about the discrepancy between individual and team performance: what is the relation between the performance of an individual player and the performance of his team as a whole?

1 Football and Science

The influence of science in the world of sports is mainly due to the increasing power of computers and the heavy competitions between athletes and teams, where details and money more and more make the difference between winning and losing. Even in football, known to be one of the most conservative sports, the interest in support from science is increasing. Especially professional football organizations from all over the world demand nowadays for the scientific support of the decisions and the analysis of players and team performances.

This paper introduces two systems, both developed by experts in 'quantitative logistics' and 'return on capital investment' at the University of Groningen, the Netherlands, in cooperation with the company Team Support Systems, B.V.; see e.g. Boon,Sierksma[2003], Sierksma[2003], Zwols,Sierksma[2006], and Koning,Sierk-sma,Siesling[2006]. All types of organizations and companies that deal with sport teams, such as football (soccer), American football, volleyball, basketball, hockey and baseball, but also ICT teams, management teams and project teams, may profit from these systems.

The emphasis is always on measuring the effects of scouting/headhunting.new team members and of the performance of individual players in comparison with the performance of the team. So the main underlying academic question is: how to measure individual player performances in relation to team performance? Of course, the answer to this question has immediate practical consequences as well.

The two systems supplement the expertise of the technical and financial staff of the club with rational arguments and conclusions. Besides this, and it cannot be emphasized enough, computer systems only <u>support</u> the decisions: the responsibilities stay with the people.

2 Coach & Scout Assistant

Coach & Scout Assistant (C&SA) is a decision support computer system with which assessments of football players can be analyzed in an objective way, based on measurable qualities, competencies and skills. These measurable attributes are usually equal to the ones used on scouting forms. Most clubs use four types of attributes: technical qualities (such as passing, heading, shooting), physical qualities (such as speed, power), and mental qualities (such as pressure resistant, team discipline). The coach or the technical staff evaluates all players on all these qualities

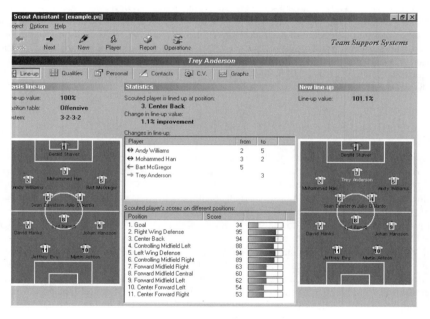

Fig. 1. C&SA Screenshot, example.

during the year on relevant moments. Also the reports of scouts can be included in the system. C&SA has the following major functionalities:
- Track and trace the development of the qualities of all players;
- Design player specific training programs;
- Determine the value-added, plus its change over time w.r.t. the team for each player on each field position;

- Determine the strongest team line up under changing circumstances (injuries, redcards, etc.);
- Determine the value-added for the team, plus an indication of the salary of new and scouted players;
- Store data and analysis of players in the easy accessible data base.

As soon as all relevant players are included in C&SA, the above formulated activities can be started. The CV of a player can be printed. The report of a scouted player, including his value added on each position (this may show, among

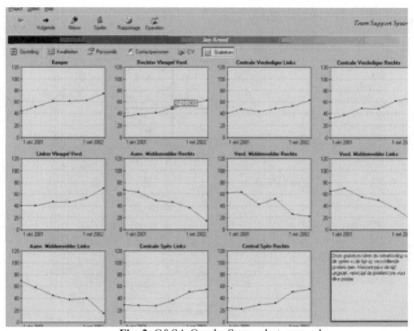

Fig. 2. C&SA Graphs Screenshot, example

others, how multi-usable a player is), can be printed. The value-added on a position is shown on the computer display by dragging the icon of the player to a desired field position. This is done on the right hand side field, shown in Fig. 1; the left hand field in this figure shows the basis (starting) line up.

Furthermore the system shows on one screen the development curves of a player on all eleven field positions. So not only the curve of the development of the player on his position in the basis line up, but also on the other field positions, are depicted. Fig. 2 shows the development of a fictitious player on all eleven field positions based on six observations. Moreover, C&SA gives for each non-basis player the qualities he needs to improve in order to become competitive with a basis player.

At the moment C&SA is used by a number of football clubs. In one situation all satellite clubs have the system on their computers and are directly connected to the computer of the main club. So this club uses the system in a network setting. This enables the club to determine at any time the value-added for the own team of (youth) players that play for a satellite club.

3 Effectivity *in* Action

In the analysis room of Team Support Systems the match Czech Republic against Holland was shown on a large TV screen. Four touch screen computer displays are connected to the main computer. Two persons for Czech Rep. and two for Holland register the activities of the teams. The photo of the player that is in ball possession is touched. Then the type of action and the score on the range of 1 to 5 are clicked.

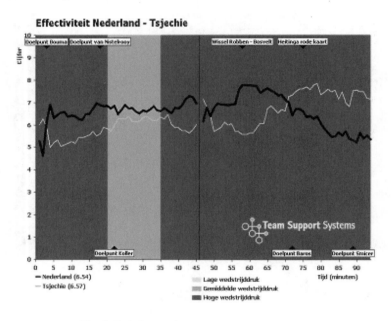

Fig. 3. E*i*A Screenshot (translation see text)

The scores 1 to 5 are well defined (this is the usual so-called Likerd scale, used for most inquiry forms). This scale does not ask for a lot of practising to use it a consistent way. The main computer calculates at any time during the match the current effectivity of both teams and draws the two effectivity graphs; see Fig. 3. Therefore, at any time of the match, the progress of the effectivity of the teams is immediately visible. As soon as the match is finished both teams receive an overall effectivity score. Also for each player an effectivity score is calculated and presented.

Fig. 3 shows the well known dramatic substitution of Arjen Robben: the effectivity of the Dutch team was still increasing while the Czechs went down when the substitution took place. After the substitution the Dutch team collapsed and the Czechs arose from the grave: Holland did not make it to the final!

Effectivity *in* Action (E*i*A) is applied at the moment in various settings. One club in Holland is testing the system for displaying the graphs on the big screen in the stadium. Coaches use the printed results, because the graphs show at one glance the effectivity during the complete match. Newspapers use good looking representations of the graphs for their readers.

4 Discussion

Both above described systems use advanced mathematics, which will not be explained here. Roughly speaking, C&SA calculates optimal matchings between demanded qualities on the positions and supplied qualities from the selected players. So the system puts the right person on the right place, based on the at that time measured qualities. The solutions are calculated by means of linear programming techniques; see e.g. Sierksma[2002].

E*i*A uses analytical formulas that apply corrections on, among others, the fact that goalkeepers are fewer times in ball possession than midfielders (in general). These formulas also translate the 1-5 scale into the more common 1-10 scale. Moreover 'weighted' averages are calculated for each consecutive five minutes (running five minutes), and a smoothing technique is applied to obtain the graphs as in Fig. 3. The back ground colors red if there is high tension in the game, orange in case of low tension, and yellow if the game is played without excitement.

References

G. Sierksma [2002], Integer and Linear Programming; Theory and Practice, Marcel Dekker, Inc., New York.

B.H. Boon and G. Sierksma [2003], Team Formation: Matching Quality Supply and Quality Demand, European Journal of Operational Research 148, pp. 277-292.

G. Sierksma, Computer Support for Team formation [2003], International Journal of Computer Science in Sport 2/1, pp. 177.

Y. Zwols and G. Sierksma [2006], Multi-event Training Support, submitted.

R.H. Koning, G. Sierksma, D.S. Siesling[2006], Soccer Salary and the Value Added of Football Players, in preparation.

www.teamsupportsystems.com

www.rug.nl/economics/sierksmag

The Effects of Multimedia Computer-Assisted Instruction on Middle School Students' Volleyball Performance

Nicholas Vernadakis, Eleni Zetou, Andreas Avgerinos, Maria Giannousi and Efthimis Kioumourtzoglou

Democritus University of Thrace, Department of Physical Education and Sport Science
nvps@otenet.gr

Abstract. The purpose of this study was to examine the effect of multimedia computer - assisted instruction (MCAI), traditional instruction (TI), and mixed instruction (MI) methods on learning the skill of setting in Volleyball. Forty-eight middle school students of seventh and eighth grade were randomly assigned into three teaching method groups: TI, MCAI and MI. Each group received ten 40-min periods of instruction divided into 3 sections: a) 5 min warm-up, b) 30-min main practice time and c) 5-min cool dawn and review. TI group participants experienced the setting ability through a series of progressive skills accompanied with drill and repetition of practice presented by an instructor. The MCAI group experienced the setting ability through a series of progressive skills accompanied with drill and repetition of practice presented by a multimedia program. The mixed group experienced the same procedure, practicing through the multimedia program as well as through traditional instruction. All students completed pre-, post-, and retention skill test. Two-way analysis of variances (ANOVA), with repeated measures on the last factor, were conducted to determine effect of method groups (TI, MCAI, MI) and measures (pre-test, post-test, re-test) on skill test. Post-test results indicated no significant differences between the groups concerning the skill test. Retention test results showed that groups retained the skill learning. However, the mixed method of instruction tended to be the most effective for skill development.

1 Introduction

Research on multimedia and related instructional technologies over the past years has been characterized by inconsistent findings about their effects on learning. There are many studies reporting that MCAI can have a positive impact on learning. A meta-analysis by Liao (1998), for example, examined 35 studies and concluded that MCAI is superior to traditional instruction. However, it is notable that 10 of these 35 studies showed the opposite, namely, that traditional instruction is superior to multimedia. A subsequent meta-analysis of 46 studies (Liao, 1999) confirmed the overall positive effect of MCAI on student achievement, but found that it largely depends on what type of instruction it is being compared with. Similar results have been reported by Wiemeyer, (2003), who reviewed nine meta-analyses of earlier and different multimedia issues and suggested that multimedia learning can be more effective and efficient than traditional learning. But this effect depends on many factors like the features of the learners, the teachers, the learning stuff, the type of learning, the features of the study, etc. Further, a meta-analysis of 167 studies (Bernard, Abrami, Lou, Borokhovski, Wade, Wozney, Wallet, Fiset and Binru Huang, 2004) concluded that a very weak learning advantage for multimedia in empirical studies is based on uncontrolled instructional methods.

Since most of the reported research findings comparing MCAI with the TI approaches have demonstrated increased learning in the cognitive domain. Only few studies have been conducted to determine if MCAI improve motor learning in the area of physical education. Vernadakis, Zetou, Antoniou, and Kioumourtzoglou (2002) reported that MCAI is a functional method in teaching the skill of volleyball setting to children aged 12 – 14 year old and is as effective as traditional teaching. In another study Wilkinson, Hillier, Padfield, and Harrison (1999) found that junior high school girls in both classes (TI and MCAI) significantly improved in their knowledge of volleyball rules and in performance of most skills (pass, set, and underhand serve) during the 16-day unit.

The effectiveness of physical education software on student achievement has yet to be determined especially through the use of the newer multimedia programs. In this study an attempt was made to compare three different instructional methods by means of the skill test scores, obtained from three groups of middle school students. The tests assessed the learning of the setting skill in volleyball. The research questions of this study was the following: a. Do students, on average, report differently on skill test using the TI, the MCAI and the MI teaching approaches? b. Do students, on average, report differently on skill test for the pre-test, post-test and re-test measurements? c. Do the differences in means for skill test between the TI, MCAI and MI teaching method groups vary between the pre-test, post-test and re-test measurements?

2 Method

2.1 Participants

Forty-eight (n=48) middle school students (25 girls and 23 boys) of seventh and eighth grade, 12-14 years of age (M=13, S.D. =1.01), selected for this study by random sampling method, enrolled in the volleyball course. Participants were randomly assigned to one of the three different teaching methods: TI (9 girls and 7 boys), MI (8 girls and 8 boys) and MCAI (8 girls and 8 boys) creating three independent groups of 16 students.

2.2 Software

A multimedia program was created and programmed in Asymetrix ToolBook to administer experimental events including 163 screens; 5 screens were introductory, 1 was main menu, 48 were information, 40 were practice, 57 were feedback, and 12 were help. Material for the multimedia application was taken from a volleyball coaching textbook (American Sport Education Program, 2001) and modified for this study. The application consisted of four sections: a) history, b) rules, court and player's position, c) skill fundamentals, and d) skill exercises. Two choices menus, one for the termination of the program and one for help, were also included at the bottom of the screen and were always available. The help menu contained a description of the active picture-buttons and suggestions for the program use. The program started with an introductive video of international volleyball federation (FIVB). The main menu with four active picture-buttons which serve as links to the other sections of the program followed.

The first two major sections addressed basic knowledge of the volleyball game pertaining to vocabulary used, history of the game, rules, court dimensions, names of positions, and rotating positions. The skill fundamental and skill exercise sections introduced basic setting skills and exercises for practical work in terms and levels that were appropriated for beginning volleyball players. A step by step instructional format that was accompanied by an exceptional graphic and video simulation depicting proper form of setting skill at different stage was included in these sections. A discussion of possible errors, what causes these errors, and what may be done to correct these errors was provided in the description of setting skill. When the user had seen enough of the setting skill, he could supplement a short quiz (multiple choices, true/false) regarding the technique and concepts that were presented. Audio was used to explain each action and give execution cues to help focus the attention of the user. The user navigated through the sections from the menu that appeared on each screen. At the end of the program, a screen with the title of the program, the names of the author and the institution were presented.

2.3 Skill Test

The AAHPER volleyball skills test (Strand and Wilson 1993) was used to evaluate setting ability in volleyball. Scorer, timer, tosser for passing and setting, and student assistants to retrieve and set ball were needed for the successful completion of the test. Testing stations were prepared as shown in figure 1.

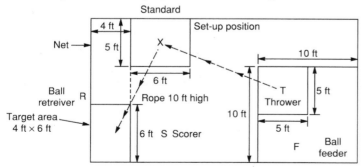

Fig. 1. Court markings for the AAHPER Volleyball Setting Test.

To begin, a thrower in a tossing area, tossed high passes to a student in the set-up zone. The receiver sets the ball over the rope into the scoring zone. Ten trials were given to the right and 10 to the left. Balls that touch the rope, the net, or hit outside the scoring zone earned no point. One point was awarded for each set that lands in the marked scoring zones. The final score was the total points for 20 trials.

2.4 Procedure

When the multimedia program was developed, the researchers gave it to an instructional technology specialist, a subject-area expert, and three subject-area teachers for evaluation. Researchers revised and improved the multimedia application according to the feedback received from those experts.

The skill test was administered on the first day of the experiment to measure participant's baseline performance on the selected setting skill. On the second day, ten 1.8 MHz Pentium IIII class computers equipped with a 17-inch color monitor, CD-ROM, soundcard and small headset, running Windows 2000 were set up in a blocked-off hallway adjacent to the gymnasium. Each computer had a volleyball skill CD-ROM created by the researchers. Before the experiment started, the MCAI and MI groups were given a 40-minute introductory session on how to use the multimedia application prepared for this study. Then the physical education instructor gave a 40-minute lecture to all participants introducing the unit of volleyball. Instruction, practice, and testing for this study were held on ten separate and successive weeks. Each group received ten 40-min periods of instruction divided into 3 sections: a) 5 min warm-up, b) 30-min main practice time and c) 5-min cool dawn and review.

The TI group received a series of progressive skills, performed in drill format, accompanied by verbal feedback in the form of positive reinforcement. They were allowed to work alone or with a partner. The physical education instructor gave verbal instruction before every drill and knowledge performance every five trials during the 30-minute of practice time. Participants in the MCAI group were asked to learn the setting ability via the multimedia program. They were allowed to work independently or with a partner. The instructor was present for organization and management supervision only. No verbal or visual reinforcement of any kind was offered by the instructor. The MI group followed the same procedure, while implementing both the multimedia program and the traditional instruction. In the first five weeks the students participated with the traditional method group, and the remaining weeks with the MCAI method group.

At the end of the treatment, the skill test that previously served as a pre-test was given to students as a post-test. One week later, the same procedure was repeated on the re-test to measure the level of retention in the selected setting skill. During the experiment, each group received an equal amount of instructional time and was provided with the same instructional materials and assignments.

2.5 Design

The experiment was a factorial design with teaching method groups (TI, MCAI and MI) and repeated measurements (pre-test, post-test and re-test) as independent variables, and skill learning performance as dependent variable.

3 Results

Two-way analysis of variances (ANOVA), with repeated measures on the last factor, were conducted to determine effect of method groups (TI, MCAI, MI) and measures (pre-test, post-test, re-test) on skill test. A significant main effect was noted for the measurements, $F(1,45) = 17.343, p<0.05$.

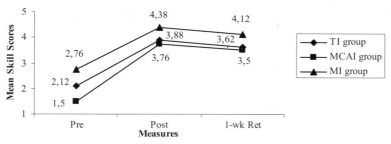

Fig. 2. The significant main effect for the measure on Skill Test

Difference and repeated contrasts were conducted to follow up the significant measurements main effect. Differences in mean rating of skill test in TI group were significantly different between pre-test and post-test, $F(1,15) = 6.622$, $p<.05$ and between pre-test and re-test, $F(1,15) = 6.818$, $p<0,05$. Differences in mean rating of skill test in MCAI group were significantly different between pre-test and post-test, $F(1,15) = 6.106$, $p<.05$ and between pre-test and re-test, $F(1,15) = 7.136$, $p<0,05$. Finally differences in mean rating of skill test in MI group were significantly different between pre-test and post-test, $F(1,15) = 4.884$, $p<.05$ and between pre-test and re-test, $F(1,15) = 4.233$, $p<0,05$. As shown in "Fig. 2", the post-test and re-test skill scores were remarkably greater than pre-test skill scores for the three groups.

4 Discussion

Research on the effects of teaching in the skill test showed a significant increase in the performance for the three instructional method groups. This increase indicates that all the instructional methods used improved the skill execution of students, regarding the setting ability in volleyball. Previous studies report equal improvement in learning with TI, MCAI (Summers, Rinehart, Simpson, and Redlich 1999; Vernadakis et al. 2002) and MI method (Vernadakis, Antoniou, Zetou, and Kioumourtzoglou 2004). The re-test measurement used to evaluate the maintenance of learning one week after the end of the educational process, showed a reduction of performance in the skill test for the three instructional method groups, which was not significant. This result is in agreement with Antoniou, Derri, Kioumourtzoglou, and Mouroutsos (2003), who supports that in a multimedia program, the combination of video, text, sound and graphics facilitate the retention ability of information.

The effectiveness of MCAI method in teaching a motor skill may have been due to a number of factors. The first factor might be that the students could practice immediately after seeing the skill on the CD-ROM. The researchers can only speculate that the worksheets provided for the MCAI group succeeded to bridge the images from the computer to the different environment. Other factors may be that students would physically imitate skills they saw on the screen. This ability to see skills over and over at the students' leisure seemed to be an advantage. Also, having the students work in pairs at the computer seemed to be helpful. Gros (2001) indicated social skill benefits occur from collaborative assignment to computer-based tasks.

In conclusion, according to the results of the present study, the MI method as a teaching aid tended to be the most effective teaching method for skill development of the setting ability in volleyball. However, these conclusions are limited for students aged 12 – 14 years old. More studies should be conducted to investigate the effect of MCAI in different ages and for various sport activities. Also, it is critical to continue researching into how students learn in different technological environments, since the researchers have only begun to explore the uses and practicality of MCAI.

References

American Sport Education Program (2001). *Coaching youth volleyball* (3rd ed). Human Kinetics, Champaign IL.

Antoniou, P., Derri, V., Kioumourtzoglou, E. and Mouroutsos, S. (2003). Applying multimedia computer-assisted instruction to enhance physical education students' knowledge of basketball rules. European Journal of Physical Education, 8(1), 78-90.

Bernard, R., Abrami, P., Lou, Y., Borokhovski, E., Wade, A., Wozney, L., Wallet, P.A., Fiset, M., and Binru Huang. (2004). How Does Distance Education Compare With Classroom Instruction? A Meta-Analysis of the Empirical Literature. Review of Educational Research, 74(3), 379 – 439.

Gros, B. (2001). Instructional design for Computer-Supported Collaborative Learning in primary and secondary school. Computers in Human Behavior, 17, 439-451.

Liao, Y. (1999). Effects of hypermedia on students' achievement: A meta-analysis. Journal of Educational Multimedia and Hypermedia, 8(3), 255-277.

Liao, Y. (1998). Effects of hypermedia versus traditional instruction onstudents' achievement: A meta-analysis. Journal of Research on Computing in Education, 30(4), 341-360.

Strand, B.N. and Wilson, R. (1993). *Assesing Sport Skills*. Human Kinetics, Champaign IL.

Summers, A.N., Rinehart, G.C., Simpson, D. and Redlich, P.N. (1999). Acquisition of surgical skills: a randomized trial of didactic, videotape, and computer-based training. Surgery. 126(2), 330-6.

Vernadakis, N., Zetou, E., Antoniou, P. and Kioumourtzoglou, E. (2002). The Effectiveness of Computer – Assisted Instruction on Teaching the Skill of Setting in Volleyball. Journal of Human Movement Studies, 43, 151-164.

Vernadakis, N., Antoniou, P., Zetou, E. and Kioumourtzoglou, E. (2004). Comparison of Three Different Instructional Methods on Teaching the Skill of Shooting in Basketball. Journal of Human Movement Studies, 46, 421-440.

Wilkinson C., Hillier, R., Padfield, G. and Harrison, J. (1999). The effects of volleyball software on female junior high school students' volleyball performance. Physical Educator, 56(4), 202-209.

Wiemeyer, J. (2003). Learning with Multimedia - More Promise than Practice? International Journal of Computer Science in Sport, 2(1), 102-116.

6 Human Factors

Synopsis of Current Developments: Human Factors

Monika Fikus

University Bremen, Germany, mfikus@uni-bremen.de

Themes and Methods

In summary the presented field of human factors is spread over a wide range of topics. They cover the analysis of sports techniques as shown in the contribution of Okuba and Hubbard who investigate the flight of a basketball under different conditions or with Marmo, Buckingham, and Blackford who explore sweeping styles in curling with an ergometer brush. The questions of three further articles deal with the needs, intentions, and requirements of people using sports equipment. Meyer wants to find out differences between cyclists in Germany and Japan, Strehler, Hasenkopf and Moritz investigate wishes and requests of older people in the field of sports and physical activity and Carré and McHutchon explore the perception of qualitative and physical properties of hockey sticks.

Three contributions deal with innovative methods of investigation. Memmert and Perl analyse game creativity with the help of neuronal networks, Kawamura, Takihara, Minamoto and Md.Zahid discuss the modelling of the interaction between sports equipment and the human body. They use tennis players as an example. The contact between the human body and natural turf investigated in the laboratory is the topic of the contribution of Stiles, Dixon and James.

The used methods are innovative mathematical and dynamic modelling that match the complexity of the objects of intervention. Further emphasis lies on qualitative methods that are used for questioning of subjective impression and individual meaning. Some investigations combine quantitative and qualitative methods which seems to benefit the analysis of the mutual interaction between humans and sport technology.

Actual and Further Developments

Obviously the role of sports engineering goes beyond optimisation of tools, the search for solutions for practical problems in sports and the investigation of the interaction between humans respectively human bodies in motion and technical equipment and devices. New dynamic modelling offers the tools necessary for the complexity of these processes. Additional sports engineering takes over the responsibility of well being, health protection, and avoidance of injuries in sports. The increasing amount of using qualitative methods takes also into account subjective meanings and individual needs.

In overtaking this part sports engineering will become a topic of growing importance in science and society. One example is the contribution of James and Haake in which demonstrates that sports technology is an appropriate issue to educate young people in physical science.

Human Movement as a Culturally Shaped Act

Imke K. Meyer

University of Bremen, Department of Sport Science, ikm@uni-bremen.de

Abstract. This research project intends to show that human movement is realised in accordance to the cultural context, and that individual meaning can be identified in human movement itself. The strategy of this research combines qualitative and quantitative methods to illustrate the inner and outer characteristics of each participant. The presentation will show the results of the main inquiry held in Bremen, Germany and Tokyo, Japan.

1 Introduction

Sport is an element of culture. Watching, for example, citybike-cyclists compared to mountainbike-cyclists in Germany or Japan (or anywhere else in the world) you can observe that they move differently, even though both cyclists are simply pedalling.

Fig. 1 and 2: Cyclists in Bremen, Germany and Tokyo, Japan·

Cycling, as an example of a sports movement or a human movement, could therefore be culturally "formatted" (Fikus and Schürmann, 2004), which means that this movement could be realised differently depending on the surrounding cultural context (however this context is defined).

2 The problem area and occurring questions

There is a large mutual interrelation between culture and sport or human movement. Mauss described in his research – already at the beginning of last century (see Mauss, 1935) – that daily acts, such as walking or using a spade, are carried out differently according to the "cultural origins". In addition, Moritz explains the strong interrelation between culture, sport and technology (Moritz, 2003), which includes the connection between human beings, cycling and the bicycle itself. Independent of place and time, human beings never seem to move or act in "amorphous movement worlds" of free arbitrariness. Instead people move or act in culturally (pre-)formed social environments, which have been marked by peculiar movement techniques (Gebauer and Wulf, 1998).

Using a social (or cultural) scientific perspective Movement Science (Sports Science) deals amongst other things with the following question: Which meaning or senses in sports can be shown, and therefore, which cultural processes can be cleared by human beings dealing with their body? Sports and human movement are elements of culture, as we represent ourselves and perceive through the body. Sport and human movement matters besides text, symbols or language especially within perception, within representation and within the change of culture. Every act – realised through the body – matters in the personal and social presentation. Sport and human movement is more than just 'sporting motor activity'.

Human beings sense their environment, their surrounding cultural context, with their body. Perception (visual perception) occurs through the body. Hence, every interaction needs the physiological process of the body eventually (Fikus, 2003) to interact mutually with the surroundings (see Gibson's Ecological Psychology, 1979). Human movement is a cultural phenomenon. Geertz (1983) points out that each movement (even a wink with the eyelid) is formatted by a social code. He claims that there is a social/(sub-)cultural code which emphasises structures of meaning.

In the setting of these thoughts or considerations the following questions arise:
- Does the surrounding cultural context cause differences
  (a) at the level of individual meaning?
  (b) at the level of movement?
- If (a) or (b) is true – what exactly are the differences?

At first sight, it seems to be trivial to claim that wherever you are in the world, you have to pedal to move your bicycle – and that this movement might be coded by the social/(sub-)cultural context. But the empirical data to prove these assumptions are still missing and will be established as part of this research project.

3 Theoretical foundations

Culture has usually been seen as 'text', as a structured interrelation from singled elements. The interest which occurs recently considers culture as something 'in progress' with the core in acting, change and dynamics. The 'new-placed' social-cultural theories of practice (e.g. Pierre Bourdieu) give the social aspect a different position in the discussions on the body-mind-dualism, as the location of sociology is now the knowing of social practices, seen as the knowing of behaviour, the 'incorporated' knowledge.

'Incorporation' means, that the surrounding environment, in the form of acting, moving, speaking, feeling, even taste, is incorporated into the body (Bourdieu, 1982, 2001). The "Entwurf einer Theorie der Inkorporierung (Draft of a Theory of Incorporation)" (Jäger, 2004) will provide another basis to prove these assumptions and will also attempt to show that individual and social meaning can be recognised in movement itself.

As mentioned above, human beings and therefore human movement as well, interact mutually with the environment. Therefore, it is very useful to integrate a theoretical foundation concerning this subject. The ecological psychology in combination with the system-dynamic approach (Williams, Davids and Williams, 1999; Gibson, 1979) – together with the ideas of Bourdieu – provides an ideal basis for the empirical proofs given within this research project.

4 Aims and benefits of this research

This research project promises sport scientific as well as cultural scientific impetus with regard to the empirical results and the theoretical (further) development. Hence, it will be of high importance for the sports engineering community.

In the course of 'globalisation' it is and will be still important and significant to "take care of cultures" (Moritz, 2003, 19) since the main aspect is the resolution and discovery of the mutual relation between human movement and social/cultural context. There are a few sport scientific assumptions (e.g. Fikus and Schürmann, 2004), which presume these differences, but as of now there exist no empirical data. This project therefore intends to provide these empirical proofs. It will help to understand movement and demonstrate whether and how a human movement such as daily cycling is realised differently depending on the social/(sub-)cultural context.

This project attempts to create methods that record human movements' inherent meaning and different solutions of one movement task (e.g. cycling) to decipher the included or inherent (sub-)cultural code. The results will be of much interest to understand human movement in general and subsequently for modeling processes in sport. Additionally, it will be attractive for researchers in the field of region-orientated development and future development in sport equipment design and in the area of non-motorized traffic. Looking at the same problem or working on the same task (e.g. cycling and the bicycle) on the basis of different (cultural) conditions, may open up a whole new dimension. For engineers it is important to know what the local needs actually are. If human movement is realised differently in dependence of the cultural context (as assumed) the empirical results will assist the technical developments to be effective in the long run. A culture-sensitive innovation considers the special needs of the local-living people.

5 Research Design

The strategy of this research is the combination of qualitative and quantitative methods, like (field) observation, questionnaire, interview and movement analysis (see Tab.1, p.5) to record and examine the inner meaning and the outer characteristics of each test person regarding his or her own movement.

5.1 Hypothesis and the operationalisation of the variables

In accordance to the formerly expressed research questions (see section 2), my hypothesis states first that the cultural context causes differences in the individual meaning of a movement. Secondly my hypothesis states that the differences are incorporated to such an extent, that you can actually 'recognise' the differences inside the movement itself.

To verify a hypothesis concerning a cultural subject it is of importance to choose a global widespread sport movement, nearly seen as a 'neutral discipline'. Cycling as a human movement can be described as such, as it is known worldwide – wherever or whoever you are, or however or whenever you cycle. It is therefore an ideal candidate for this type of research.

The operationalisation of the independent variable (IV) is based on the latest cultural scientific discussions on inter-, multi- and transculturality (see in use: Krüger and Meyer, 2005). Against the background of these discourses, the IV is first of all divided into two 'cycling subcultures': citybike-cyclists and mountainbike-cyclists. At a second level this IV will be split into two further groups: Bremen, Germany and Tokyo, Japan.

The operationalisation of the dependant variable (DV) contains the way human movement can be modelled adequately. The *understanding* of human movement is in the foreground of this research (rather than the *explanation* of human movement, see furthermore: Fikus, 2001). On the one hand, the dependent variable (DV) will be recorded with the "semantic differential", in which individual meaning to the person's own movement are captured. On the other hand, the DV will be set by the observation criteria, recorded through the video movement analysis to show the exterior view of the person's movement. This analysis is additionally attested by cycling and sport scientific experts.

The combined record of the inner and outer characteristics of each test person will show the effects on the dependant variable (DV) through the independent variable (IV).

5.2 Methods

The random sample consists of 10-14 German and 10-14 Japanese cyclists – citybike- and mountainbike-user –, who use their bicycle almost daily in everyday life. In the following table you will see a detailed line-up of the contents of this research. It shows further information on the methods, used to record and analyse the data:

No.	(method)	record of data	use of data	analysis of data
1	qualitative	…field) **observation** (participant and direct observation) …picture/video	…study the sub-cultures of citybike-cyclists and moun-tainbike-cyclists in general …record/categorise different characteris-tics of cyclists	…thick description (Geertz, 1987), if possible …movement analysis (experts) …field diary
2	qualtitative and quantita-tive	**interview** … picture-confrontation … (half structured) (Witzel, 2000)	…let participants describe the cyclist-models and let them assign to one of them …record individual meanings in cycling	…guide for coding …analyse inter-view according to Witzel, 2000
3	qualitative and quantita-tive	**questionnaire** (Se-mantic Differential)	…record individual meaning with a standardised method	…descriptive statistics, …presentation of individual profiles …factor analysis
4	qualitative and quantita-tive	video-**movement – analysis**	…verify that individ-ual meanings can be recognised in move-ment itself (record of the body techniques and connect them to the results of the interview and the questionnaire)	…descriptive statistics …factor analysis …analysis of variance …produce neural networks to show individual mean-ings in movement

Tab.1   Detailed line-up of employed research methods

6 Results and Prospects

The short lecture, which will be presented at the ISEA 2006, will show the results of the main inquiry held in August and September 2005 in Bremen, Germany and To-kyo, Japan and will show whether the assumptions can be proved and thus whether the hypothesis can be confirmed. The field observation has already shown some differences and similarities in the use of the bicycle and the cycling movement itself. The interviews also show that citybike-cyclists in Tokyo attach a different impor-tance to their cycling movement than citybike-cyclists in Bremen. In Tokyo you cycle rather slowly on the sidewalk, the saddle is in a low position, in order to be able to place the feet on the ground at any times. The knees are very much bended during the pedalling movement (see Fig.2, p.1) and might be rather uncomfortable. Including often mentioned local needs (e.g. a bike-basket) and conditions (see furthermore Moritz, 2003) it calls for cultural-sensitive innovations and improvements in the area of non-motorized traffic, described here for the special situation in Tokyo.

As the results aim a theoretical impetus and development – based on empirical data – they will be interesting not only for Tokyo or in the area of bicycle improvement, but in times of globalisation for any cultural-orientated innovations as well.

Acknowledgements

For the realisation concerning this research project, I express my special thanks to Prof. Dr. Tetsunari Nishiyama and his Human Performance Laboratory with T. Sato, K. Watanabe, and T. Nobukuni as well as Ms. T. Yasukawa and Ms. W. Kurosawa at the Nippon Sport Science University, Tokyo, Japan. (この日本での研究実現にあたっては、特に日本体育大学の 西山哲成 教授をはじめ同大学身体動作学研究室 の 佐藤孝之さん、渡辺船平さん、延国毅さん、安川 貴子さん、黒澤和歌子さんに多大なご協力を賜りました。ここに心より感謝の意を表します)

References

Bourdieu, P. (1982). Die feinen Unterschiede. Frankfurt/Main: Suhrkamp.

Bourdieu, P. (2001). Meditationen. Frankfurt/Main: Suhrkamp

Fikus, M. (2001). Bewegungskonzeptionen in der Sportwissenschaft. In V. Schürmann (Ed.), Menschliche Körper in Bewegung. Frankfurt/Main: Campus.

Fikus, M. (2003). The role of body and movement in culture(s), and implications on the design of research on culture and technology. In E.F. Moritz (Ed.), Sports, Culture, and Technology. An Introductory Reader. Sottrum: artefact.

Fikus, M. and Schürmann, V. (Eds.) (2004). Sportwissenschaft als Kulturwissenschaft. Bielefeld: transcript.

Gebauer, G. and Wulf, C. (1998). Spiel – Ritual – Geste. Mimetisches Handeln in der sozialen Welt. Reinbek: rororo.

Geertz, C. (1983). Dichte Beschreibung. Frankfurt/Main: Suhrkamp.

Gibson, J. J. (1979). The ecological approach to visual perception. Boston: Houghton Mufflin.

Jäger, U. (2004). Der Körper, der Leib und die Soziologie. Entwurf einer Theorie der Inkorporierung. Königstein/Taunus: Ulrike Helmer.

Krüger, P. und Meyer, I. K. (Eds.) (2005). Transcultural Studies. Interdisziplinarität trifft Transkulturalität. Bremen: Universitätsdruckerei.

Mauss, M. (1935). Soziologie und Anthropologie, Band 2. Frankfurt/Main: Suhrkamp.

Moritz, E. F. (2003). Sports, Culture, and Technology. An Introductory Reader. Sottrum: artefact.

Williams, A. M., Davids, K. and Williams, J. G. (1999). Visual perception and action in sport. Routledge: E & FN Spon.

Witzel, A. (2000). Das problemzentrierte Interview. FQS 1(1), 9.

Understanding Human Perception of Field Hockey Stick Performance

Matt Carré and Mark McHutchon

Sports Engineering Research Group, University of Sheffield, m.j.carre@shef.ac.uk

Abstract. This study applies the semantic differential approach to measure hockey players' emotional responses to different hockey sticks, based on factors relating to performance (e.g. comfort, power, control) as well as physical properties (e.g. weight, stiffness). Each player was given a set of disguised sticks and asked to carry out a set of ball-strikes, before rating the sticks using a semantic differential questionnaire. The hockey sticks varied in their physical properties and results indicated that although players were unable to perceive stick stiffness directly, they were able to perceive the effect of stiffness through their rating of comfort and power. Perception of stick 'weight' was strongly linked to the reaction torque felt at the hand when the stick was held in equilibrium, prior to a hit. Sticks judged in a positive light were also reported to be comfortable and powerful and had relatively high rigidity. Results from this study have been incorporated into an all-encompassing design methodology for sports equipment, including analytical and empirical models of equipment performance.

1 Introduction

Today's hockey players can choose from a wide range of hockey sticks that differ in terms of their physical parameters (e.g. weight, stiffness), as well as performance (e.g. comfort, power, control). However, the extent to which players actually perceive these differences remains unclear. This study sought to investigate the perception of a range of hockey sticks, and compare this to physical properties.

Previous authors have interpreted the 'feel' of sports equipment in different ways. When 'feel' is used with regards to the physical sensation of holding sports equipment, it is the combined psychological interpretation and physiological response, to the transmission of forces and vibration to the human body. For tennis rackets, the characteristics that affect feel are numerous and include weight, 'swing weight', frame and string-bed stiffness, head size and shape, handle grip and shock absorbing properties (Brody, Cross and Lindsey 2002). A study on golf clubs investigated 'feel' in a broader sense (Hocknell, Jones and Rothberg 1996). Three main components of feel were identified: the trajectory of the ball flight, the sensation in the hands and the sound of the impact. It was found that a reduction in perceived softness of the sensation in the golfer's hands corresponded to an increase in low frequency vibration energy for certain club-ball combinations.

One approach to investigating human perception of product performance is known as "Kansei engineering" (Nagamachi 1995). This technique aims to quantify consumer feelings towards products and identify correlations between these feelings and product properties and/or design elements. Semantic pairs (e.g. good/bad) are used to aid subjects in describing their emotions. The methodology can be broken down into several stages, including definition of the study, collection of an appropriate list of semantic pairs, reduction of the list and design of a suitable subjective questionnaire (Schütte and Eklund 2001). Such an approach was used in this study.

2 Methodology

An initial list of 103 semantic differential pairs was compiled by collecting descriptive words from relevant magazines, and informal interviews with hockey players. The list was then reduced through consultation with twenty-five experienced hockey players who chose the six most important pairs to describe the sticks. These related to comfort, power, weight, stiffness, control and overall performance. The six pairs were used to form a questionnaire using a seven-point scale.

Five sticks were chosen for evaluation from a commercially available range. Table 1 shows the mechanical and physical properties of the sticks, including mass, length, moment of inertia about the swing axis (MOI), centre of mass distances from the handle (COM) and modulus of rigidity, EI - a stiffness measurement based on a three point bend test. All of the sticks were wrapped in black tape to disguise the make and model and hence to eliminate any pre-held views of performance. The stick heads and handles were left uncovered in order that the normal stick performance would not be altered. Each stick was labelled with a different symbol for purposes of identification (e.g. diamond, circle etc.).

Table 1. Physical properties of the sticks.

Stick	Design	Mass, kg	Length, m	MOI, kgm^2	COM, m	EI, Nm^2
1	Fully composite	0.62	0.928	0.0356	0.556	1041
2	Fully composite	0.58	0.926	0.0356	0.578	754
3	Reinforced wood	0.61	0.930	0.0354	0.551	435
4	Reinforced wood	0.64	0.929	0.0344	0.555	556
5	Reinforced wood	0.59	0.927	0.0353	0.542	308

Twenty-six university-level hockey players volunteered to participate in the study. The mean and standard deviations for age and height of the two gender subgroups (13 male and 13 female) were 20.5 ± 1.9 years, 1.74 ± 0.1 m and 20.2 ± 1.0 years, 1.64 ± 0.1m respectively. All of the subjects considered themselves to be of average to high athletic ability, with more than three years of playing experience. Players were asked to strike five stationary balls into a goal 3 m away using a double-handed shot (both hands close together at the top of a handle). They were requested to hit the balls at near maximum power, whilst remaining comfortable and using their normal posture. The sticks were tested in random order each time. The subjects were asked to record their responses to each stick after each exercise, using the questionnaire. Finally, the players were then asked to choose their favourite and least favourite, based solely on performance.

3 Results and Discussion

3.1 Power, Stiffness and Weight

The subjects' responses on each semantic scale were averaged for each stick, resulting in the semantic profile shown in Fig. 1. One-way ANOVA tests were conducted for each of the scales and the null hypothesis was not retained at the 0.01 criterion for any of the scales except *flexible-rigid*. Hence, players were able to consistently perceive a difference between the sticks in terms of control, power, weight, control and overall performance, but they were unable to directly distinguish any difference in stiffness. This result was somewhat surprising as stiffness was the physical stick parameter that varied most (see Table 1).

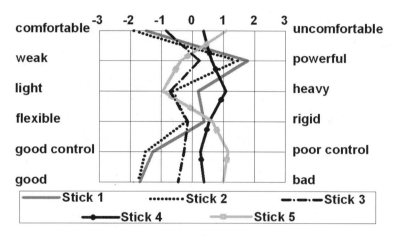

Fig. 1. The semantic profile plot of the mean perceived ratings on the SD scales.

However, players were able to perceive the *effect* of increased stiffness, through their responses on the power scale. Fig. 2a shows perceived power plotted against modulus of rigidity (stiffness), displaying a significant positive trend between the two variables ($R^2 = 0.94$). A previous study showed that increasing hockey stick stiffness by around 100% can increase the post-impact ball velocity by approximately 10 % (McHutchon, Curtis and Carré 2004). Hitting power is also affected by mass and moment of inertia, due to mass distribution although no significant correlations were found with these properties.

(a)

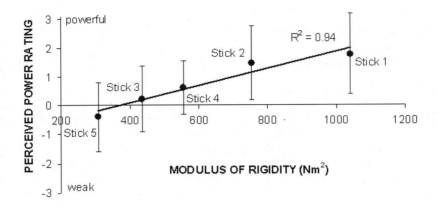

(b)

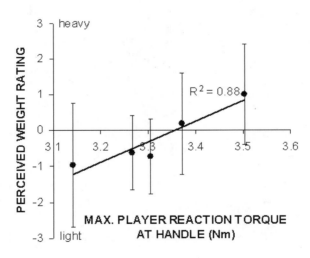

Fig. 2. The mean perceived power ratings (± 1sd) plotted against modulus of rigidity (a) and the mean perceived weight ratings (± 1sd) plotted against max. reaction torque (b).

There were significant trends between perceived weight and both the reaction force (effectively the weight of the stick) and the maximum reaction torque at the handle required to hold the stick in equilibrium (caused by the moment due to the weight and the position of the centre of mass). The strongest trend was with reaction torque, as shown in Fig. 2b. Hence, the perception of stick weight is enhanced by a large stick mass and/or a mass distribution that serves to concentrate more mass towards the head end of the stick.

3.2 Comfort and Control

No significant, strong trends were identified between perceived comfort and any of the separate stick properties. However, there was a weak, positive correlation ($R^2 = 0.58$) between perceived comfort and stiffness. Studies on tennis (Brody *et al.* 2002) have found that high levels of vibration in flexible tennis rackets add to the total shock, or jarring sensation experienced during impact. As racket stiffness increases, the vibration frequency increases and amplitude decreases, reducing discomfort. Therefore it is possible that increasing the stiffness of a hockey stick, would also lead to reduce the amplitudes of vibration which should make the stick more comfortable.

During the exercises there was little scope to judge stick control, since the hitting mechanism was relatively straight forward and unlikely to result in many miss-hits. A positive trend was found between perceived control and modulus of rigidity ($R^2 = 0.71$), but it is thought that players found it difficult to respond to this question accurately and simply considered *'good control-poor control'* to be synonymous with *'good-bad'*.

3.3 Overall performance

Unsurprisingly, there were strong correlations between overall performance (*good-bad*) ratings and the other performance-related scales of comfort, power and control ($R^2=0.96$, 0.82, 0.98 respectively). These responses were in line with the players' choice of stick preference, with Sticks 1 and 2 fairing best (sharing 25 from 26 'favourite' votes) and Stick 5 fairing worst (voted 'least favourite' by 18 subjects).

No trends could be found between overall performance and the physical properties of mass or moment of inertia. A weak correlation was found between overall performance and centre of mass (COM) position ($R^2 = 0.57$) suggesting that players may have a slight preference for weight to be centred near the stick head. However, it is possible that the COM data was skewed due to Stick 2 having a COM significantly closer to the head than the other sticks.

There was no correlation between players' perceptions of overall performance and stick weight ($R^2 = 0.0001$). Preference for a particular 'weight' of stick is likely to be influenced by the player's strength, playing position and playing style. This was confirmed by comments from certain players who chose Stick 4, the heaviest stick, as their most or least favourite stick based on weight alone. However, it is reasonable to expect that there will be a desired weight range within which all sticks should lie, neither being 'too light' or 'too heavy'. There was a positive trend between overall performance rating and stiffness ($R^2 = 0.73$), thought to be due to a combination of the positive responses to stiff sticks, through comfort and power ratings, as previously discussed.

4 Conclusions

This study found that the Kansei approach to understanding human perception of products could be applied successfully to some sports equipment. Although players were unable to perceive stick stiffness directly, they were able to perceive the effect of stiffness through rating comfort and power, based on responses made after repeatedly striking a ball. Stick 'weight' was found to be most strongly linked to a reaction torque being felt at the hand when the stick was held in equilibrium, prior to a hit. It is thought that the exercises the players were asked to perform did not give them adequate opportunity to judge control and that this rating was aligned with overall performance. Sticks judged in a positive light were also reported to be comfortable and powerful and had relatively high rigidity. This approach can now be applied more readily by sports equipment manufacturers, as well as other industries (e.g. medical devices) to gain an insight into player's emotional responses to design parameters such as stiffness, mass and centre of mass (COM). Data from this study has already been used to form an evolutionary design methodology for field hockey sticks that also includes an analytical model of hitting power and an empirical model for dribbling performance.

References

Brody, H., Cross, R. and Lindsey, C. (2002). *The Physics and Technology of Tennis.* Solana Beach, CA: Racquet Tech Publishing.

Cross, R.C. (1998). The sweet spots of a tennis racket. Sports Engineering, 1, 63-78.

Hocknell, A., Jones, R. and Rothberg, S. (1996). Engineering 'feel' in the design of golf clubs. In: S.J Haake (Ed.), *The Engineering of Sport: Proceedings of the 1ˢᵗ International Conference on the Engineering of Sport* Rotterdam: Balkema, pp. 333 - 338.

McHutchon, M.A., Curtis, D. and Carré, M.J. (2004). Parametric design of field hockey sticks. In: M. Hubbard, R. Mehta and J. Pallis (Eds.), *The Engineering of Sport 5, Volume 1: Proceedings of the 5ᵗʰ International Conference on the Engineering of Sport*, ISEA, UK, pp. 284-290.

Nagamachi, M. (1995). Kansei Engineering: a new ergonomic customer-oriented technology for product development, International Journal of Industrial Ergonomics, 15, 3-11.

Schütte, S. and Eklund, J. (2001). An approach to Kansei Engineering - Methods and a Case Study on Design Identity. In: *Proceedings of the International Conference on Affective Human Factors Design, Singapore.* London: Asean Academic Press, pp. 27-29.

Strategies for Bank Shots and Direct Shots in Basketball

Hiroki Okubo[1] and Mont Hubbard[2]

[1] National Defense Academy, Japan, ookubo@nda.ac.jp
[2] University of California, Davis, USA

Abstract. A dynamic model is used to analyze the release conditions of direct and bank basketball shots: horizontal and vertical distance from the hoop center, release velocity, release
angle, lateral deviation angle, and backspin angular velocity. The dynamic model includes
radial ball compliance and damping and contains six sub-models: ball-rim, ball-bridge, ball-
board, ball-bridge-board, ball-rim-board, and pure gravitational flight. Each ball-contact sub-
model has slipping and non-slipping interactions. Margins in release angle and velocity are
defined and used to characterize robustness of capture to perturbations in these variables.
Based on these margins we suggest strategies for bank and direct shots depending on the court
position.

1 Introduction

Basketball shots are of two types: direct shots and bank shots. A direct shot is one in
which the ball first contacts the rim; a swish is a successful shot that contacts neither
the rim nor backboard. A bank shot bounces first off the backboard. Basketball players sometimes use the bank shot, deciding between the direct or bank shot depending
on their court position. The bank shot is rarely seen in three point shots, free throws,
and jump shots which are attempted from positions where the initial ball path is
nearly parallel to the backboard surface. On the other hand, players often try bank
shots when they shoot near the hoop and around 45 degrees from the end line.

There have been some previous analyses of bank shots. Shibukawa (1975) analyzed clean bank shots in free throws using a two dimensional algebraic impact
model. With a more complex, algebraic impact model allowing ball-rim and ball-
board contacts, Hamilton and Reinschmidt (1997) calculated release conditions for
successful free throws. Huston and Grau (2003) expanded this to a three dimensional
algebraic impact model. Recently, Okubo and Hubbard (2002; 2003; 2004a; 2004b;
2006) and Silverberg et al. (2003) presented general dynamic models for ball motion
with radial ball compliance. The model of Okubo and Hubbard (2004b) calculates
the ball-rim, ball-bridge, ball-backboard, and ball-bridge-board interactions with
slipping and non-slipping ball-contact.

In a general direct or bank shot, the basketball may contact the rim, board, bridge,
bridge and board, or rim and board simultaneously with a slipping or non-slipping
interaction. We investigate release conditions leading to capture for these two types

of shots in terms of horizontal and vertical distance from the hoop center, release

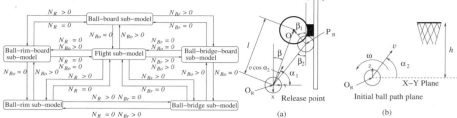

Fig. 1. Overall model.   **Fig. 2.** Geometry of field shots. (a) top and (b) side views.

velocity, release angle, deviation angle, and release backspin angular velocity.

2 Dynamic Model

2.1 Overall model

Our overall model for basketball shots has six sub-models; pure gravitational flight, and ball-rim (Okubo and Hubbard 2002; 2004a), ball-board (Okubo and Hubbard 2003) and ball-bridge-board (Okubo and Hubbard 2004b; 2006) and ball-rim-board contacts as shown in Fig. 1. Each contact sub-model has possible slipping and non-slipping motions. We switch between the sub-models depending on the reaction forces at the contact point. The slipping or non-slipping motion of each sub-model is used if the normal force is positive. Switching between slipping and non-slipping is based on the contact velocity and friction forces.

2.2 Geometry of field shots

A right-handed coordinate system (Fig. 2) with the origin O at the center of the hoop has its X axis parallel to the backboard, Y axis perpendicular to the board and Z axis vertical. The release position O_R is $(l\cos\beta,-l\sin\beta,-h)$ where β is the angle between the XY projection of line OO_R and the XZ plane, and l and h are horizontal and vertical distances from the hoop center, respectively. The release velocity is characterized by four input parameters; velocity v, angular velocity ω, lateral angle from the YZ plane α_1, and release angle α_2. In general, shots are attempted with $\alpha_1 = 90 - \beta$ deg for direct shots, and with $\alpha_1 = 90 - \beta_2$ deg and $\beta_1 \approx \beta_2$ for bank shots, where β_2 is the projected angle of incidence, β_1 is the angle between the board surface and the XY projection of line OP_B, and P_B is the location of the ball center when the ball first contacts the board. If the ball were modeled as a frictionless particle, $\beta_1 = \beta_2$ would be the projected angle of reflection. We assume, however, that the ball has non-zero rotational inertia, radial compliance, and friction, and possibly slips at the contact point; in general, therefore, the angle of reflection $\beta_1 \neq \beta_2$.

Table 1. Numerical values of simulation parameters.

m kg	R_b m	I kgm^2	k KN/m	c Ns/m	R_h m	R_r M	μ	μ_{Bo}
0.6	0.12	0.0057	47	19	0.234	0.009	0.5	0.7

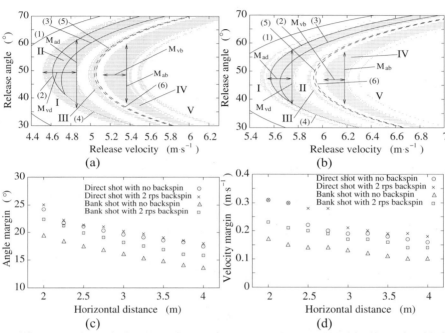

Fig. 3. Shots for $\alpha_1 = 0$, $\beta=90$ deg, and $h=0.2$ m. Capture combinations for (a) $l=2$ m and (b) $l=3$ m with $\omega = 4\pi$ rad/s, (c) margins of angle, and (d) margins of velocity.

3 Numerical Simulations

3.1 Simulation parameters

In the simulations described below we choose the parameters shown in Table 1: m, I, and R_b are the basketball mass, central mass moment of inertia, and undeflected ball radius, respectively; k and c are equivalent stiffness and damping coefficients; R_h and R_r are major and minor radii of the toroidal rim; and μ and μ_{Bo} are the coefficients of friction between the ball and rim/bridge, and board, respectively.

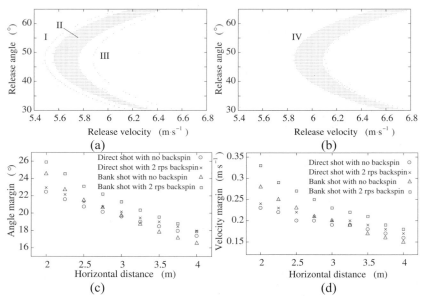

Fig. 4. Shots with $\beta = 33.5$ deg and $h = 0.2$ m. Capture combinations with $l = 3$ m and $\omega = 4\pi$ rad/s for (a) direct shots with $\alpha_1 = 56.5$ deg, (b) bank shots with $\alpha_1 = 49.2$ deg, (c) margins of angle, and (d) margins of velocity.

3.2 Shots with initial ball path in saggital YZ plane

Figures 3(a) and (b) show capture conditions (shaded) in the velocity-release angle space for $\omega = 4\pi$ rad/s, $\alpha_1 = 0$ and $\beta = 90$ deg and release positions 2 and 3 m behind and 0.2 m below the hoop center (typical jump shot conditions). Almost all capture conditions lie in five regions, with similar crescent shapes to those of longer free throws (Hamilton and Reinschmidt, 1997). Shots first hit the front of the front rim in region I. Region II, where shots first touch inside the rim or swish, is the largest for direct shots. Swishes lie between lines (2) and (3). Shots first contact the back rim in region III and shots in region IV are successful bank shots. We define four margins M_{vd}, M_{ad}, M_{vb}, and M_{ab} (Figs. 3(a) and (b)) that characterize the width and depth of the largest capture regions for direct and bank shots, respectively. M_{vd} and M_{ad} are direct shot velocity and angle margins at the minimum velocity on the right boundary of region II. M_{vb} and M_{ab} are similar margins for bank shots in region IV.

Figures 3(c) and (d) show the margins versus horizontal distance from 2 to 4 m from the hoop center. Shots nearer the hoop of both types have larger margins. Close shots have large velocity margins because regions I and II coalesce into a single area (compare Figs. 3(a) and (b)). The margins for direct shots (region II) are larger than those for bank shots with either 0 or 4π rad/s backspin. From these results we can see no advantages of attempting bank shots for ranges from 2 to 4 m from the hoop center in the saggital YZ plane, and therefore recommend direct shots exclusively.

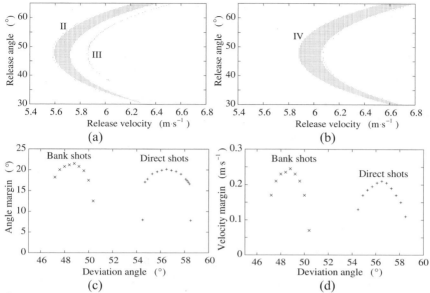

Fig. 5. Direct and bank shots with $\beta = 33.5$ deg, $l = 3$, $h = 0.2$ m, and $\omega = 4\pi$ rad/s. Capture combinations for (a) direct shots with $\alpha_1 = 58.5$ deg, (b) bank shots with $\alpha_1 = 51.2$ deg, (c) margins of angle, and (d) margins of velocity.

3.3 Shots from directions around 45 degrees from the end line

Capture combinations for direct and bank shots are plotted in Figs. 4(a) and (b), for which the initial pure backspin angular velocity is 4π rad/s, and the release position is $(3\cos33.5, -3\sin33.5, -0.2)$ m. The release angular velocity vector is assumed to be perpendicular to the vertical plane of the initial ball path. The lateral angles α_1 for the direct and bank shots are near 56.5 and 49.2 deg, respectively. In the bank shots, the ball first contacts the board with $\beta_1 \approx \beta_2$. Both direct and bank shots have one large capture region as shown in Figs. 4(a) and (b). The possibility of lucky shots with multiple bounces is very small. To be captured the bank shots need almost the same release angle as, and a slightly larger release velocity than, successful direct shots.

The margins of angle and velocity for bank shots near the hoop are larger than those for direct shots as shown in Figs. 4(c) and (d). Thus players should try bank shots near the hoop. Backspin helps capture for both types of shots. Especially of note is the fact that backspin is more effective for bank shots.

3.4 Capture conditions with perturbations in lateral deviation angle

Figures 5(a) and (b) show capture combinations with two different (non-optimal) cases of lateral angle than those shown in Fig. 4; $\alpha_1 = 58.5$ deg for direct shots and

$\alpha_1 = 51.2$ deg for bank shots. The shots have the same backspin $\omega = 4\pi$ rad/s, and release position (3cos33.5, -3sin33.5, -0.2) m. The large capture areas (region II for the direct shots and IV for bank shots) with these non-optimal lateral deviations (Figs. 5(a) and (b)) are smaller than those of shots with more optimal lateral deviations (Figs. 4(a) and (b)). The direct shots with perturbations of lateral deviation from optimal are missing region I entirely. The likelihood of lucky shots with numerous bounces is virtually zero with substantial errors in lateral deviation.

The angle and velocity margins as a function of lateral deviation angle are shown in Figs. 5(c) and (d). The direct shots have the largest margins when the vertical plane containing the initial ball path includes the hoop center. The bank shots have the largest margins with $\alpha_1 = 49.6$ deg and the angle of incidence $\beta_2 = 40.4$ deg. The projected reflection angle β_1 (= 44.8 deg) is slightly larger than β_2.

4 Conclusions

We have used our dynamic model to calculate successful release conditions for direct and bank shots that may contact the rim, bridge, and backboard. Margins in release velocity and release angle measure the robustness of capture to perturbations in these variables. These margins show that direct shots have advantages when the initial ball position is in or near the saggital YZ plane. On the other hand, bank shots are useful near the hoop from directions around 45 deg from the end line. Backspin generally helps capture after bouncing off the rim and board.

References

Hamilton, G.R. and Reinschmidt, C. (1997) Optimal trajectory for the basketball free throw. Journal of Sports Sciences. 15, 491-504.

Huston, R.L. and Grau, C.A. (2003) Basketball shooting strategies. Sports Engineering. 6, 49-63.

Okubo, H. and Hubbard, M. (2002) Dynamics of basketball-rim interactions. In: S. Ujihashi and S. J. Haake (Eds.), The Engineering of Sport 4. Blackwell Science, Oxford, UK, pp. 660-666.

Okubo, H. and Hubbard, M. (2003) Dynamics of basketball-backboard interactions. In: A. Subic, P. Trivailo, and F. Alam (Eds.), Sports Dynamics: Discovery and Application. RMIT University, Melbourne, Australia, pp. 30-35.

Okubo, H. and Hubbard, M. (2004a) Dynamics of basketball-rim interactions. Sports Engineering. 7, 15-29.

Okubo, H. and Hubbard, M. (2004a) Effect of basketball free throw release conditions using a dynamic model. In: M. Hubbard, R.D. Mehta and J.M. Pallis (Eds.), The Engineering of Sport 5. ISEA, Sheffield, UK, pp. 372-378.

Okubo, H. and Hubbard, M. (2006) Dynamics of the basketball shot with application to the free throw. Journal of Sports Sciences, in press.

Shibukawa, K., C. (1975) Velocity conditions of basketball shooting. Bulletin of the Institute of Sport Science, The Faculty of Physical Education. 13, 59-64.

Silverberg, L., Tran, C. and Adcock, K. (2003) Numerical analysis of the basketball shot. ASME Journal of Dynamic Systems, Measurement, and Control. 125, 531-540.

Optimising Sweeping Techniques for Olympic Curlers

Brett A. Marmo, Mark-Paul Buckingham & Jane R. Blackford

University of Edinburgh, brett.marmo@ed.ac.uk

Abstract. In the sport of curling players sweep the ice in the front of curling stones to increase the distance that the projectiles slide. Their vigorous sweeping raises the surface temperature of the ice thereby reducing its coefficient of friction. The change in ice temperature is dependent on the velocity that curlers sweep the ice, the downward force they apply and the pattern that is swept. The forces and velocities applied by Olympic level curlers were recorded on an instrumented brush. A numerical model was used to determine optimal sweeping pattern based on the curlers sweep force and velocity profiles.

1 Introduction

The Winter Olympic sport of curling is the only target-based sport where a projectile can have its trajectory corrected once it has left a player's hand or delivery device. This is done by sweeping the ice in front of an approaching stone to modify the coefficient of friction of ice resulting in metre scale variations of the distance the stone travels. Sweeping techniques and the athlete's fitness can provide the crucial difference between winning and losing a game of curling at both club and Olympic levels. Despite its importance, sweeping methods and equipment have developed over centuries of curling in a qualitative and anecdotal manner.

A *sweep ergometer* was developed to quantify the brush head velocities and forces achieved by Olympic level curlers (Buckingham, Marmo and Blackford 2006). Initially, the sweep ergometer was developed to monitor the fitness of athletes. However, results from the sweep ergometer also provide us with information about the dynamics of the sliding interface between the nylon brush head and the ice. A three-dimensional thermo-mechanical numerical model has been developed based on these results. The model determines the amount of heat generated by curlers and can be used to determine the optimal style of sweeping.

2 Ice Friction and Curling

In curling two teams of four players alternatively slide 19kg granite stones across 28m of ice to a target area known as the *house*. Teams score points by having the stone(s) closest to the centre of the house. Once a stone has been released players sweep in front of the stone to modify its trajectory.

Ice friction is central to the sport. Friction on ice depends on a number of parameters including the velocity, thermal properties and surface roughness of the sliding object and on the morphology and temperature of the ice (Akkok, Ettles, and Calabrese 1987). At ice rink temperatures (-5 °C) and over almost all the velocity range experienced in curling frictional heating is sufficiently high to melt the ice surface and provide a lubricating film of liquid water (Bowden and Hughes 1939; Bowden and Tabor 1950; Evans, Nye and Cheeseman 1976). Lubricated sliding is therefore the dominant friction mechanism in curling. For a given load, the frictional heating and the thickness of the fluid film increase with velocity, resulting in a non-linear reduction of friction with velocity ($\mu \propto v^{-1/2}$) (Evans et al. 1950, Stiffler 1984). It is this inverse relationship that is responsible for the curved trajectory of curling stones (Marmo and Blackford 2004).

Friction on ice is strongly dependent on temperature. As ice approaches its melting point less thermal energy is required to melt its surface. With higher temperature more lubricating melt is produced and μ decreases. Friction is increases with the sum surface roughness of the counter-facing surfaces (Hutchison 1992).

Sweeping reduces μ by polishing the ice and raising its temperature. Polishing the ice of relatively smooth Olympic standard curling ice has a negligible effect on the sum roughness of the counter-facing surfaces. Raising the temperature of the ice by frictional heating via sweeping has the greatest effect on μ and is the focus of the following study.

3 Sweeping dynamics

A curling brush was equipped with a series of strain gauges and a tri-axial accelerometer to produce a sweep ergometer that measures the forces and velocity applied by curlers (Buckingham et al. 2006). A typical sweeping profile of an Olympic level male curler is shown in Fig. 1. Elite curlers sweep with a frequency ~4.5 Hz with peak velocity of ~2.5 ms^{-1} and a peak downward force of ~450 N (Figs 1a & b). The peak velocity achieved by each curler is generally close to the centre of their stroke (Fig. 1c). The peak downward force does not coincide with the peak velocity (Fig. 1a & b), but occurs at the point in the stroke where the brush head is closest to the player's feet (Fig. 1d). When the brush head is closest to the player the horizontal moment arm from the curler's centre of mass is reduced to a minimum, increasing the vertical force exerted on the brush head.

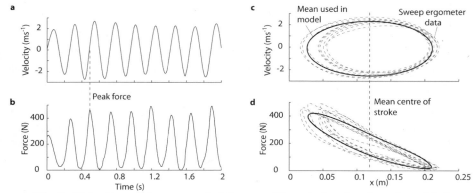

Fig. 2. Velocity and force recorded from an Olympic level male curler with the sweep ergometer. a) Horizontal velocity-time history. b) Vertical force-time history. Peak velocity and force do not coincide. c) Variation of velocity with position for a curler who's centre of mass is close to the origin. d) Variation of vertical force with position. Solid line shows the mean used in the numerical model (see Fig. 4).

4 Thermal Model of Sweeping

A three-dimensional thermal model has been developed based on recorded sweeping dynamics. The model is used to determine which sweeping styles and patterns produce the greatest temperature increase in the surface temperature of ice. The model determines the heat produced by rubbing at the interface between the brush head and the ice and solves thermal conduction equations to determine how the ice temperature varies both temporally and spatially.

The model employs two 3-dimensional cubic meshes that are configured to represent the brush head and ice. The ice surface lies in the x-y plane (at $z=0$) of a right-hand Cartesian co-ordinate system where y is parallel to the sliding direction of the curling stone and z is vertical. The mesh representing the brush has is based at $z=0$ and can have any position and orientation in the x-y plane so long as it lies within the outer boundary of the ice mesh. In the x and y directions the mesh elements have equal side lengths $\Delta x = \Delta y = 0.1$ mm. Most heat flow occurs in the z-direction so elements have shorter side lengths in this direction ($\Delta z = 0.01$ mm). The mesh representing the brush is configured to represent the brush head of the popular *Performance Curling Brush*, which has side lengths of 0.22 m and 0.07m. Elements in the ice and brush meshes are assigned the appropriate physical and thermal properties (see Table 1). The ice and brush are assumed to initially be in thermal equilibrium at −5°C and a numerical time step of $\Delta t = 0.001$s was used. The frictional heat Q generated by rubbing is equivalent to the work done over a finite time Δt:

$$Q = Fv\Delta t \qquad (1)$$

		Ice	Nylon
Thermal Conductivity	k (m kgs^{-3} K^{-1})	2.2	0.35
Density	ρ (kgm^{-3})	927	900
Specific Heat	c (J kg^{-1} K^{-1})	2090	2000
Thermal Diffusivity	κ (m^{2}s^{-1})	1.14x10^{-6}	1.95x10^{-7}

Table 1. Thermal and physical properties used in the numerical model.

where v is the sliding velocity and F is friction. This is descretised by re-writing Eq.1 as the heat per surface area of each mesh element, which allows the incorporation of μ and the load per unit area σ.

$$\frac{Q}{\Delta x \Delta y} = \mu \sigma v \Delta t \qquad (2)$$

The coefficient of friction is dependent upon the velocity of the slider. Based on a least squares fit of velocity and force data from the sweep ergometer μ=0.13+0.01$v^{-0.5}$. Both the ice and nylon absorb heat and increase in temperature. Heat is partitioned between each material based on their relative thermal conductivity k to maintain thermal equilibrium. The model determines which elements in the ice surface are in contact with elements in the surface of the nylon brush head and their temperature increased according to:

$$\Delta T = \frac{Q}{c\rho \Delta x \Delta y \Delta z} \qquad (3)$$

where c and ρ are the specific heat and density of the relevant material. Heat then conducts through each material according to the 3-dimension thermal conduction equation:

$$\frac{dT}{dt} = \kappa \left(\frac{\partial^2 T}{\partial x^2} \frac{\partial^2 T}{\partial y^2} \frac{\partial^2 T}{\partial z^2} \right) \qquad (4)$$

where κ is the thermal conductance of the relevant material. Heat conduction also occurs across the sliding interface. For simplicity it is assumed that the counter-facing surfaces are in perfect thermal contact, no thermal energy is lost to the surrounding air and energy lost to over come the latent heat of fusion is negligible.

5 Thermal Footprint from Sweeping

At each numerical time step the velocity and vertical forces recorded with the sweep ergometer and substituted into Eq. 2, the heat of friction calculated and allowed to conduct through the system (Eqs. 3 & 4). The model produces avi format (audio video interleave) movies showing the thermal footprint (Fig. 2) left by sweeping, which can be shown to players to augment their training program.

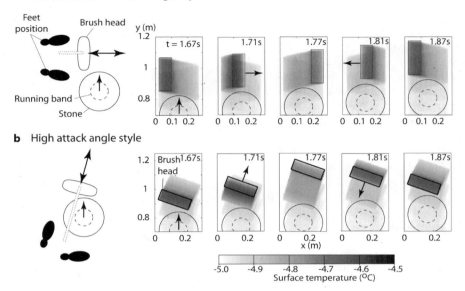

Fig. 3. Thermal footprint produced by two popular sweeping styles with schematic representation feet and brush position. a) A conventional low attack angle with frames showing modeled ice surface temperature for once complete sweeping cycle. b) A high attack angle style of sweeping. Each successive stroke overlaps the previous producing elevated ice surface temperatures.

The model can also be used to determine the optimal pattern for sweeping the ice. Figure 2 shows the effect of sweeping using two popular styles. To satisfactorily compare sweeping patterns the force and velocity history needs to be more consistent than that of an actual player so a velocity-force history based on mean sweep ergometer results was used (Fig. 1c & d). Conventionally, curlers sweep across the path of the approaching curling stone with a *low attack angle* with respect to the x-axis (Fig. 2a). This style of sweeping leaves a sinusoidal thermal footprint as the curler brushes back and forth in front of the while progressively moving along the ice in front of the sliding stone (*y*-direction). When curlers use a conventional style the asymmetry in downward force (Fig. 1d) results in higher surface temperatures on the side of the stone that the curler sweeps from. The greatest heat is generated as the player begins to sweep away from himself (Fig 1a, frame 2). Highest temperatures occur where successive strokes overlap each other, where temperature increases of 0.5 °C are produced.

A *high attack angle* style is a popular variation of the conventional style (Fig. 2b). A significant advantage of this high angle style is that brush tends to sweep over the same piece of ice several times adding heat each time to produce surface temperatures ~0.2°C higher than for an equivalent conventional sweeping style. Again the maximum heat generation occurs closest to the players feet, which lies in a band directly in front of the stone. The thermal footprint is therefore less asymmetric than a conventional sweeping style.

6 Conclusion

A method of analysing the effects of sweeping in the sport of curling has been developed by integrating quantitative force and velocity measurements made with a sweep ergometer with numerical modeling of heat generation and conduction through both the ice and brush head. The model allows coaching staff to show curlers what effect their sweeping had on the ice. Different sweeping styles can be compared and an optimal style chosen. The advantage of using a high attack angle style over a conventional style has been demonstrated to be due to the over-lapping of successive strokes.

Acknowledgments

We would like to thank Mike Hay and all at the Scottish Institute of Sport for their advice on curling and their support. Thanks to all members of British Olympic curling squad for their patients. Support from UK Sport and funding from EPSRC is also gratefully acknowledged.

References

Akkok, M., Ettles, C.M.McC. and Calabrese, S.J. (1987) Factors affecting the kinetic friction of ice. *Transaction of the ASME, Journal of Tribology*, 109, 552-560.

Bowden, F.P. and Hughes, T.P. (1939) Mechanisms of sliding on ice and snow. *Proc. R. Soc. London A*, 172, 280-298.

Bowden, F.P. and Tabor, D. (1950) *The Friction and lubrication of solids*, Oxford University Press, Oxford

Buckingham, M-P., Marmo, B.A. and Blackford J.R. (2006) Monitoring sweeping technique in the sport of curling. *IMechE Journal of Materials: Design and Applications – Part L*.

Evans, D.C.B., Nye, J.F. and Cheeseman, K.J. (1976) The kinetic friction of ice. *Proc. R. Soc. London A*, 345, 493-512.

Hutchison, I.M. (1992) *Tribology, Friction and wear of engineering matericals.* Butterworth-Heinemann, Oxford.

Marmo, B.A. and Blackford, J.R. (2004). Friction in the sport of curling. *The 5th International Sports Engineering Conference, Davis, California, September 2004*, V1, 379-385.

Stiffler, A.K. 1984. Friction and wear with a fully melting surface. *Journal of Tribology*, 106, 416-419.

An Initial Investigation of Human-Natural Turf Interaction in the Laboratory

Victoria H. Stiles[1], Sharon J. Dixon[1] and Iain T. James[2]

[1] University of Exeter, V.H.Stiles@exeter.ac.uk
[2] Cranfield University

Abstract. It is essential to provide high quality, safe and affordable sports surfaces in order to attain the health and social benefits from sports participation. Investment, construction and research into artificial sports surfaces have increased to meet this provision (Kolitzus, 1984; Nigg & Yeadon, 1987). Full provision cannot be met without natural turf surfaces, which also have an important role as greenspaces in the built environment. For improved access to sports facilities, there needs to be a significant improvement in the durability of natural turf surfaces and thus greater understanding of the human-natural sports surface interaction. Research into human interaction with natural surfaces is complicated by integrating natural soil media and sustaining turf growth in the laboratory environment. This study describes and provides data on methodology incorporating the biomechanical assessment of natural turf in the laboratory. Practicalities of using natural turf in the laboratory were overcome by using 10 portable plastic trays (0.57 m x 0.38 m x 0.08 m), turfed with ryegrass in a sand rootzone. Trays were positioned lengthways in the laboratory on non-slip matting (6 mm thick) to form a continuous runway and cover the force plate (AMTI, 960Hz). Ground reaction force (GRF) data were collected from two subjects wearing football boots (artificial turf/hard pitch design) for running, turning, and acceleration from rest. Mean GRF values compared well with the range of magnitudes presented in the literature for similar movements (Stucke, Baudzus & Baumann, 1984; Munro, Miller and Fuglevand, 1987; Miller, 1990) demonstrating that the incorporation of natural turf in the laboratory environment has been achieved successfully. Compared to running (subject 1, -0.41 ±0.06 BW; subject 2, -0.34 - ±0.04 BW), peak horizontal force increased for turning (subject 1, -0.50 ±0.06 BW; subject 2, -0.90 ±0.01 BW) and accelerating from rest (subject 1, -0.52 ±0.05 BW; subject 2, -0.44 ±0.09 BW), reflecting greater braking and propulsive requirements for the respective movements for both subjects. Peak vertical impact forces were 1.89 BW (±0.24) and 2.01 BW (±0.26) for subjects 1 and 2 respectively during running and 1.40 BW (±0.02) and 2.57 BW (±0.37) respectively during turning. To improve human-natural turf interaction, future studies will assess multiple subjects, movements, footwear and a range of natural turf conditions using the methodology developed here.

1 Introduction

Health and social benefits derived from sports and exercise participation are well documented (Department of Health, 2004). Affordable, safe and appropriate sports facilities make an important contribution to the promotion and attainment of a healthy nation. Participation in traditional sports such as hockey, football, tennis, rugby, cricket and lacrosse at school, club or elite level provide competitive opportunities to reap the health and social benefits of sports participation. These opportunities are much enhanced if the condition and provision of sports surfaces is appropriate.

Increased use of artificial surfaces in sport, particularly in tennis, hockey and to some extent football, has provided an all year round playing surface. Artifical surfaces are also less affected by adverse weather conditions, can require lower levels of maintenance and provide a smaller and more cost-effective facility tolerating regular multi-sport use compared to a natural surface (Kolitzus, 1984; Nigg & Yeadon, 1987; Cox, 2004). A natural sports surface is highly influenced by changes in the weather, does not tolerate a frequent multi-sport usage (problems of wear and degradation) and requires a larger area of ground to rotate pitch use. Artificial sports surfaces have made an important contribution to the provision of functional sports surfaces and increased sport participation. However the importance of maintaining natural turf sports surfaces is two-fold: it is crucial to protect greenspaces and playing fields in the built environment and to maintain the fundamental playing characteristics of sports such as football, rugby, golf, cricket and lacrosse.

Modification of hockey pitches that started in the 1970's from natural turf to artificial turf surfaces resulted in certain playing skill adaptations with a loss of some surface-related skills and an enhancement of other skills together with a faster-paced game (Spencer, Lawrence, Rechichi, Bishop, Dawson and Goodman, 2004). However, sports such as football are more reluctant to adapt the characteristics of their game to an artificial turf unless the surface allows complete replication of the characteristics of play (UEFA, 2005). The change in playing characteristics and reluctance encountered when switching from a traditional natural turf surface to artificial turf highlights the importance of maintaining the availability of natural turf sports surfaces. However, advancement in the construction and sustainability of natural sports surfaces is required if their provision is to be maintained for training and competitive use in sports.

There is a scarcity of research incorporating natural soil media into the biomechanics laboratory due to the logistical complications of integrating and sustaining turf growth in an un-natural environment. This fact is highlighted in a study that assessed studded footwear designed for use on grass with participants performing on artificial surfaces in a laboratory (Morag & Johnson, 2001). However, analysis of natural turf surfaces has been performed in the field to assess grip performance while performing cutting manoeuvres (Coyles, Lake & Patritti, 1998). Plantar pressures have also been assessed during soccer-specific movements in the field (Eils, Streyl, Linnenbecker, Thorwestern, Volker & Rosenbaum, 2004). However, the incorporation of natural turf in the laboratory is required in order to utilize biomechanical equipment that is either too sensitive or practically inappropriate to be used in an outdoor environment.

Advancement of natural sports turf engineering requires increased understanding of how turf responds to variations in human movement, ideally sports specific movement. Biomechanical assessment is required to provide input characteristics regarding the loading of natural turf. The first step in achieving the overall aim of a biomechanical analysis of sports specific movements on natural turf in the laboratory is to incorporate natural turf in the laboratory and collect initial data. The present study describes and provides data on methodology incorporating the biomechanical assessment of natural turf in the laboratory environment.

2 Methods

Ten portable plastic trays (0.57 m x 0.38 m x 0.08 m) were turfed with ryegrass in a sand rootzone (Fig. 1). Trays were positioned lengthways in the laboratory on non-slip matting (6 mm thick) to form a continuous runway and cover a force plate (AMTI, Massachussetts, 960 Hz). One tray was used to cover the force plate (force plate tray). All trays remained in place for the entire data collection period. A surrounding supportive runway of rubber matting and foam covered with an acrylic top-surface was constructed on either side of the turf runway for the safety of partici-pants. Figure 2 illustrates the laboratory lay-out. Figure 3 presents a detailed picture of the laboratory set-up.

One female (Subject 1, S1) weighing 745 N and one male (Subject 2, S2) subject weighing 824 N both wearing football boots (size 10) of an artificial/hard pitch de-sign performed football specific movements along the natural turf runway in the laboratory. Ten running trials at a speed of 3.83 m.s^{-1} (±5%) were performed by each subject. Contact with the force plate was made with the right foot. A second movement involving turning, was performed on the force plate with the subject exit-ing the force plate pointing in the direction from which they had started the run-up. The third movement required the subject to start on the force plate and accelerate from rest using their preferred push-off leg. Three turning and accelerating from rest trials were performed by S1. Three turning trials and ten accelerating from rest trials were performed by S2. Approval for the collection of data from human participants was obtained from the School of Sport and Health Sciences, University of Exeter Ethics Committee.

A right-footed step was analysed for both the running and turning movements. Push-off during the accelerating from rest movement was analysed from the appro-priate leg. Ground reaction force data (GRF) from the AMTI force plate (960 Hz) was collected from respective movement steps. In the case of the running and turn-ing manoeuvres, data were analysed from steps that had been made without evidence of step alteration in order to meet the force plate. Where subjects failed to make contact with the plate in a natural style, data were discarded and the trial recollected.

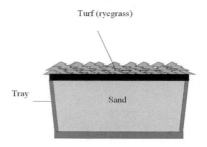

Fig. 1. Plastic tray turfed with ryegrass on a sand rootzone

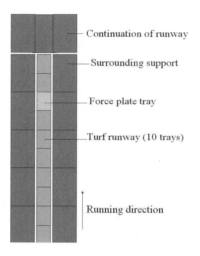

Continuation of runway

Surrounding support

Force plate tray

Turf runway (10 trays)

Running direction

Fig. 2. Laboratory lay-out

Fig. 3. Laboratory set-up

3 Results

Mean ground reaction force data for each subject performing each of the sports-specific movements (running, turning and accelerating from rest) are presented in Table 1.

Table 1. Mean ground reaction force results for subjects 1 and 2.

	Number of Trials	Peak Fz (BW)	Peak Fz Time (s)	Peak LR (BW.s^{-1})	Average LR (BW.s^{-1})	Peak Fy (BW)	Peak Fy Time (s)
Running (Peak Fy represents peak horizontal force during braking)							
S1	10	1.89 (±0.24)	0.03 (±0.01)	137.59 (±33.65)	64.85 (±17.25)	-0.41 (±0.06)	0.057 (±0.006)
S2	10	2.01 (±0.26)	0.03 (±0.01)	175.25 (±51.63)	63.97 (±21.64)	-0.34 (±0.04)	0.053 (±0.017)
Turning (Peak Fy represents peak horizontal force during braking)							
S1	3	1.40 (0.02)	0.078 (±0.018)	37.589 (±8.96)	18.653 (±5.1)	-0.50 (±0.06)	0.083 (±0.019)
S2	2	2.57 (±0.33)	0.043 (±0.004)	169.93 (±28.55)	60.02 (±1.55)	-0.90 (±0.01)	0.051 (±0.002)
Accelerating from rest (Peak Fy represents peak horizontal force during propulsion)							
S1	2	1.59 (±0.06)				-0.52 (±0.05)	
S2	10	1.45 (±0.18)				-0.44 (±0.09)	

4 Discussion

The present study collected data from two subjects performing sports specific movements on a natural turf surface in the laboratory. Peak impact forces found in the present study (1.89 BW and 2.01 BW) compared well to those presented in the literature. Running at a speed of 3.75 m.s^{-1} typically yields peak vertical impact forces of approximately 1.86 BW (Munro et al., 1987). Subjects in the present study yielded peak horizontal braking forces representing 41% and 34% of their bodyweight, magnitudes that are also supported by values presented in the literature (Miller, 1990).

In the present study although a fewer number of trials were collected for the turning movement, subjects yielded higher magnitudes of peak braking force during turning (S1, -0.50 ±0.06 BW; S2, -0.90 ±0.01 BW) compared to their running step (S1, -0.41 ±0.06 BW; -0.34 ±0.04 BW). This expected result reflects a higher braking requirement for turning compared to running. It is suggested that, since turning and accelerating movements may have implications for turf wear, ground reaction force data collected in the present and future studies may inform future turf design.

Subjects in the present study, when accelerating from rest yielded propulsive horizontal force components of approximately half their bodyweight. Time histories in the literature depicting peak horizontal propulsive forces for a similar movement on cinder and artificial turf yield higher magnitudes of approximately 0.64 BW and 0.79 BW respectively (Stucke et al., 1984). Differences in values yielded from the present study compared to the literature are likely due to the differing surface condition and the performance of the accelerating movement.

Overall, the comparison of ground reaction force data derived from the present study with values presented in the literature has demonstrated that subjects performed movements on the natural turf surface yielding expected force data. This finding suggests that the incorporation of natural soil media in the laboratory environment and the collection of typical movement data were successful.

Ongoing research will assess multiple subjects performing sports specific movements on a variety of natural turf surfaces. Biomechanical data will quantify how variations in turf conditions affect human response. Such measures will also act to quantify the input characteristics of turf wear and degradation. Turf wear and soil deformation will be measured using novel and standard techniques for natural turf sports surfaces (BS EN 12231:2003 & BS EN 14954:2005). Increased understanding of human-natural turf interaction will improve understanding of the cause and effect nature of turf wear. It is important to attain this information in order to engineer more sustainable natural turf sports surfaces in the future, improve the provision of natural turf in both the sports and built-up environment and help maintain the characteristics of some sports that would have been changed if played on an artificial surface.

Acknowledgements

The authors gratefully acknowledge the funding of this research by the Engineering and Physical Sciences Research Council, UK under project EP/C512243/1(P).

References

Department of Health (2004) At least five a week–evidence on the impact of physical activity and its relationship to health. Department of Health, Physical Activity, Health Improvement and Prevention.

Kolitzus, H.J. (1984) Functional standards for playing surfaces. In: E.C. Frederick (Ed.), *Sports Shoes and Playing Surfaces: Biomechanical Properties*. Human Kinetics Publishers Inc: Champaign, IL, pp. 98-118.

Stucke, H., Baudzus, W. and Baumann, W. (1984) On friction characteristics of playing surfaces. In: E.C. Frederick (Ed.), *Sports Shoes and Playing Surfaces: Biomechanical Properties*. Human Kinetics Publishers Inc: Champaign, IL, pp. 87-97.

Miller, D.I. (1990) Ground reaction forces in distance running. In P.R. Cavanagh (Ed.) *Biomechanics of Distance Running*. Human Kinetics: Champaign, IL, pp. 203-223.

Munro, C.F., Miller, D.I. and Fuglevand, A.J. (1987) Ground reaction forces in distance running: a reexamination. Journal of Biomechanics, 20, 147-155.

Nigg, B.M. and Yeadon, M.R. (1987) Biomechanical aspects of playing surfaces. Journal of Sports Sciences, 5, 117-145.

Spencer, M., Lawrence, S., Rechichi, S., Bishop, D., Dawson, B. and Goodman, C. (2004) Time motion analysis of elite field hockey, with special reference to repeated-sprint activity. Journal of Sports Sciences, 22, 843-850.

UEFA, Turf first in the Netherlands, (2005) Retrieved from **www.uefa.com** 06/10/05.

Analysis of Game Creativity Development by Means of Continuously Learning Neural Networks

Daniel Memmert[1] and Jürgen Perl[2]

[1] University of Heidelberg, Institute of Sports Science, Daniel.Memmert@urz.uni-hd.de
[2] University of Mainz, Institute of Computer Science, perl@informatik.uni-mainz.de

Abstract. Experts in ball games are characterized by extraordinary creative behavior. This article outlines a framework of analyzing creative performance based on neural networks. The aim of this study is to compare the potential of different kinds of training programs with the learning of game creativity in real field contexts. The training groups (soccer group, $n=20$; field hockey group, $n=17$) showed significant improvement in comparison to the control group ($n=18$) with respect to the three measuring points, although no difference could be established between the groups. As regards the development of performance, five types of learning behavior can be distinguished, the most striking ones being what we call "up-down" and "down-up". In the field hockey group in particular, an up-down fluctuation process was identified, whereby the creative performance increases initially, but at the end is worse than in the middle of the training session. The reverse down-up fluctuation process was identified mainly in the soccer group. The results are discussed with regard to recent training explanation models, such as the super-compensation theory, with a view to future investigation.

1 Introduction

Creative behavior plays an important role in the performance of sport. It is a significant feat of attentiveness—particularly in sport—for the generation of tactical response patterns and for seeking original solution ideas, when a player is also able to perceive objects that appear unexpectedly and to incorporate them into the game plan alongside his initial and most significant plan. Creative soccer players, for example, set themselves apart in situations such as the following: although they may actually intend to pass the ball to player A, they are able to perceive at the last moment that player B is suddenly unmarked and better positioned and pass the ball to him instead. In a general and more scientific context, Sternberg and Lubart (1999, p. 3) define creativity as "the ability to produce work that is both novel (i.e. original, unexpected) and appropriate (i.e. useful)."

Nowadays it is broadly accepted that creativity is a stochastic combinational process (cf. Dietrich, 2004; Simonton, 2003). This means that in such a mental state, creative thinking is characterized by unsystematic drifting and is more chaotic, permitting more loosely connected associations to emerge. Neural network-based qualitative analyses seem a good starting point for evaluating creative performance. The stochastic approach reduces the total recorded data to only a few statistical quantities and checks their significance by means of variance analysis. In contrast, neural networks—considering all available data to be high-dimensional points that correspond to neurons—can be used to extract specific striking features and qualitative trends from all original data. The field of sports seems a fruitful area in which to study complex behavior in a complex context. In particular, complex situations such as ball games enable creative performance to be analyzed in an ecologically valid way, as is recommended by many investigators (see Runco and Sakamoto, 1999; Simonton, 2003). Neural networks have not yet been used for the study of creative behavior in complex field situations such as sports.

2 Creativity and Dynamically Controlled Networks (DyCoN)

Creativity is a stochastic combinational process (see Boden, 2003; Simonton, 2003). Martindale (1995) used a wide range of examples (cf. associative hierarchies, defocused attention) to show that major theories of creativity can be translated into neural networks and modeled with connectionist theories, as both are nearly identical. For these reasons computational psychology seems a suitable way of modeling creative performance without any a priori detailed predications or explanations.

Up to now only a few psychometric measures of creativity are widely accepted in literature (cf. Plucker and Renzulli, 1999). To test the theory of creative thinking, the two conceptions of originality and flexibility identified via factor-analysis by Guilford (1967) were used and extended to sport-specific situations. A new procedure was developed and evaluated for assessing the two creativity characteristics. This is described in more detail above. An important point is that it is generally agreed that creative performance characteristics should be analyzed individually. Together with Simonton's assertion (2003) that creativity is a stochastic combinational process, neural networks seem to be a good solution for data evaluation.

DyCoN is based on the concept of Kohonen Feature Maps (KFM), whereby neurons are trained with data and so build clusters of similar input data, without needing any additional information. These clusters define *types* of input data and thereby help to recognize and identify test data after the training phase. A given test input is recognized by the net as corresponding to the cluster to which it is most similar and is therefore identified by the type (e.g. name, specification) of that cluster. The new idea behind DyCoN is that each neuron learns and offers information continuously and, more importantly, *individually* (cf. Perl, 2002). This method has been used successfully in several projects during the last years. One reason is that DyCoN not only can be used as a tool for data analysis but, because of its specific architecture, also and in particular fits as a model for description and analyses of learning processes.

In the presented project, DyCoN complements quantitative stochastic methods with regard to associative clusters, bridging neurons and semantic relevance, which enables for analyzing qualitative phenomena, such as game creativity, and therefore make the children's creativity development much more transparent.

3 Method

Participants and Procedure

90 children of around seven years of age participated in the field study. Their mean age at end of the 15-month training period was 8.4 years (s = .83) and 28% were female. Of course not all children were in one of the different treatment groups for the whole time. The number of children decreases over time for a variety of different reasons (e.g. move to another town). All children were tested before the treatment and during the treatment after six months and 15 months. Also, not all children could take part in the three measurements, leaving a total of 55 children who participated in all tests. The co-variables of age during training, time of training and participation in training have no influence on the presented results.

Treatment

Two treatment groups and a control group were compared.
Specific treatment − soccer (n = 20): Each training session was structured in the same way, namely, the soccer teams started with non-specific training exercises and played a free soccer game for 30 minutes at the end of each lesson. For the rest of the time they were taught technical skills, such as passing, stopping, blocking or dribbling. No closed relationship to specific positions or drill training existed. The children regularly took part in competitions.
Specific treatment − field hockey (n = 17): The children were given coordination exercises and different mini games involving reacting with hand, foot and hockey stick. For the remainder of the time they practiced hockey-specific skills, such as how to handle the stick or the ball. They also learnt how to dribble, push-pass and control the ball. At the end of the training session they often played a field hockey game with modified rules. Also, the children regularly took part in competitions.
Control group (n = 18): The children in the control group only participated in the regular school sports program twice a week. No child was a member of a special club.

Instrumentation

Game-test situations were constructed as an instrument for data recording (cf. Memmert & Roth, 2003). This instrument contains a context-dependent real world setting which can provoke creative solutions directly in ecologically-valid situations. In recurring comparable situations, this game-test situation has evoked creative behavior in using gaps. The game-test situations involved three different kinds of skill (hand, foot and hockey stick) in a system where the players take turns (two rounds for each person). Game idea, number of players, rules and environmental conditions were given. This game-test situation was tested for objectivity, reliability and validity in many preliminary studies. For instance, the internal consistency coefficients for creative thinking in the game-test situations are .72 and therefore in a similar area to usual measurements of creativity.

The children's performance in the game-test situation was recorded on videotape and judged using a subsequent *concept-oriented expert rating*. This means that first the experts were given exact observation criteria according to Guilford (1967). The main evaluation criterion concerned the originality or unusualness (scale 10 to 1) and the number of unusual solutions they came up with (= flexibility; scale 10 to 1). *Originality* denotes the unusualness, innovativeness, or even uniqueness of the children's ideas and decision making. It is defined as the statistical rareness of solutions to the game test situations. *Flexibility* also shows close connections with unusualness, innovativeness, etc. and characterizes the ease with which someone changes between thought levels, uses other systems of reference, generates different hypotheses, and modifies information. It is defined as the number of different solutions to the game-test situations. Secondly, the experts were trained using special videotapes. Thirdly, the experts underwent a final video-based test to check their expert quality. Only experts showing a high reliability as measured against a golden standard of ball games experts were chosen. The children in the game-test situations were each judged by three ball games experts. Because of the two player rotations, each child received two creative values from each of the three coders. All of the intra class correlation coefficients (inter-rater reliability) were above the crucial limit of .80.

4 Results

Each child was analyzed individually with neural networks and a qualitative analysis was conducted. Figure 1 shows trajectories that represent individual training processes. Among the 20 soccer players, the subjects' game creativity with the "hand" motor function shows different types of development over the 15 months. In 5 out of 20 cases (25%) the creative performance increases initially, but at the end, at the yellow square step, is worse than in the middle of the training session (up-down fluctuation process). The reverse development was identified in 30% of the children (down-up fluctuation process). In 25% the performance increases and in 10% it decreases. Only in a few cases does no development occur (10%).

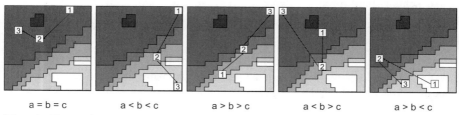

| a = b = c | a < b < c | a > b > c | a < b > c | a > b < c |

Fig. 1. Exemplary representation of the trajectories of the subjects specifically trained in field hockey. The learning process starts with "1" (first test) end ends with "3" (third test). The levels of creativity quality stretch from excellent (white) to very poor (dark grey). "a", "b", and "c" mean the levels of creativity at the regarding steps.

Table 1 presents the overall results from the two trained groups and the control group, and also from the three motor functions. Interestingly, the field hockey group shows a different results pattern than the soccer group, independently of the motor functions. Only a comparison of both fluctuation processes' Chi2-statistic shows a significant effect between the field hockey and the soccer group ($\chi 2(1) = 18.942$; $p < .001$, $n = 52$). The children of the field hockey trained group showed a stronger up-down fluctuation than the soccer group, and vice versa. No significant differences occur between the soccer and control group in either fluctuation process. Nor were any important differences found in the three other learning behaviors—stable, increase, or decrease of game creativity—between the three groups. Such a process-oriented presentation of the result might help to detect problems and find reasons. In this particular example, this could be due to the fact that the whole training process was underlined by different kinds of learning behavior according to the training groups.

Tab. 1. Summary of the results of all trajectories of the three treatment groups. The five different types of learning behavior are presented. The number of each learning behavior in the three motor functions (hand, foot, hockey stick) is given in brackets.

		Field hockey group	Soccer group	Control group
a=b=c	————	3 (1/0/2)	5 (2/3/0)	4 (1/3/0)
a≤b≤c	/	10 (1/5/4)	11 (5/2/4)	11 (7/2/2)
a≥b≥c	\	12 (5/4/3)	8 (2/1/5)	17 (4/6/7)
a<b>c	/\	24 (9/7/8)	11 (5/0/6)	11 (4/4/3)
a>b<c	\/	1 (1/0/0)	17 (6/9/2)	6 (2/3/1)

5 Discussion

The intention of this study was to analyze whether and how—indicated by individual learning behavior—game creativity could be improved by following a structured field training program. A second aim was to evaluate neural networks by studying creative performance. The soccer and the field hockey group considerably improved their creative performances over time. In contrast, no significant improvement in performance was evident for participants in the control group over the three measuring points. It was surprising to discover that five main types of learning behavior exist, depending on the training group. Only two of them show a consistent profile with constant or improved performance over time, as expected (see Tab. 1, first and second rows), while a third type shows a continuous deterioration (Tab. 1, row 3) and two types—down-up and up-down fluctuation processes—show strange behavior with alternately increasing and decreasing performance (Tab. 1, rows 4 and 5). Thus, a practice-oriented training program can lead to very different kinds of learning performances.

We analyzed this phenomenon by means of the Performance Potential Metamodel (PerPot, cf. Perl, 2002), which was originally developed in order to analyze physiological adaptation processes. The use of PerPot here is based on the idea that creative behavior can also be understood as a dynamic adaptation process (e.g. Boden, 2003; Simonton, 2003). Typical results of PerPot-analysis are the delay values: In the presented case the ratio between the delay of learning (DL) and the delay of unlearning

(DU) characterizes the way the learning process is adapted to the training load stimulus. If DU is smaller than DL then unlearning becomes effective first, otherwise the learning effect is faster (down-up fluctuation process). Therefore, as is demonstrated in Tab. 2, five types of learning behavior can be characterized theoretically, simulated by means of PerPot and observed from original data.

Tab. 2. Five types of observed learning behavior compared to the corresponding delay values taken from PerPot simulation. The delay values of the "strange" types of behavior are highlighted, demonstrating the change of delay ratio.

		DS: delay of unlearning	DR: delay of learning
a=b=c	——————	1.5	1.2
a≤b≤c	⟋	7.0	3.5
a≥b≥c	⟍	4.7	7.5
a<b>c	⟨	10.5	6.5
a>b<c	⟩	3.3	3.5

The results show that adaptation is not a smooth, continuous process, and that at least 2 or 3 additional measurement times are required if a deeper understanding is to be gained. One main question could be: Are there individual learning types, characterized for instance by particular ratios of delays, that can be recognized and then help to individually optimizing learning strategies?

References

Boden, M.A. (2003) Computer Models of Creativity. In: R.J. Sternberg (Ed.), *Handbook of Creativity*. Cambridge University Press, Cambridge, pp. 351-372.

Dietrich, A. (2004) Neurocognitive mechanisms underlying the experience of flow. *Consciousness and Cognition 13*, 746-761.

Guilford, J.P. (1967) *The nature of human intelligence*. McGraw Hill, New York.

Kohonen, T. (1995) *Self-Organizing Maps*. Springer, Berlin - Heidelberg - New-York.

Martindale, C. (1995) Creativity and connectionism. In: S.M. Smith, T.B. Ward and R.A. Finke (Eds.), *The creative cognition approach*. MA: MIT Press, Cambridge, pp. 249-268.

Memmert, D. and Roth, K. (2003) Individualtaktische Leistungsdiagnostik im Sportspiel. *Spektrum der Sportwissenschaft 15(1)*, 44-70.

Perl, J. (2002) Adaptation, Antagonism, and System Dynamics. In: G. Ghent, D. Kluka and D. Jones (Eds.), *Perspectives – The Multidisciplinary Series of Physical Education and Sport Science*. Meyer & Meyer Sport, Oxford, pp. 105-125.

Plucker, J.A. and Renzulli, J.S. (1999) Psychometric Approaches to the Study of Human Creativity. In: R.J. Sternberg (Ed.), *Handbook of Creativity*. Cambridge University Press, Cambridge, pp. 35-61.

Runco, M.A. and Sakamoto, S.O. (1999) Experimental Studies of Creativity. In: R.J. Sternberg (Ed.), *Handbook of Creativity*. Cambridge University Press, Cambridge, pp. 62-92.

Simonton, D.K. (2003) Scientific creativity as constrained stochastic behavior: The integration of product, person and process. *Psychological Bulletin 129*, 475-494.

Sternberg, R.J. and Lubart, T.I. (1999) The concept of creativity: Prospects and paradigms. In: R.J. Sternberg (Ed.), *Handbook of creativity*. Cambridge University Press, Cambridge, pp. 3-15.

Influence of Sports Equipments on Human Arm

S.Kawamura[1], H.Takihara[2], H.Minamoto[3] and Hossain Md.Zahid[4]

Toyohashi University of Technology, Department of Mechanical Engineering,
[1] kawamura@mech.tut.ac.jp   [2] takihara@dynamics.mech.tut.ac.jp
[3] minamoto@mech.tut.ac.jp   [4] zahid@mech.tut.ac.jp

Abstract. Impacting on muscles is a normal phenomenon when a ball strikes a racket or a bat during playing tennis or baseball. It causes vibration, which propagates to human body, especially, hand and arm. Vibration is not desired because it may be a cause of the tennis elbow injury or fracture. So, designing of sports equipments is very important to avoid these injuries of players. In this study, the influence of sports equipments on the muscles and skeleton of human arm during impact is focused. A case is considered where a ball strikes against a tennis racket, which is fixed by hand and arm. A multi-degree-of-freedom model of the ball and racket is constructed by finite element method, and human hand, lower arm and upper arm are considered as an individually rigid linked model. The joint moments are calculated when a ball strikes against a racket using modal analysis technique. Then, some muscles are adopted and a musculoskeletal model is constructed. The muscular forces are identified such as to satisfy the joint moments by the optimization method. Finally, the influence of property of racket on the muscles is investigated.

1 Introduction

Today, tennis is played all over the world as a sport which everyone can easily enjoy. However, an injury called tennis elbow occurs so frequently that design of tennis racket is very important to avoid these injuries. There are many studies about vibration of a racket and restitution of a gut and a ball (Iwatsubo, Kawazoe, Matsuhisa). The other hand, injuries are received on the human body so that it is important to analyze both sports equipment and human body simultaneously. When playing tennis is considered, the racket vibration by impact effects the human body, especially, hand and arm.

In this study, the influence of sports equipments on the muscles and skeleton of human arm during impact is focused. A case is considered where a ball is struck against a tennis racket, which is fixed by hand and arm. A multi-degree-of-freedom model of the ball and racket is constructed by finite element method, and human hand, lower arm and upper arm are considered as an individually rigid linked model. The joint moments are calculated when a ball is struck against a racket using modal analysis technique. Then, some muscles are adopted and a musculoskeletal model is constructed. The muscular forces are identified such as to satisfy the joint moments by the least squares method. The muscular forces are derived for the various striking conditions, such as property of racket.

2 Dynamics model

When a player plays tennis, he/she always takes a racket but a huge muscle arm appears at the impact between a ball and racket. Tough the muscle forces during swing have effects on the phenomenon at the impact, the modeling and analysis are only focused at the impact in this study.

The musculoskeletal structure of human arms is modeled as three rigid links with muscles as shown in Fig.1. And there many kinds of muscle in human arms, but, in this study, we gather them as eight muscle groups from the viewpoint of muscle function in the musculoskeletal model.

A racket, a ball, and a clasp of a palm are modeled a multi-degree-of-freedom model by two dimensions as shown in Fig.2. The racket is treated as a uniform beam and modeled by the finite element method. Furthermore, a racket and a wall are connected with the rotational spring and the translational spring that shows the clasp of the palm. Then, there assumed to be a viscous damping in each spring in parallel.

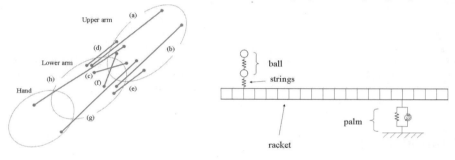

Fig.1. Upper limb model   Fig.2. Ball-racket-palm model

3 Equation of the motion

The equations of the racket, ball, strings and clasp of the palm are combined, and we obtain the equation of the complete system. However this equation is only applicable during the contact between the ball and racket. When the ball leaves from the racket, the strain of string changes from compression to tension. After that time, the parameters of the ball and strings are neglected in analysis.

4 Governing equation of the upper limb

The equilibrium of forces and moments on the n-th link are constructed as shown in Fig.3. $\mathbf{F}_{n-1,n}$ is the force acting from the link $n-1$, $\mathbf{F}_{n,n+1}$ is the force acting from the link $n+1$, as well as $\mathbf{M}_{n-1,n}$ and $\mathbf{M}_{n,n+1}$ mean the moments acting from the link n and $n-1$ respectively. The mass of link n is m_n, the vector of the acceleration of gravity is $\mathbf{g}$, the force acting on link n is expressed

$$F_n = F_{n-1,n} - F_{n,n+1} + m_n g. \tag{1}$$

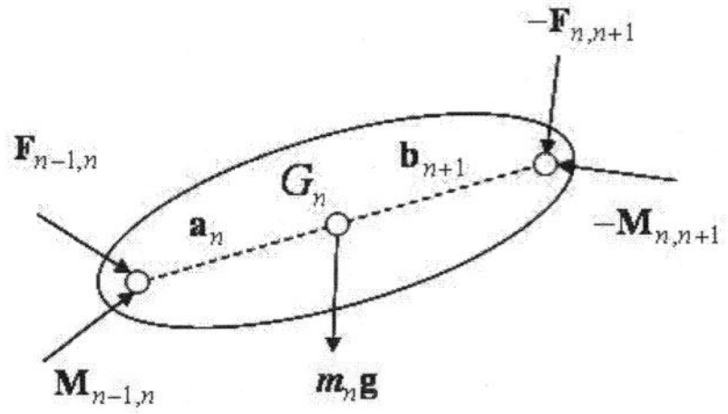

Fig.3. the link model

Then, the vector of the center of the link n and $n+1$ from the center of gravity of the link n is a_n and b_{n+1}, the moment acting on the link n is expressed

$$M_n = M_{n-1,n} - M_{n,n+1} + a_n \times F_{n-1,n} - b_n \times F_{n,n+1}. \tag{2}$$

Here, the force and moment at the end of link $F_{n,n+1}, M_{n,n+1}$ are external force and moment acting on the hand. Furthermore, the joint moment of each joint can sequentially be calculated from the hand by giving properties and condition of arm and hand.

When the muscular force is f_j and the moment arm is R_{ij} for the link n, the reaction force of joint $F_{n-1,n}$ and the reaction moment of joint $M_{n-1,n}$ are expressed

$$F_{n-1,n} - F_{n,n+1} = \Sigma f_j, \tag{3}$$

$$M_{n-1,n} - M_{n,n+1} = \Sigma f_j \times R_{ij}. \tag{4}$$

Using the attached position of muscles and the vector of attached position X_j, muscular force is expressed

$$f_j = \frac{X_{aj} - X_{bj}}{|X_{aj} - X_{bj}|} \cdot f_j, \tag{5}$$

where a and b denote the origin and insertion position, respectively. Substituting Eq. (5) into Eq. (3),(4), we obtain the equation with respect to the magnitude of muscular force.

In this research, eight muscle groups are considered shown in Fig.1. The groups (a) and (d) are attached on the front side, the groups (b) and (e) are attached on the back side, the group (c) is attached on the inside, and the group (d) is attached on the outside. The groups (g) and (h) are attached to the hand, (g) is back side, and (h) is front side. Because eight muscle groups are considered, the number of unknown parameters, that are forces of muscle groups, is fewer than the number of equation, so that the least squared method can be used. The weighting function isn't used in the least squared method. Its effect will be considered in the future study.

Table.1. Values of parameters

arm	l_u	0.28 m	racket	l	0.7 m
	l_l	0.3 m		m	5.0 kg
	l_h	0.18 m		ρA	0.518 kg/m
	m_u	1.944 kg		EI	121.0 Nm^2
	m_l	1.152 kg	strings	k_s	4.15×10^4 N/m
	m_h	0.432 kg	palm	k_{p1}	1.0×10^8 N/m
ball	m_{b1}	0.028 kg		k_{p2}	1.0×10^6 N/m
	m_{b2}	0.028 kg		c_{p1}	29.8 Ns/m
	k_b	7.98×10^4 N/m		c_{p2}	2.64×10^{-2} Nms/rad

5 Results and discussion

A reaction force and moment acting on a fixed edge corresponding to the hand by the modal analysis are calculated using the date of Table.1 (Matsùhisa). The results are shown in Fig.4. Here, the initial velocity of ball is 30[m/s]. Furthermore, the racket has a slight dumping. In this study, we pay attention to the behavior at the beginning of dynamic response.

Then the joint reaction force and moment in the upper limb model are calculated against the impact force and moment, which are the reaction force and moment on the fixed end of racket. After that, using Eq.(3) and (4), the muscular force of each of muscle groups is calculated. The results are shown in Fig.5. It is recognized that the muscular forces of the group (b), (c) and (e) are great. When a ball is struck by a racket with forehand style, these muscle groups work hard. Here some of muscular force are great than 10^6 N. Tough it is very huge, it is considered that the pre-stressed force is included and the precise analysis is need.

In the case of tennis elbow, a patient has pain on the back side of the elbow. Therefore, it is thought that this result tends similarly to an actual symptom. Furthermore, the activity of the muscle group (a) is especially small because the moment arm of the muscle group (a) is larger than the other one. Moreover, it is thought that a symmetrical result is obtained because there are only two kinds of the muscle group (g), (h) concerning the hand, and both moment arms are also almost equal.

In the results, the muscle forces are often negative because some muscles are only adopted as representative to express the moment. The muscles are antagonist one so that the positive muscular force of a muscle is greater than the result because the other muscle has the same effect by its negative value.

Next, the property of the racket, Young's modulus, is changed with 80% and 120% to show the effect of racket property on the muscular forces. The results are also shown in Fig.5. It is shown that the frequency of each muscular force is due to the external force, and the forces increase with the stiffness of the racket, and the periods of vibration decrease with one.

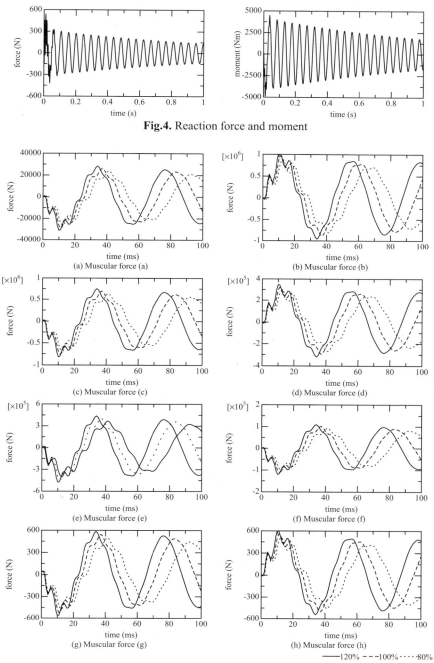

Fig.4. Reaction force and moment

(a) Muscular force (a)

(b) Muscular force (b)

(c) Muscular force (c)

(d) Muscular force (d)

(e) Muscular force (e)

(f) Muscular force (f)

(g) Muscular force (g)

(h) Muscular force (h)

—— 120% - - - 100% ····· 80%

Fig.5. Muscular forces

6 Conclusions

In this paper, the muscular forces are identified when a ball is struck by a racket. As a result, it is shown that the muscular forces of muscle group of back side and inside are great. And when the property of racket is changed, the frequency of each muscular force is due to the external force, and the forces increase with the stiffness of the racket, and the periods of vibration decrease with one.

References

Iwatsubo, T., et al. (2002) A finite element vibration analysis of a tennis racket with an impact shock protection system. The engineering of Sport 4, pp.273-280.
Kawazoe, Y., et al. (2002) Prediction of yours factors associated with tennis impact: Effects of large ball and string tension. The engineering of Sport 4, pp.176-184.
Matsuhisa, H., et al. (1995) Design of a Tennis Racket for Reducing Impulsive Force of Arm.Trans JSME. 61-591, pp.4429-4435.(in Japanese)

Using Sport to Educate and Enthuse Young People About Engineering and the Physical Sciences

David M James and Stephen J Haake

Sports Engineering, Sheffield Hallam University, d.james@shu.ac.uk

Abstract. Over the past decade, the United Kingdom has experienced a decline in the proportion of young people continuing with their education in the physical sciences through to university level. In general, young people believe these subjects to be 'boring' and irrelevant to their everyday lives. The image of the physical sciences is a serious concern to the UK Government since in order to maintain the UK's position as a technological leader in the world economy, a continual flow of high calibre graduates is essential. Numerous initiatives have been undertaken to encourage young people to study the physical sciences with varying degrees of success. This paper discusses one successful initiative coordinated and delivered by the Sports Engineering Research Group from the University of Sheffield. A series of interactive lectures and workshops were devised to explore how modern science and technology plays an ever increasing role in sport. In just two years, more than 13,000 young people took part in the initiative during 120 events across the UK. The impact of the initiative was monitored and evaluated throughout its duration by collating questionnaire data. Responses from the questionnaires, as well as other forms of feedback, showed the use of sport to be highly effective in terms of engaging young people's interest. The project demonstrated that the physical sciences need not be taught in a manner that disengages young people; by using examples and demonstrations from the world of sport, its perception can be one of interest and fascination.

1 Science Education in the UK

The UK Government recently announced its ten year strategy for science and innovation (Science and innovation framework 2004-2014). At the heart of the strategy is the recognition that science and innovation are key drivers of economic well-being and the quality of life. In order to maintain its position in the global economy, the Government has set itself the challenging task of raising national investment in research to 2.5% of GDP. The strategy's primary objective is to place the UK as a key knowledge hub for global industries, and to build on its reputation for scientific and technological discovery. In order to achieve these aims, the UK requires a highly skilled population that can readily engage in scientific issues. Furthermore, a great many young scientists need to be nurtured such that they will be able to develop the 'big ideas' of the future.

It is in this backdrop that science and engineering are facing a crisis in UK education. Since 1996, the number of A-level candidates in physics has declined by over 13 per cent, in chemistry by over 7 per cent and in mathematics by over 13 per cent, and this at a time when overall A-level numbers have increased by around 9 per cent. The UK is not alone in this respect – in the United States, student numbers in engineering, physical sciences and mathematics have reduced by 24, 12 and 7 per cent respectively over the last twenty years. The significant drop in the number of students studying the physical sciences at A-level has led to numerous high profile closures of university departments, and an associated drop in the percentage of students studying these subjects at degree level. According to the Higher Education Statistics Agency (www.hesa.co.uk), the total number of students in UK higher education rose from 1.75 million to 2.25 million between 1996 and 2004, but the percentage of students studying engineering dropped from 7.6% to 6%, and the percentage of students studying the physical sciences dropped from 4.3% to 3.3%. These trends are shown in Fig. 1 and are clearly at odds with the UK Governments strategy for science and innovation.

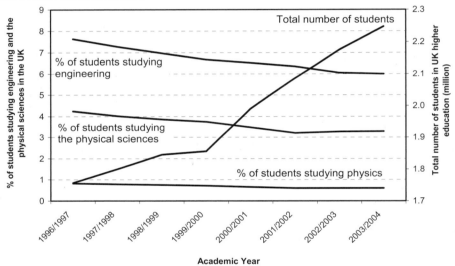

Fig. 1. Figures complied by the Higher Education Statistics Agency showing the rapid rise in UK student numbers, but the decline in engineering and the physical sciences.

It has become apparent that young people in the UK are becoming increasingly disengaged from science and engineering, and that if the Government's ambitious strategy is to be achieved; a dramatic change in student perceptions is required. It is difficult to pin point a single reason for the demise in the popularity of the physical sciences, but a compelling explanation lies in the absence of specialist physics teachers in schools. According to a recent independent report (Smithers and Robinson 2005) almost a quarter of all 11-16 UK schools had not one single teacher who had studied physics to any level at university. Science should be the most exciting subject on the school curriculum, but if schools do not have staff with the knowledge to bring the subject to life, then its perception of being 'boring' and 'difficult' is assured. A great many scientists from industry and academia are now involved in a myriad of projects that aim to bring science to life in the classroom; and the sports engineer has a clear role to play. Good examples of the input that industry and academia can have in schools are the multimedia resources developed by SportScience and Engineering in Education (*SEE*). These resources were developed to enrich the student learning experience and have achieved considerable success (Cooke and Bryce Taylor 2002; Cooke and Taylor 2004).

2 Taking sports engineering into schools

Sport has a near universal appeal to all people and particularly to the young. It is quite justified to say that no other topic can stimulate more public interest. Countless examples can be presented to support this statement, however one need look no further than to the fact that an estimated 3.7 billion people (more than half the worlds population) are believed to have watched the Sydney 2000 Olympic Games. Sport offers a fabulous context to teach many elements of the science curriculum, particularly in physics. Topics such as forces, acceleration and motion; energy transfer; work and power can all be taught by using examples from the world of sport. Sport makes these topics instantly accessible for young people because it is a subject that relates to their everyday lives and aspirations.

Since its formation, the University of Sheffield's Sports Engineering Research Group have worked with schools. At the start of 2004 this activity became a major focus for the research group with the commencement of a high profile nationwide schools lecture tour. With support from the Institute of Physics and the Engineering and Technology Board, an interactive lecture was devised and delivered to more than 8,000 young people during 55 events. The target audience for the lectures were 11 to 18 year olds. The hour long lecture focused on the technology of Olympic sports to coincide with the Olympic Games in Athens. The tour was well publicised and attracted considerable media attention. Needless to say, that with an average audience size of more than 145, the lectures proved to be very popular with school groups.

Riding on the success of the nationwide schools lecture tour, a second major project with schools was launched in the autumn of 2004. This second project was supported by the local education authorities in South Yorkshire and consisted of an extensive programme of workshops to be delivered across the region. More than 4,400 secondary school students were able to take part in the project during events at 44 different schools. The target audience for the programme of workshops were 11 to 18 year olds.

In addition to these two projects, the research group have also delivered numerous 'one off' events for young people all over the UK and the world. In just two years, the group have become established as a well respected provider of curriculum enrichment for secondary schools and have worked with well over 13,000 young people.

3 Developing an engaging interactive lecture

Sport provides a perfect vehicle to communicate topics in science that young people may find somewhat impenetrable. However, one can not simply use the 'hook' of sport and expect young people to be engaged with its underlying science. During the development of the aforementioned sports engineering lectures and workshops, a continuous process of formative evaluation was used to assure that the end product was of the highest standard. During the delivery of each lecture and workshop, school teachers completed a questionnaire that provided feedback on the scientific topics covered during the hour and the enjoyment of the students. Teachers were also asked to make comments about the style of presentation and for ways in which the lecture could be improved. This feedback was used to constantly refine and develop the lectures throughout the tours.

In essence, the lectures explored the role that technology plays in sport by presenting examples from four or five different sports. An additional theme of the lectures was the ethical question of whether it is cheating to use new technologies in sport. The workshops followed the same themes as the lectures but did so in a more 'hands on' format incorporating many experiments with specialist apparatus.

Over the course of the projects, lots of new lecture and workshop material was developed for an increasing number of different sports. The material for each sport was written such that it formed a stand alone unit that could be picked for inclusion during the lecture or left for another day. Before each event, different units were chosen based on the expected sporting interests and academic level of the audience.

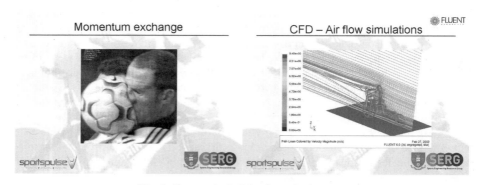

Fig. 1. Two typical slides from the lectures.

The lectures were delivered with the aid of presentation software. The slides were carefully designed and contained a myriad of videos, animations, diagrams and pictures to keep the audience engaged. Figure 1 shows two typical slides from the presentation. The slides demonstrate the deliberate method of using a minimum amount of text; this enabled the lecturer to talk in a non-scripted manner and provided room for improvisation and discussion with the audience.

During the lectures, every effort was made to interact with audience through the use of live experiments, demonstrations and discussions. Lots of sporting equipment was taken to the lectures for use as props as well as numerous pieces of bespoke experimental apparatus. The workshop events used so much specialist equipment that a large van was needed to transport it all to the schools. Audience sizes were deliberately restricted during the workshop sessions such that all participants could be engaged in 'hands on' demonstrations, experiments and discussions. An example of a popular experiment that was used in both the lectures and the workshops showed the effect and need for pronounced seams in sports balls. A volunteer was invited to throw a very smooth ball across the lecture theatre and would subsequently experience serious difficulty in making the ball fly straight or far. By simply pulling a couple of elastic bands over the ball, its flight was shown to be significantly enhanced due to the new surface roughness. During the workshops the students were also able to film themselves kicking footballs using high-speed video, use artificial bowling machines to explore the effects of side spin on flight and use accelerometers to find the 'sweet-spot' on tennis rackets to name just a few of the activities.

4 Evaluation

Essential to any activity of this kind is a process of evaluation; firstly to assure that the product being delivered to schools is appropriate, and secondly to ascertain the impact of the activity on students attitudes towards science. Teachers completed questionnaires during every lecture and workshop and their feedback was overwhelmingly positive. Based on responses from 117 questionnaires, teachers attitudes towards the lectures and workshops can be summarised as follows:

How well did the lecture/workshop fit into the National Curriculum / how appropriate was its content?

The majority of teachers (50%) rated the lecture and workshops as fitting into the National Curriculum very well. A smaller group of teachers (29%) said that the lecture fitted into the National Curriculum reasonably well, and 21% of the teachers said that the lecture fitted loosely to National Curriculum. It is evident that lectures and workshops could have been designed to fit with the learning objectives of the National Curriculum better. However, it is also worthy to note that part of the appeal of specialist lectures is that they are able to discuss exciting areas of science that are not normally covered in the school curriculum.

How interesting do you think the students found the lectures and workshops?

The overriding majority (68%) of teachers believed their students to find the lecture very interesting. A smaller group of teachers (31%) believed their students to find the lecture fairly interesting, and only a handful of teachers (1%) believed their students to find the lecture of average interest.

During the regional tour of sports engineering workshops, an independent evaluation of the activity was conducted by the University of Sheffield's School of Education. The evaluation was based on a workshop delivered to a single sex, female class of mixed ability (11 to 12 years old). The impact of the workshop was assessed by determining the student's attitudes towards science both before and after the visit. The assessments were made by interpreting responses from questionnaires.

The results from the questionnaire prior to the workshop were very disheartening. A massive 67% of the students rated their normal science lessons as not being enjoyable. The teacher of the class described the group as being very unmotivated and uninterested in science. Two weeks after the workshop was delivered, the class's attitudes towards their science lessons were re-evaluated in a second questionnaire. The class reported an increase in general levels of enjoyment in relation to their normal science lessons. A pleasing 58% of the class reported science as being enjoyable in a surprising but very welcome change of attitude. The students also made comments about the workshop such as: 'it was fun and different' and 'it was an interesting topic, I'd like to know more'. Of course not every student was happy and comments such as, 'I learnt something but it was a bit boring' were also made.

It is accepted that this project's evaluation is less than exhaustive, but the results do suggest that the activities were generally successful. It is intended that future projects will incorporate a more detailed process of evaluation including long term tracking of student choices.

5 Conclusions

Engaging young people in science and engineering is essential for the economic prosperity of nations such as the UK. The universal appeal of sport makes it an ideal context to discuss ideas in science and engineering that young people may feel are 'irrelevant' or 'boring'. The provision of exciting and engaging lectures and workshops has been shown to be a highly valuable activity, eagerly welcomed by schools. In just two years, more than 13,000 young people have taken part in sports engineering activities during 120 events. The direct contact between young people and a practitioner from the world of science challenges their preconceptions about who technologists are and sort of work that they do. With UK schools struggling to attract teachers who have studied physics to any level at university, this direct contact is clearly valuable. There is no 'quick fix' to the issues facing science education in the UK, but projects such as those described in this paper can only help to encourage young people to consider career pathways in science and engineering.

References

Cooke, A.J. and Bryce Taylor, A.M. (2002) Sportscience and engineering in education: a multimedia methodology to attract young engineers. In: S.Ujihashi and S.J. Haake (Eds.), *The Engineering of Sport 4.* Blackwell Science, Oxford, pp 40-45.
Cooke, A.J. and Taylor, A.B. (2004) Using an interactive, computer-based resource on the science of tennis for creative science education in the UK. In: M. Hubbard, R.D. Mehta, J.M. Pallis (Eds.), *The Engineering of Sport 5, volume 2.* ISEA, Sheffield, pp 582-588.
Higher Education Statistics Agency website www.HESA.co.uk.
HM Treasury, Department for Education and Skills and Department for Trade and Industry (2004) Science and innovation investment framework 2004-2014. Her Majesty's Stationary Office.
Smithers, A. and Robinson, R. (2005) Physics in schools and colleges; teacher deployment and student outcomes. Centre for Education and Employment Research, University of Buckingham.

Movement, Health and Quality of Life for Senior Citizens - Prerequisites, Approaches, Concepts

Martin Strehler[1], Andreas Hasenknopf[2] and Eckehard Fozzy Moritz[1]

[1] SportKreativWerkstatt GmbH, München, Germany, ms@SportKreativWerkstatt.de
[2] MLD

Abstract. In most of industrialized countries societies are aging. Today already more than 35% of the German population is over 50 years old and the average age is still increasing. As the percentage of seniors in our society is growing, the financial load on the working part of the society also increases due to the pension fund contributions and due to the high medical expenditures for elderly people. The costs for medication could drastically be reduced by regular physical exercises. To promote the positive effects of exercises for elderly people appropriate devices for training are required. Therefore we will present some research about what sports devices should be like that are especially designed for the needs of senior citizens, and on this basis introduce a few concepts of innovative developments that could serve as pilot projects paving the way for more physical activities of senior citizens.

1 Background

According to the Bavarian Ministry of Environment, Public Health and Consumer Protection in Germany nowadays 80 billion euros is spent on diseases that could be prevented e.g. by physical exercise. But while there are lots of sports concepts and devices for young people and adults, there are nearly no sports devices specially designed for the needs of older people. By the year 2030 about 34% of the German population will be over 60 years old and for old people health is the topic priority number one, ranked even higher than family and friends (Denk and Pache 1995).

Health can be restored, maintained or even improved by physical exercises even in old age. Sports devices are an aid for this target group to find appropriate, healthy and joyful ways to move. However these purposes can only be fulfilled by the sports devices, if they are developed to match the requirements of senior citizens.

Considering that the population over 50 years in Germany also represents 50% of the overall purchasing power; new sports devices for seniors could also be a very lucrative market segment.

2 Senior Citizens and Exercise

At the age of around 20 years the human body is at its highest fitness level. Under steady conditions by the age of 25 a physiological decrease can start. With older age visible changes on the body become recognizable, the body grows weaker and also gets more easily affected by disease. The psychological changes are more difficult to perceive and many seniors are not aware of their problems.

The loss of muscle strength with increased age has major effects on the human body. Muscles do not only have the task to perform activities with strength but also to hold the skeleton. If the muscle mass decreases or the muscles weaken the spine and the joints suffer from increased loads and start to wear and tear. Very often muscular deficits in the back, in the abdomen and in the bottom are the causes for diseases of the spine, for example the well known pain in the lower back, arthritis and intervertebral disc injuries. Weak muscles in old age also increase the risk of falling and fractures, which have a deep impact on the life and the status of the seniors. The loss of muscle mass often results in a decreased independence and self-consciousness. The weak feeling of the body transfers to the mind and the seniors get into a kind of vicious circle which they can only escape by regular exercise (Meusel 1996). Finally, exercise and physical activity keep the senior citizens mobile and independent. Independence, mobility and health lead to higher quality in the life of elderly people.

3 Requirements for Senior Sports Devices

To identify essential requirements for equipment suitable for senior citizens, a group of 55 elderly people was questioned. The participants were between 50 and 86 years old (average 64 years) and three quoters of them were female. A pre-study gave the framework for a questionnaire that was developed after the guidelines of Mummendey (1995) with the assistance of sports psychologists of the Technische Universität München. During three workshops the questionnaires were filled in. After each workshop needs and existing concepts for seniors were discussed more deeply. All data was recorded and quantitatively and qualitatively analyzed. Some of the results are displayed below.

About 2/3 of the participants stated that they are interested or very interested in sporting activities. More than 65% said that they would be interested or very interested in suitable sports devices. On the contrary, only 10% stated that they already use sports devices. This shows that there is a great demand for new equipment.

Most of the seniors (66%) like to participate in physical activity together with friends and their family. Only 10% do not like to be surrounded by others while doing sports. These figures show how important other people are during exercises. Only one third of the participants liked to do sports alone from time to time. Interestingly, the family and the partners are not as often companions as friends are.

Computer supported sports devices are in our days not interesting for elderly people. Most likely this attitude will change with the next generations of old people who already grow up with computers and accept them as everyday companions. Competition is not very attractive for senior citizens any more. The need for a direct comparison with others to confirm their fitness that can be found among young ambitious people is vanishing with age. On the contrary the probability of a disability

grows with age. For a relatively high percentage of the participants it would be interesting if the sport device could compensate handicaps. Either the participants want to
continue training even in case of an injury, or they liked the idea to integrate handicapped friends and family members in their training.

Almost half of the participants were interested in new devices that have to be
moved for training. They appreciate it even more if they would be moved by the
sports devices as in cycling or cross country skiing.

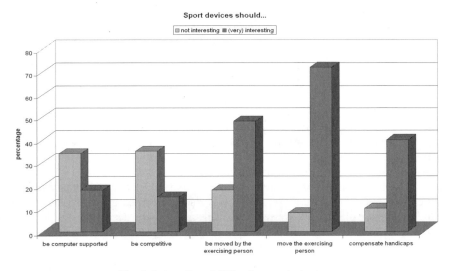

Fig. 1. Interesting abilities for sports devices

Nearly all of the participants train with specific purposes. The improvement of
flexibility (94%) and endurance (92%) are their main aims of training. Also very
important was the enhancement of coordination (84%). As health reasons dominate
the motivation of seniors and competition is less important than for young persons,
strength training (64%) is the last but still quite strong motive for exercising.

	Endurance	Strength	Flexibility	Coordination
(very) important	92 %	64 %	94 %	84 %

Fig.2. Purposes for training

Among the participants the color of the equipment was a rather negligible design
feature. Regarding shape, the tendencies show that round and small, handy shapes
are preferred and big or edgy devices are disliked.

Half of the participants stated that they liked wood as a material. Therefore new
sports products for senior citizens should be made out of wood or at least have wood

as interface between sports device and human being. Wood is followed as preferred material by cloth (30%), plastic (26%), metal (24%) and rubber (16%).

The most relevant factors for the decision to buy a new sports device are so called quality-factors. Device quality includes factors like durability (90%), security (96%) and efficiency (92%). Relevant for the buying decision are the price (86%) and good service (94%). After the purchase, senior citizens lay emphasis on a good and illustrated manual (80%) and customer support (84%). Only one third of the participants said that they would take the design as a purchase criterion.

As an answer to the question, what the preferred training surrounding would be the most common answer was "in the nature" (82%). Still 60% of the participants also like to exercise at home and 40% said that they even would go to a gym. These statements stand in a strong contrast to actual figures that show that a very low percentage of senior citizens actually train in gyms. Perhaps this gap could be reduced by appropriate new training devices for senior citizens and an adequate surrounding.

Not only the strength but also the effectiveness of the senses decreases with age. That means that new sport concepts also have to meet these new requirements by e.g. using slower and bigger balls or by making it easier to keep the balance on the devices. Moreover senior citizens do not like action and sudden changes anymore and prefer "round", continuous movements like cycling and swimming.

One of the most important requirements for senior sports device is the security aspect. People become more scared of injuries and accidents with increasing age. The senior devices must be more stable, less risky and must give the user a secure feeling. However old people are not only concerned about accidents but also about the wrong kind of training or too much training. Good instructions, support and new monitoring devices like a pulse watch and displays can contribute a big part to the acceptance, if they are easy enough to be handled by senior citizens.

It seems to be difficult to activate people to start special and complex sports like tennis or soccer in old age. For senior citizens who have been inactive for years it is important to start at a low level with the possibility to advance in little steps and not to ask too much of them. Senior citizens like it simple. It is more important to move and to exercise than to learn techniques and rules. Their needs are quite similar to the needs of children. They both want to be encouraged to move but they do not want to fulfill strict patterns and rules (Blaumeister and Wappelshammer 2000).

To give them the chance to be more active and to change the attitude of being sedentary there is the need to promote an active life style. If older people are to do more exercise the existence of sports equipment is surely not sufficient. The awareness of the positive effects of physical activity must be promoted e.g. through their doctors (Denk and Pache 1995). At the same time new sports devices and concepts must be available for the senior citizens to enhance successful ageing within our society. The study showed that senior citizens were absolutely aware of the positive effects of exercise. Nearly everyone had the opinion that exercise is an appropriate prevention tool (92%) and that exercise keeps people healthy (98%). So why do not all elderly people actually do physical exercise? The answer surely is not easy and must consider many aspects like the social acceptance of seniors participating in sports and the lack of intrinsic motivation. But at least a certain contribution could be delivered by appropriate sports devices which are motivating, simple and promoted in a greater setting.

4 Equipment Innovations for Senior Sport

The SportKreativWerkstatt in Munich addresses to questions such as the one raised above. How can the life quality of senior citizens be enhanced by new and motivating sports devices and concepts? Based on several own studies (see above), literature and a good network in workshop and seminars some innovative concepts for sports equipment for seniors were developed and tested. Three of the most interesting and most promising approaches are displayed below.

4.1 Multi-person Bikes

One of the most common and widely accepted sports devices in our society and especially among elderly people is the bicycle. As a natural consequence bicycles are adapted to senior citizens. The most common approach to make bikes more suitable for elderly people is to prevent them from tipping over by improving their stability (3 or 4 wheels). Everybody wants to get old but no one wants to be old. This means that senior citizens do not accept obvious "senior bikes". Our approach is, to give the bike a new purpose that can only be realized by a stable, non-tilting construction: The multi-person bike. Therefore, to the already existing purposes - transportation and perhaps fitness – new features like social contact are added. On the multi-person bike weaker people can ride together with healthier ones. Another benefit is that handicapped people can be reintegrated by alternative drives.

4.2 Best Age Basketball

Reaching a certain age, different kinds of movements are not very healthy any more. These include sudden stop and go movements like those needed in the classic basketball game. The "Best Age Basketball" concept builds on the already common acceptance of this game. Certain changes could make basketball more suitable for older people but still keep it attractive for the younger. This can be accomplished by having only one central basket that can be thrown at from every side and new rules. As a result coordination and throwing abilities are still challenged but the sprint phases are reduced. Moving around remains a key feature in the new basketball concept but older people can also play more with tactics and experience.

4.3 The Dynamic Garden

The Dynamic Gardens are planned to be places in public space accessible at all times, that motivate people of different target groups to move, play, and communicate. Thereby they effectively support fitness, prevention, and health.

Fig.3. The Dynamic Garden

The objective of the project is in first place the development of new system solutions for these Dynamic Gardens. With the term system solution we express our conviction that it is not enough to design individual pieces of equipment only. Much more than that, a comprehensive conceptualization and development of all installations of such a Dynamic Garden, plus the integration of these elements among each other and into a balanced environment is necessary. Furthermore, the question of financing, administration, maintenance and access has to be addressed; new solutions have to be developed.

Consequently we will try to find solutions how these Dynamic Gardens can be integrated into the everyday life of people. Here we address questions like which political measures will support the use of the Dynamic Gardens for the sake of improving health and well-being of the people, at which places instalments can best be positioned, and how continuous improvement and further development can be ensured.

References

Blaumeister, H and Wappelshammer, E. (2000) Mit Senioren die Zukunft gestalten. Deutscher Seniorentag 2000. Nürnberg.

Denk, H. and Pache, D. (1995). Die gesundheitliche Bedeutung von Bewegungs- und Sportaktivitäten in der Sicht der Älteren. BAGSO-Nachrichten 3.

Meusel, H. (1996) Bewegung Sport und Gesundheit im Alter. Quelle & Meyer, Wiesbaden.

Mummendey, H.D. (1995) Die Fragebogen-Methode. Hogrefe, Göttingen.

7 Performance Sports

Synopsis of Current Developments: Performance Sports

Chikara Miyaji
Japan Institute of Sports Sciences, Japan, Chikara.MIYAJI@jiss.naash.go.jp

Resulting the best performance on world class competitions requires improved movement, innovative equipment, and new training method. Research and engineering should help and support such activities. This section introduces several articles which explain such activities.

Current Developments about Performance Sports

Innovative equipment is one of the big contribution of the Sports engineering on performance sports. Klap-skate is one typical result, and also many other sports have similar results such as spike shoes on track and field, pole of pole jump, ski boots and so on.

To improve performance with new equipment, it needs new skill to use the equipment and new training method to exploit the capability. To improve skill and performance, videos are the tool frequently used in various sports. But there are some limitations on videos; it can not capture internal force and faster movement.

Also Sports engineering contributes such part. Sensor and data monitoring tools are the field which evolve drastically in recent years. Many sensors are miniaturized: for example IC-Accelerometer and IC-gyro, and can attached on human body easier, All signals are digitalized and easier to transmit on wireless. The combination of videos and sensors will compensate each other.

It is important to record the history of training day by day to monitor the improvement. If the videos and sensors data are captured and accumulated day by day, it will become good tool to improve ones skill. And there will be good opportunity for Sports engineering to provide such integrated system.

The Application of Inertial Sensors in Elite Sports Monitoring .

Daniel A. James
Griffith University, Australia, d.james@griffith.edu.au

Abstract. Arguably the performance of elite athletes today has almost as much to do with science, as it does with training. Traditionally the measurement of elite athlete performance is commonly done in a laboratory environment where rigorous testing of biomechanics and physiology can take place. Laboratory testing however places limits on how the athlete performs, as the environment is sufficiently different to the training environment. In addition, performance characteristics are further augmented during competition when compared to regular training. By better understanding athlete performance during the competition and training environment coaches can more effectively work with athletes to improve their performance. The testing and monitoring of elite athletes in their natural training environment is a relatively new area of development that has been facilitated by advancements in microelectronics and other micro technologies. Whilst it is a logical progression to take laboratory equipment and miniaturize it for the training and competition environment, it introduces a number of considerations that need to be addressed. In this paper the use and application of inertial devices for elite and sub-elite sporting activities are discussed. The capacity of accelerometers and gyroscopes to measure human motion thousands of times per second in multiple axis and at multiple points on the body is well established. However interpretation of this data into well-known metrics suitable for use by sport scientists, coaches and athletes is something of a challenge. Traditional brute force techniques such as achieving dead reckoning position and velocity by multiple integration are generally regarded as an almost impossible task. However novel derivative measures of performance such as energy expenditure, pattern recognition of specific activities and characterisation of activities into specific phases of motion have achieved greater success interpreting sensor data.

1 Introduction

Athletic and clinical testing for performance analysis and enhancement has traditionally been performed in the laboratory where the required instrumentation is available and environmental conditions can be easily controlled. In this environment dynamic characteristics of athletes are assessed using treadmills, rowing and cycling machines and even flumes for swimmers. In general these machines allow for the monitoring of athletes using instrumentation that cannot be used in the training environment but instead requires the athlete to remain quasi static thus enabling a constant field of view for optical devices and relatively constant proximity for tethered electronic sensors, breath gas analysis etc. Today however by taking advantage of the advancements in microelectronics and other micro technologies it is possible to build instrumentation that is small enough to be unobtrusive for a number of sporting and clinical applications (James, Davey and Rice 2004). One such technology that has seen rapid development in recent years is in the area of inertial sensors. These sensors respond to minute changes in inertia in the linear and radial directions. These are known as accelerometers and rate gyroscopes respectively. This work will focus on the use of accelerometers, though in recent years rate gyroscopes are becoming more popular as they achieve mass-market penetration, thus increasing availability and decreasing cost and device size.

Accelerometers have in recent years shrunk dramatically in size as well as in cost (~$US20). This has been due chiefly to the adoption by industries such as the automobile industry where they are deployed in airbag systems to detect crashes. Micro electromechanical systems (MEMS) based accelerometers like the ADXLxxx series from Analog Devices (Weinberg, 1999) are today widely available at low cost. The use of accelerometers to measure activity levels for sporting (Montoye, Washburn, Smais and Ertl 1983), health and for gait analysis (Moe-Nilssen, Nene and Veltink 2004) is emerging as a popular method of biomechanical quantification of health and sporting activity and set to become more so with the availability of portable computing, storage and battery power available due to the development of consumer products like cell phones, portable music players etc.

2 Inertial Sensors

Accelerometers measure acceleration felt at the sensor itself and typically in a single axis only (uni-axial). Using MEMS technology these sensors are now millimeters or smaller in size. In general a suspended mass is created in the design and has at least one degree of freedom. The suspended inertial mass is thus susceptible to displacement in at least one plane of movement. These displacements arise from changes in inertia and thus any acceleration in this direction. Construction of these devices vary but typically use a suspended silicon mass on the end of a silicon arm that has been acid etched away from the main body of silicon. The force on the silicon arm can be measured with piezoresistive elements embedded in the arm. In recent years multiple accelerometers have been packaged together orthogonally to offer multi-axis accelerometry.

Accelerometers measure the time derivative of velocity and velocity is the time derivative of position. Thus accelerometers can measure the dynamics of motion and potentially position as well. It is well understood though that the determination of position from acceleration alone is a difficult and complex task and something of a holy grail in the field (Davey and James 2003).

Instead, accelerometers are often used for short-term navigation and the detection of fine movement signatures and features (such as limb movement). Accelerometers can be used to determine orientation with respect to the earth's gravity as components of gravity are aligned orthogonal to the accelerometer axis. In the dynamic sports environment, complex physical parameters are measured and observed in relation to running and stride characteristics (Herren, Sparti, Aminian and Shultz 1999), and in the determination of gait (Williamson and Andrews 2001).

Researchers have also used accelerometers for determining physical activity and effort undertaken by subjects. These kinematic systems have been able to offer comparable results to expensive optical based systems (Mayagoitia, Nene and Veltink 2002). Rate gyroscopes, a close relative of the accelerometer, measure angular acceleration about a single axis and are also used to determine orientation in an angular co-ordinate system, although these suffer from not being able to determine angular position in the same way accelerometers have trouble with absolute position. Additionally many physical movements, such as lower limb movement in sprinting, exceed the maximum specifications in commercially available units that are sufficiently small and inexpensive for such applications.

2.1 Inertial navigation

If we consider the simplest case of motion in a single direction (x) with constant acceleration (a_x) double integration with respect to time should enable displacement (s_x) to be calculated. Thus the equations for velocity (v_x) and displacement (s_x) become:-

$$\text{Velocity}_x \qquad v_x = \int a_x \,.\, dt = v_x + c_1 \qquad\qquad (1)$$

$$\text{Displacement}_x \qquad s_x = \int (v_x + c_1).dt = s_x + c_1.t + c_2 \qquad (2)$$

$$\text{Offset}_x \qquad = c_1.t + c_2 \qquad\qquad (3)$$

In each integration step additional constants (C_1, C_2) emerge. Thus for constant acceleration a sloped straight line offset appears after double integration (Eq. 3). Clearly for a well-bounded problem these constants can be removed, but if there is any error in the boundaries the solution becomes increasingly unbounded. Additionally because the integration is numerical then the sampling considerations can contribute significantly. In practice this is much more complex as acceleration is nonlinear.

Simple test applications involving displacements over just a few meters have resulted in positional errors of many meters after just a few seconds of double integration. In addition accelerometers are usually sampled discretely giving rise to errors from the sampling and digitization process. One technique is to apply known boundary elements to the data at regular intervals; because sporting movement is very often cyclic in nature a 'zero' point can then be applied at regular intervals. This has been applied successfully to monitoring the motion of an exposed heart (Hoff, Elle, Grimes, Halvorsen, Alker and Fosse 2004). In applications such as running gait, motion in the vertical direction is for flat terrain self normalising. Wixted (Wixted, Thiel, James, Hahn, Gore and Pyne 2005) has shown that a simple low-pass filter can be used to extract variations in axial tilt to keep the co-ordinate system aligned orthogonal to the ground. Recent work (Channells, Purcell, Barrett and James 2005) shows that lower limb shank rotation can be calculated using two accelerometers. Gyroscopes, in general are unsuitable for this application due to the high rates of rotation of the limb during running. By monitoring the movement of individual limbs during the limb stride it may be possible to reconstruct the entire gait cycle. Applying anthropomorphic measures and measuring angles of takeoff for the flight phase potentially allow the displacement covered by the athlete between steps and the total distance traveled could be calculated.

2.2 Sampling Considerations

Accelerometer output is typically recorded digitally. Depending on the application, sample rate and resolution are important factors. Thus careful adherence to sampling theory is required to ensure that aliasing due to under sampling does not occur (Cutmore and James 1999). This is particularly relevant to studies based on human motion. Whilst the peak of human activity occurs below 20 Hz, sampling at much higher rates is required to capture the full detail of the motion, or if short term navigation is sought as there are significant higher order terms in the data that contain important information. Typically high rates such as 1000 Hz and 12-bit resolution preserves data quality for most applications. Of course sampling at these rates for multi-axial devices generates large data sets very quickly. Recent research (Lai, James, Hayes and Harvey 2003) suggests that in many cases the resolution can be reduced to 8 or 6 bits without loss of information, though preserving the high data rate is important. It is argued that error minimsisation through shaping the measurement error in time and space, mindful of the noise characteristics of the device such as the noise floor of the sampling circuit and the bandwidth response of the devices themselves are more important than the number of bits used in sampling.

2.3 Sensor artifact

The mechanical coupling of inertial sensors to the region of interest requires a number of considerations. Firstly within the musculoskeletal system it is usually the dynamics of joint and bone segments that are of primary interest. However, it is rarely possible to mount these sensors directly to these segments. Instead sensors are mounted directly onto the skin surface and the tissue beneath it necessarily separates itself from the bone of interest in usually non-linear coupling that is susceptible to stretching and distortion generated during movement. Secondly subject comfort and the neutrality of the sensor is important; if the subject is uncomfortable in any way the performance characteristics of their movement may be subtly different. Finally, attachment of the acceleration sensors also has the potential to affect the very dynamics of what is to be measured by the addition of additional mass and surface area on the site of interest. For example, introducing added mass creates drag in monitoring swimmers, and skin based artifact when monitoring leg or shank movement.

Fortunately accelerometers today weigh only a few grams and can be packaged simply using epoxy or similar compounds. The advent of body hugging clothing and suits in many sports allows for the incorporation of sufficiently small sensors and accompanying instrumentation relatively easily using sewn pouches for example.

3 Applications

A number of sporting activities have been investigated using accelerometers. Central to these investigations has been the development of a modular sensor system that can be customized for the intended sport. This system (James et al. 2004) was initially applied to both rowing and swimming, before being trialed in a number of other sports. In each case the system was packaged separately and allowed additional modules to be added as required to facilitate data storage, RF communications in near real-time and post event IR data download and sealing for aquatic applications.

Rowing was trialed initially because of a number of advantages. Firstly rowing is predominantly a 1-D activity thus it was hoped that inertial navigation might be possible. Secondly the monitoring equipment could be mounted to the scull rather than the athlete simplifying packaging constraints. Thirdly, and almost most importantly, rowing is a technical sport and already uses a number of technologies in the training environment, thus there was little cultural change required by athletes and coaches to adopt a newer technology. Very quickly it was noticed that there was a lot of high frequency accelerations, perhaps due to mechanical artifacts such as oarlock movement and the seat movement. Data from the accelerometers proved useful in identifying stroke phase characteristics such as the catch, drive and recovery phases, something which is difficult to do even with a video system. From these basic measures such as stroke rates and counts could easily be extracted and combined together with GPS and other data such as heart rate to analyse performance and further develop race strategy.

The system was then encased in water proofing material and applied to trunk movement in swimming. Visually examining tri-axial traces enabled the identification of stroke type, stroke counts, lap times and tumble turns. A model of body roll dynamics was constructed allowing the development of automated algorithms to extract performance characteristics and stroke identification routines (Davey, and James 2003). These produced data that exceeded hand timed and counted data and was comparable with underwater video and touch pad equipped pools.

Uni-axial accelerometers have been used successfully for measuring walking gait and energy expenditure, however the correlation is less clear for running and other physical activities such as in the various football codes. Energy expenditure can be calculated with greater accuracy using tri-axial accelerometers (Wixted et al. 2005). By extracting and examining body tilt other activity types can also be recognized as distinct signatures. Thus by determining activity type and intensity an estimate of energy expenditure can be determined, based on historical and laboratory based calibration. Furthermore by doing correlation studies with in-sole pressure sensors and high speed video it has been found that trunk mounted accelerometers alone can reliably determine gait cycle characteristics such as heel down and toe off as well.

Implements used in sports have also been instrumented (James, Gibson and Uroda 2005; Ohgi, Baba and Sakaguchi 2005) showing that key characteristics of swing can be measured and extracted, and that they and correlate well with athlete skill. Recently the device has been applied to winter sports to detect and quantise aerial activity time and type in ski and snowboard events.

4 Discussion

A purely technology based approach using accelerometers for sporting applications has yielded little success, whereas informed signal processing of the data through the use of sport specific knowledge and involvement of sport scientists has allowed the extraction key features in the data which can then be interpreted in a useful manner. Critical to the success of this work has been to ensure that the development of the technology has been in partnership with key stakeholders including athletes, coaches and sport scientists. Keeping the technology development and interpretation firmly grounded on providing useful outputs that benefit athletes has been critical to enhancing sporting activity.

References

Channells, J., Purcell, B., Barrett R. and James, D.(2005), Determination of rotational kinematics of the lower leg during sprint running using accelerometers, SPIE Bio MEMS and Nanotechnology, December, Brisbane Australia

Cutmore, T.R.H., and James, D.A. (1999). Identifying and reducing noise in psychophysiological recordings. Int J Psychophys., 32, 129-150.

Davey, N. and James, D.A. (2003). Signal analysis of accelerometry data using gravity based modelling, Proceedings of SPIE Vol. 5274, SPIE, Bellingham, WA, 362-70.

Herren, R., Sparti, A., Aminian, K., and Schutz, Y. (1999). The prediction of speed and incline in outdoor running in humans using accelerometry. Med Sci Sports Exerc, 31, 1053-1059.

Hoff, L., Elle, O., Grimnes, M., Halvorsen, S., Alker, H. and Fosse, E. (2004). Measurements of Heart Motion Using Accelerometers. IEEE Sensors, Vienna

James, D., Gibson, T. and Uroda, W.(2005), Dynamics of a swing: A study of classical Japanese swordsmanship using accelerometers, *The Impact of Technology on Sport*, Subic. A., Ujihashi, S. ASTA, 355-60

James, D.A, Davey, N. and Rice, T.(2004), An Accelerometer Based Sensor Platform for Insitu Elite Athlete Performance Analysis, IEEE, Sensors, Vienna

Lai, A., James, D.A., Hayes, J.P. and Harvey, E.C. (2003). Application of triaxial accelerometers in rowing kinematics measurement, Proceedings of SPIE 5274, (SPIE, Bellingham, WA) 531-42

Mayagoitia, R., Nene, A. and Veltink, P. (2002). Accelerometer and rate gyroscope measurement of kinematics: an inexpensive alternative to optical motion analysis systems, Journal of Biomechanics 35, 537-542.

Moe-Nilssen, R., and Helbostad, J. L.,(2004), Estimation of gait cycle characteristics by trunk accelerometry, J Biomech, 37(1), 121-126

Montoye, H., Washburn, R., Smais, S., Ertl, A., Webster, J.G., and Nagle, F.J., (1983), Estimation of energy expenditure by a portable accelerometer, Med. Sci. Sports Exerc., 15, 403 (1983)

Ohgi, Y., Baba, T. and Sakaguchi, I., (2005) measurement of Deceleration of Golfers Sway and uncock timing in driver swing motion, *The Impact of Technology on Sport*, Subic. A., Ujihashi, S. ASTA, 349-54

Weinberg, H. (1999). Dual Axis, Low g, Fully Integrated Accelerometers, Analogue Dialogue 33, Analogue Devices (publisher).

Williamson, R. and Andrews, B. J. (2001). Detecting absolute human knee angle and angular velocity using accelerometers and rate gyroscopes. Med Biol Eng Comput, 39, 294-302.

Wixted, A., Thiel, D., James, D., Hahn, A., Gore, C. and Pyne, D. (2005), Signal Processing for Estimating Energy Expenditure of Elite Athletes using Triaxial Accelerometers, IEEE Sensors 2005, Irvine, USA

Requirements and Solution Concepts for the Development of Sport-Specific Measuring Units in High Performance Sports

Klaus Knoll and Klaus Wagner

Institute for Applied Training Science Leipzig (IAT), wagner@iat.uni-leipzig.de

Abstract. In the design and development of measuring units in elite sport the request for high objectivity, relialibilty and validity of the results creates high requirements for the construction of the specific devices which usually are not covered by comemercially available systems. But only a sufficiently exact simulation of the competition conditions enables the researcher to evaluate the results with respect to the performance prerequisites of the sport in question and the actual training state resp. athletic shape of the athlete. This refers to the sport specific movement patterns as well as to the robustness caused by the interaction between athlete and measuring unit. The paper presents solutions of selected measuring units.

1 Introduction

For years measuring units have been an indispensable basis for the determination of physical abilities of elite athletes (diagnostic). However, they also increasingly applied in training (feedback training) in particular for learning and correcting of motions (technique training), as well as for the optimization of competition forming e. g. in the endurance sports (race optimization). In spite of the intense evolution in the field of both the software simulation systems and the measurement technology for the analysis of competition performance, measuring units will keep their high importance as a component of the training control too further.

Utilization of the measuring units in high-performance sports leads to high demands on these systems. Special requirements are e. g.

- The resistance against the extreme force actions and the movement frequencies to be encountered in the high performance range
- The protection of the reliability of the measurement results (comparability of the results, if several identical measuring units are used; robustness against environmental impacts as temperature or humidity etc.)
- Sufficiently precise simulation of the real conditions, so that the won results can be interpreted in meaning of physical abilities (validity). That concerns both the guarantee of the sport-specific movement and the robustness against the actions of the athlete at the measuring unit.

It has to be taken into account that the individual requirements dependent from each other in most cases.

2 Sport-Specific Measuring and Information Systems (MIS)

We will concentrate our representation on systems which simulate competition like conditions, since it is not possible to deal with all the different requirements mentioned above in this paper. This goal contains a high standard of objectivity, reliability and validity of the results gained. The following proprietary developments of the last years were selected as examples.

2.1 Starting Block Dynamometer for Competitive Swim

In swimming the start is of great importance, above all for the short distances. A comprehensive analysis of the start technique includes the measurement of the forces occurring during the take-off. Based on the fact that the parallel take-off was the only start technique in the past, starting block dynamometers were applied that monitor summaries of the take-off force (e. g. Härting, Zschocke, Alvermann, Wecker and Schattke 1982; Kibele 2004).

The development of the grab and trac start technique (Fig. 1) requires a separate measurement of the reaction forces of both feet and of the hands. Investigations of an Australian team (Benjanuvatra, Lyttle, Blanksby und Larkin 2004) that used a divided starting block with two 2D force platforms were the first published internationally.

Fig. 1. Grab and track start

With the new starting block dynamometer the supporting forces are measured separately for every leg in vertical and horizontal direction (jump direction) and a summary for the hands in vertical direction is monitored too. The take-off areas are provided with an anti-slip coating and separated by a gap of 7 mm (Fig. 2). The total area has the dimensions of 50 cm × 50 cm with a 5° inclination towards the front as the standard starting block.

For the construction the following maximum load, based on previous measurements and by estimate, formed the basis: 1700 N for the vertical and 1000 N for the horizontal take-off force for each side as well as 1000 N for the drag of the hands. Hottinger Baldwin Messtechnik load cells served as measuring sensors. By installation of mechanical gliding elements, the number of load sensors could be limited to five.

Fig. 2. Starting block dynamometer (left: without hood)

The force-time curves in Fig. 3 clearly show the expected different course of the curve for the take-off forces of the left (front) and right (rear) leg in the grab and track start.

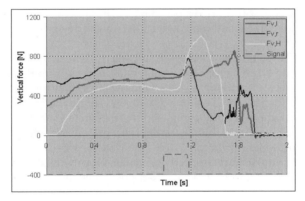

Fig. 3. Vertical force-time graphs of a grab and track start
(Fv,l - left (front) leg; Fv,r - right (rear) leg; Fv,H - Hands)

2.2 Canoe Ergometer for Flatwater Canoeing and Kayaking

Canoe ergometers have been introduced as a training apparatus and for the diagnosis of specific endurance abilities in flatwater canoeing. In our case, the double-bladed paddle is connected via two ropes with a load resistor system, to the single-bladed paddle over an individual rope, in order to simulate the reaction force occurring during stroke. The quality of the simulation depends on the design of the load device.

Our canoe ergometer is a further development of Heinze's ergometer (1978). The mass of boat and athlete that has to be accelerated is represented by equivalent rotating masses. The hydrodynamic resistance of the boat is simulated by an eddy-current brake. As a result, it is possible to convert the strength effort respectively the athletes's performance into the theoretically covered distance of the canoe. Thus it becomes possible to simulate the course of races or to arrange feedback training.

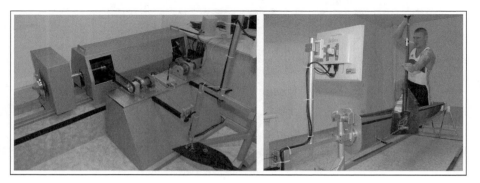

Fig. 4. Canoe ergometer

The drag ropes are coiled by a retrieval device (drum with free-wheel clutch and spiral spring). The retrieval device, a further development of the institute FES Berlin, allows frequencies up to 160 strokes per minute. These frequencies are clearly above the previous stroke frequency and correspond to the recent top performances in canoe racing sport.

In tests with latest boat models in the towing tank of the shipbuilding research station Potsdam flow resistances required for the braking torque were determined. The experiments were carried out with varying towing velocity and load.

The eddy current brake was calibrated with an electric motor primed for a torque measuring. The leading control parameter for the voltage to be supplied by the computer to the eddy current brake was the rope velocity which corresponds to the boat speed in good approximation.

In accordance with his driven speed the canoeist is presented an amount that is an expression of the braking force. On a monitor positioned in front of him (Fig. 4, right picture) relevant parameters as boat speed, covered distance and paddle force are presented numerically and graphically. The force is measured by sensors attached to the paddle and transmitted by radio.

2.3 Measuring Unit Throw/Shot Put

In most cases investigations into the sports technique in the throw and shot disciplines of track and field are carried out with photogrammetric procedures. Measurements of ground reaction forces while throwing and shot putting offer a deeper insight into the movement structure. However, this possibility only consists in sports halls where the sports apparatus is thrown into a net.

In the test hall at the Institute for Applied Training Science Leipzig (IAT), four 3D force platforms were fit into the flat ground for discus throwing and shot put. The platforms rest on flute steel rails which were embedded into the concrete foundation. Their spatial arrangement covers the area which has been used as support points of the feet (cp. Fig. 5). A non-skid wooden throwing ring which was cut out at the platform edges was attached on the platforms. The discuses were thrown into a catch net and in shot put the balls landed on mats.

The measuring unit is completed by corresponding measuring amplifiers and low-pass filter for 16 measuring channels as well as by synchronized video recording for 3D photogrammetry.

Fig. 5. Measuring unit throw/shot put - assembly of the force platforms P1-4

2.4 Measuring and information system ski-jump

The MIS ski jump "Fichtelbergschanze" has been designed for feedback training as well as to get a deeper insight into performance structure under competition-like conditions. However, it can also be employed directly in competitions (Dickwach and Wagner 2004). The essential technological basics for it are a chain of dynamometric platforms and the digital video recording synchronized with measuring.

By monitoring the force curve and the corresponding video sequence directly before and during the take off it becomes possible to generate the most important take-off information as direct feedback, i. g. the immediate display of essential biomechanical parameters and the synchronous presentation of the force curve and the video sequence (as kinegram or video clip) after the jump (Fig. 6).

The chain of platforms consists of 12 one-dimensional force platforms which are fit in a concrete tub of the final 12 m of the run-up (ski jump radius, take-off area). Any platform is equipped with four load cells which are designed for a nominal force of 2 kN. For measurements during winter time, a separator device can be mounted on the ski jump table. With this device a slit is built to prevent traction between the snow coating of the measuring area (run-up lane) and the surrounding area. To guarantee a high quality of the measurement the platform superstructure has to enable an independent measurement of any platform.

The determination of the radial force acting on the athlete when moving through the ski jump radius has been a special challenge. The radial force is required in order to calculate the take off force of the athlete (lower curve) from the measured total

force (upper curve). The calculation from the ski jump radius did not supply results
of sufficient quality. Therefore in an additional experimental study a rigid mass was
mounted on a sledge. The results which we gained in this study could be verified by
the computer simulation system Alaska.

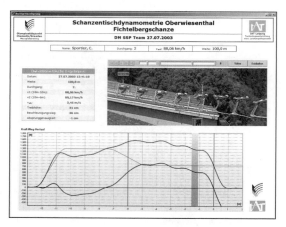

Fig. 6. Kinegram and force-time diagram of a ski jump take off

3 Conclusion

Application of measuring units in high performance sports puts high requests on
these systems which are not usually covered by commercially available products.
Therefore in most cases the introduced approximately fifty measuring units at the
IAT represent an advancement of commercial systems which have been customized,
custom-made products of specialized companies or proprietary developments.

References

Benjanuvatra, N., Lyttle, A., Blanksby, B. and Larkin, D. (2004). Force development profile
    of the lower limbs in the grab and track start in swimming. Proceedings of the ISBS
    Symposium. Ottawa.
Dickwach, H. and Wagner, K. (2004). Neue Möglichkeiten der Analyse und Technikkorrektur
    im Skispringen durch die Kopplung visueller Informationen mit Kraftverläufen. *Leistungs-
    sport, 34 (1)*, 12-17.
Härting, B., Zschocke, M., Alvermann, F., Wecker, U. and Schattke, U. (1982). Plattform zur
    Messung mehrerer orthogonaler Kraftkomponenten. Berlin: Amt für Erfindungs- und
    Patentwesen, Patentschrift 209519.
Heinze, M. (1978). Entwicklung einer Methode der sportartspezifischen Ergometrie für den
    Kanusport. Diss., Technische Hochschule Otto von Guericke Magdeburg.
Kibele, A. (2004). Abschlußbericht zum BISp-Projekt „Mobiler Mess-Startblock" (VF
    0407150102). Universität Kassel.

Analysis of Fulcrum Skate Performance

Kim B. Blair, Kieran F. Culligan and David Walfisch

MIT, Center for Sports Innovation, blairk@mit.edu

Abstract. The winning and losing times in long-track speed skating competitions can differ by only a few hundredths of a second, so even marginal improvements in technology can give skaters a significant advantage. The fulcrum skate incorporates a new hinge mechanism for use in long-track speed skating. Klapskates, used by virtually all skaters today, provide only rotational motion at the hinge, but the fulcrum skate's hinge mechanism introduces both translational and rotational motion. Previous skating research shows that increases in ankle-opening angle, hinge-opening angle and blade-to-ice contact time enabled by the klapskates can be correlated to faster skating speeds. Using a slide-board, the skating motion while using both fulcrum and klapskates was analyzed in an off-ice laboratory setting. Nine test skaters were used for the experiment. Each skater performed eight runs each on the fulcrum skate and klapskate. High-speed video recorded the skating motion. Maximum ankle-opening angle, maximum hinge-opening angle, and blade-to-ice contact time were measured from the high-speed video. Our results showed that, on average, compared to the klapskate, the fulcrum skate offers over 12% increase in ankle-opening angle, 33% increase in hinge-opening angle, and 23% increase in blade-to-ice contact time. These results show that using the fulcrum skate offers skaters increased contact time with the ice, which could equate to faster on-track skating performances.

1 Introduction

In the 1980's, the introduction of the klapskate into long-track speed skating resulted in record-breaking performances. While traditional skates have the blade fixed to the boot, the klapskate incorporates a hinge located between the blade and the boot near the front of the skate. The heel of the boot is not attached to the blade, allowing the blade to rotate about the pivot point and stay in contact with the ice longer on each skating stroke, allowing a longer application of force during each stroke.

Okolo Sports Technologies has developed the fulcrum skate, incorporating a new hinge mechanism that has both translational and rotational motion (Fig. 1). The additional translational motion of the hinge mechanism will allow for a further increase of blade-to-ice contact time, which will result in increased skating speed. The goals of this skate design can be validated if quantitative, statistically significant differences in the skating stroke, using a fulcrum skate compared to a klapskate, can be demonstrated.

Previous research comparing the performance of klapskates and traditional skates showed a significant increase in skating speed for klapskates over traditional skates. Further results from biomechanical and exercise physiological testing, as well as computer simulation data have not been able to explain why klapskates produce such a significant increase in speed. (Koning, Houdijk, and Ingen Schenau 1997; Koning, Houdijk, Groot, and Bobbert 2000; Koning, Houdijk, Groot, and Bobbert 2000a; Koning, Houdijk, Groot and Bobbert 2002).

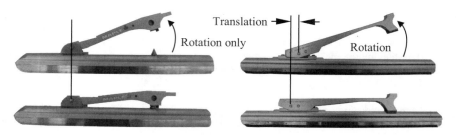

Fig. 1. Comparison of the hinge motion of the klapskate (left) and the fulcrum skate (right).

Using a high-speed video camera, Van Horne (2003) compared the performance of traditional skates and klapskates with three different pivot positions, 55%, 62.5%, and 70% of the total boot length. It was found that klapskates, by providing the skater with increased blade-to-ice contact time, allow the skater to apply force for a longer period during each stroke and therefore go faster. Further, it was concluded that different skaters require pivot points at varying locations based on their physical size and skating style.

As the fulcrum skate is new, this is the first reported research on the performance of this skate design. The present study compares the performance of the klapskate with the new fulcrum skate design. The testing environment used herein follows from the work of Van Horne (2003), using high-speed video to measure the attributes of the skating stroke in an off-ice, controlled laboratory environment.

2 Methods

The objective of this experiment was to analyze the speed skating stroke in a laboratory setting in order to compare the performance of the new fulcrum skate design and the existing klapskate design. Skaters were asked to test both the fulcrum and klapskate designs. In order to alleviate the variability associated with on-ice testing, skaters were asked to take single strokes on a slide-board. Blade-to-ice contact time, maximum ankle-opening angle, and maximum hinge-opening angle were measured from high-speed video images of the skating strokes.

Nine test subjects were used in this experiment. The subjects were selected from competitive speed skaters living near Boston, MA USA. The test skaters were all male with ages between 17 and 67, with an average age of 37 years. Because human test subjects were involved in the project, approval from MIT's Committee on the Use of Humans as Experimental Subjects (COUHES) was required. COUHES approval number for the experiment is 0502001092. All skaters signed a consent form, and skater confidentiality was maintained before, during, and after the actual testing.

The slide-board used during the test (Fig. 2) consists of a large piece of melamine board. The skater wears the test skate on his right foot, and a shoe covered with a slick cloth on the left foot. The skater assumes a pre-stride position placing the skate blade firmly against a wood starting block fixed to the slide-board. When ready, the skater pushes off with his right foot using his regular skating stroke and slides across the board on his left foot. The experiment is then repeated as required for both skate types. A high-speed video camera, set so that the focal plane is parallel to the skate/blade plane, with a frame rate of 250/sec, records the skate and lower-leg motion. Circular white dots were placed at six locations on the skate and leg of test skaters. These dots served as reference points during video post-processing for angle measurements. It was determined that a camera misalignment error of 5% would result in a measurement error of less than 1%.

Fig. 2. Experimental set-up showing the skater taking a stroke on the melamine slide-board. The skate is braced against the starting block.

2.1 Skater Testing

The testing was completed over the course of three weekends to accommodate the schedules of our test skaters. One skater with extensive experience using both the klapskate and the fulcrum skate came to the test area and served as a sample test subject to assist with any de-bugging required for the experimental set-up.

For the data collection phase, the experienced skater demonstrated the test for each test subject so that the skaters had a clear understanding of what was required of them for successful data collection. In order to account for the possible presence of order affects, test subjects were randomly selected to start with a particular skate type. Half of the skaters used the klapskate first, and the others started with the fulcrum skate. The skater donned the assigned skate and completed a minimum of ten practice strokes to become accustomed to skating on the slide-board. The video camera was turned on, and the skater made eight identical skating strokes on the slide-board. Then recording was stopped and the skater switched to the other type of skate. The warm-up and testing was repeated with the other skate. Each of the nine test skaters completed eight test runs with each type of skate, totaling 144 skating strokes.

2.2 Data Reduction

The 144 skating strokes were recorded on high-speed video. Each video segment includes the push-off part of the stroke. For each of the 144 strokes, the video was examined frame by frame, and two frames were selected from each stroke. The first frame was when the hinge mechanism begins opening; the second frame was when the blade began to leave the wood block. Figure 3 illustrates examples the two time frames. The time between these frames is defined as the blade-to-ice contact time. The x-y coordinates for six points, defined by the optical markers were selected on each frame and then used to calculate initial and final angles of the ankle and the hinge. Thus, for each skating stroke, the blade-to-ice contact time, the maximum ankle-opening angle and maximum hinge-opening angle was recorded. The values for each skater, as well as the aggregate for all skaters, for each type of skate were averaged. The percent difference in each of the analyzed values was compared between the fulcrum skate and the traditional klapskate. Statistical significance of the performance difference between the skates was evaluated using t-tests ($\alpha = 0.05$).

| Time 1 | Time 2 |
| Hinge begins to open | Skate leaves wood block |

Fig. 3. Sample initial and final time frames from one trial.

3 Results

Figure 4 shows the average values and percent differences for the ankle-opening angle. The top graph shows the mean for each skater for each type of skate, the other shows the percent difference between the traditional and fulcrum skate. The average ankle-opening angle was between 18-39°. Three of the subjects saw a statistically significant increase in ankle-opening angle using the fulcrum skates. The average increase in ankle-opening angle for all skaters was statistically significant at 13%.

Figure 5 shows the average values and percent differences for the hinge-opening angle. For all skaters, the hinge-opening angle was between 9-22°. Seven of the subjects saw a statistically significant increase in hinge-opening angle using the fulcrum skates. The average increase in hinge-opening angle for all skaters was statistically significant at 33%.

Figure 6 shows the average values and percent differences for the blade-to-ice contact time. For all skaters, the blade-to-ice contact time was between 43-80 ms. Six of the subjects saw a statistically significant increase in blade-to-ice contact time using the fulcrum skates. The average increase in blade-to-ice contact time for all skaters was statistically significant at 23%.

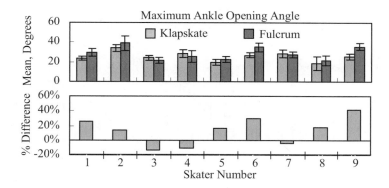

Fig. 4. Ankle-opening angle measurement shown for each skater.

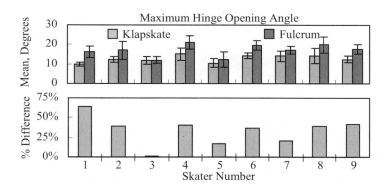

Fig. 5. Hinge-opening angle measurement shown for each skater.

4 Discussion/Conclusion

While individual skater results vary, the overall average results show a significant difference in all measured parameters comparing the fulcrum skate to the traditional klapskate. The effect of training and experience may have some effect on the measured results. One subject spent much of his training time on a slide-board like device resulting in minimal variation between individual test strokes. Further, subjects with the most fulcrum skate experience showed larger increases in all measured values compared less experienced test subjects. Consequently, a skater might expect even larger performance gains with continued training and use of the fulcrum skates.

An increase in each of the measured values is considered advantageous in improving on-ice skating speed. Thus, by switching from klapskates to fulcrum skates, a skater can expect a significant improvement of the critical performance parameters of blade-to-ice contact time. Just as klapskates increased blade-to-ice contact time resulting in on-ice performance improvements in the 1980's, the present study indicates that the fulcrum skate may improve performance in the 2000's.

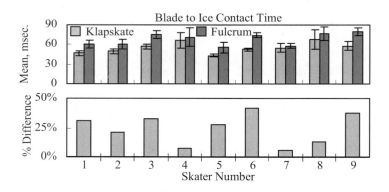

Fig. 6. Blade-to-ice contact time measurement shown for each skater.

References

Koning, J.J., Houdijk, and H., Ingen Schenau, G. J. (1997) "Background to the klapskate," Dept. Kinesiology, Faculty of Human Movement Science, Free University, Amsterdam, The Netherlands.

Koning, J.J., Houdijk, H., Groot, G., and Bobbert, M.F. (2000) "Physiological responses that account for the increased power output in speed skating using klapskates", *European Journal of Applied Physiology*, Vol. 83, No. 2-3, pp. 283-288.

Koning, J.J., Houdijk, H., Groot, G., and Bobbert, M.F. (2000a) "From Biomechanical theory to application in top sports: the Klapskate story," *Journal of Biomechanics,* Vol. 33, pp. 1225-1229.

Koning, J.J., Houdijk, H., Groot, G., and Bobbert, M.F. (2002) "How klapskate hinge position affects push-off mechanics in speed skating," *Journal of Applied Biomechanics*, Vol. 18, No. 3, pp. 292-305

Van Horne, S. (2003) "Mechanical and Performance Effects of a Modified Point of Foot Rotation During the Speed Skating Push" Masters Thesis, University of Calgary.

The Klap-Trap: A Device to Monitor Dynamics of the Klap Mechanism in Speed Skating

Sean Maw, Clifton R. Johnston and Allen Yuen
University of Calgary, Canada, smaw@ucalgary.ca

Abstract. An electromechanical system has been designed and built to measure the kinematics of the klapskate mechanics in speed skating. A string potentiometer was used to determine the opening and closing dynamics of a klap mechanism during actual speed skating at the Calgary Olympic Oval. Custom designed and fabricated clamps were used to attach the potentiometer to the klapskate mechanism. A preliminary study was conducted on four speed skaters. Klap dynamics were distinguishable based on skater fatigue levels, the location of the skater on the track and effort level of the subject. Parameters analyzed were maximum klap opening, stride frequency, and klap opening and closing timing. The opening angle of the klap mechanism was found to increase with increased effort and was higher in straights than corners. The result indicated that the concept proposed can measure the dynamics of the klapskate and may be an important training tool.

1 Introduction

Sport technology has led to important improvements in performances in speed skating. Since the late 1990s, perhaps no more important advancement in speed skating has been the klapskate (Versluis, 2005). Most, if not all, Olympic caliber skaters have employed the klap mechanism in their skating. The klap offers an increase in the biomechanical efficiency of the push-off. However, different skaters open and close their klap mechanism to different extents and at different rates. These parameters may vary depending on their fatigue level and where they are on the skating oval (straights or corners).

At present, we know of no system to measure the actual dynamics of the klap mechanism. Several studies have examined the positioning of the klap pivot in power production (Van Horne and Stefanyshyn, 2005; Houdijk, Bobbert, de Koning and de Groot, 2003; Allinger and Motl, 2000) as well as changes in skating mechanics with klapskates (Houdjik, de Koning, de Groot, Bobbert and Schenau 2000). However, none of these studies has examined the opening-closing klap dynamics as part of their investigation. The dynamics of the klap mechanism provides critical information in understanding the power improvements possible with the klapskate.

The current study examines the possible methods that could be used to measure the klap dynamics and to prove the feasibility of the concept on actual skaters. The method selected for measuring the movement of the klap mechanism was a string potentiometer attached to the klap spine and blade with custom built clamps. The potentiometer would allow the relative position of a point on each of the spine and blade to be measured. Those two relative positions would, in turn, allow the opening angle of the klap mechanism to be calculated.

A device was manufactured in the Department of Mechanical & Manufacturing Engineering at the University of Calgary and was tested on four speed skaters.

2 Data Collection System

As no known system for measuring the opening-closing dynamics of the klapskate existed, we first had to examine possible technologies and determine the most feasible method for implementation. Any system developed for this work would require three main components: position transducer, skate mounting system and data acquisition.

2.1 Position Transducer

Technologies that were initially considered for measuring the klap position were the rotary potentiometer, LVDT, optical transducer and string potentiometer. Of these possible approaches, a 75 mm string potentiometer (SpaceAge Control, Palmdale, CA) was chosen because of its small size (28g), accuracy and linear response (0.5% over full scale).

Several tests were conducted on the potentiometer to confirm the specified responses. The linearity of the transducer was experimentally examined. The linearity was tested using a Mitutoyo digital micrometer. The linearity was found to fall well within the manufacturer's specifications.

The next issue to be addressed was the positioning of the potentiometer on the klap mechanism. The string had a maximum extension length of 75mm, thus becoming the limit on the placement of the attachment points. Using a simple geometric model, the string displacement corresponding to a range of attachment points was determined. In order to preserve a safety factor, the maximum displacement of the string should be no more the 40 mm.

It was found that the change in length of the string potentiometer was more sensitive to the position of the attachment point on the blade than to the attachment point on the spine. Therefore, if the string was attached 25 mm from the hinge on the blade, the attachment on the spine would be unrestricted while still limiting the extension of the string to 40 mm and the speed of the string movement to 1 m/s.

The last issue examined was the frequency response of the transducer. The string potentiometer would be required to follow the opening and closing speed of the klap mechanism. The maximum closing speed of the klap mechanism was experimentally found to be 8.3 m/s. However, the placement of the transducer on the spine and blade limited the maximum speed to 1 m/s. Therefore, subjecting the transducer to a sinusoidal motion that would produce an equivalent 1 m/s maximum velocity validated the frequency response within the required range.

2.2 Klapskate Mounting Brackets

One of the requirements governing the design of this device was that it be easily attached and removed from a wide range of klap mechanisms. The final design could not be permanently attached to any one skate. Therefore, a reusable, non-marring, adjustable mounting bracket was designed.

Two brackets were designed and built: a bracket for the spine and a bracket for the blade of the klapskate. The top bracket was to be temporarily fastened to the spine, while also providing the attachment point for the string potentiometer. The bottom bracket would attach to the blade (fitting over the tube) and also provide the attachment for the end of the string of the potentiometer. The final realization of the two brackets are shown in Fig. 1.

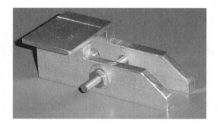

Fig. 1. Final design of top and bottom brackets

The top clamp, made from aluminum, consisted of two parts: the potentiometer attachment and a clamping face. A bolt passing through the two pieces provided the clamping force. The bolt was located to take advantage of a common through hole in a majority of klapskate spines. A flange on the top of the potentiometer mount bracket ensured that the bracket was evenly attached on both sides of the spine. Holes were tapped into the bracket to allow the string potentiometer to be fastened.

The bottom bracket was manufactured from a single block of aluminum and was designed to rest on the blade tube. The bottom bracket was fixed to the blade by three setscrews on the side of the bracket. Another threaded hole was created so that a large bolt could be used as the attachment point for the potentiometer string.

2.3 Data Collection

The voltage output of the potentiometer was passed to a National Instruments USB-9215 data acquisition module. This system was connected to a laptop carried in a backpack by the skater. The USB port of the computer powers the data acquisition module. The potentiometer and klapskate were calibrated prior to testing by measuring the closed klap voltage and the maximum-open klap voltage. Data was collected at 100 Hz and no filtering of the signal was deemed necessary.

An on ice test session consisted of 4 laps at 40% effort, 2 laps at 60% and 1 lap at 80% effort. This sequence was repeated in reverse to give two sets for each effort level. The sensor was attached to the right skate to minimize the probability of the brackets or sensor hitting the ice. A total of four subjects participated in test sessions and each provided informed consent to participate as per the Conjoint Health Research Ethics Board approval. One of the subjects had to be excluded due to timing errors. During each test one of the investigators was positioned in an observation gallery. The role of this person was to record the time the subject entered and exited corners. This information was used to validate the potentiometer data. Each test session was also video taped for future reference. Each subject was also asked to complete a post-test questionnaire. The subjects' responses provide demographic information as well as general comments about their impressions of the test apparatus.

3 Data Analysis

The data collected was analyzed in three steps using Microsoft Excel. Future applications of this will include the development of more automated data reduction.

The first step in data analysis involved the calibration of the raw voltages collected from the potentiometer into klap opening angles. Using linear interpolation between the lengths and voltages of the open and closed calibrations the displacement of the string potentiometer was calculated for each sample taken during the test. The known geometry between the spine and blade brackets then allowed the relative opening angle of the klap to be determined.

The second step in the data analysis involved the characterization of the peaks of the klap angle opening and closing curves determined in the first step. A minimum angle threshold was applied to determine the start of an opening/closing cycle. Any event that did not fall between the minimum and a corresponding maximum threshold was excluded. For example, a foot extension that only slightly opened the klap but was not a full should not be included in the analysis. Data was then searched to identify the time and angle of maximum klap opening and the associated duration of klap opening and closing.

The final step in data analysis was used to determine the mean and standard deviation for each of "time for the klap to open", "time for the klap to close", "total klap open time", and "ration of opening to closing time". Stride frequency was also determined using the time between each peak. This information was then used to determine the means and standard deviations for each parameter while the skater was in a straight, inner corner or outer corner.

4 Results

The determination of the transition from straight to corner does not associate itself to any exact marker on the track or in the data collected. As such, the moment of these transitions was judged visually. However, we found that we were, indeed, able to characterize these moments of transition to within 1 or 2 klap events. As these transition events did not fit definitively in either the straight or subsequent corner we have excluded these events in the reporting of the data.

The maximum angle of opening of the klap was determined for each of the three skating conditions (straight, inner corner, outer corner) and for each effort level (40%, 60%, 80%). An example of the maximum clap opening angle for one of the subjects is shown in Fig. 1.

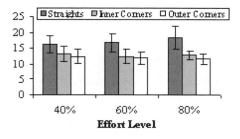

Fig 1. Maximum angle of klap opening.

Figure 1 illustrates that as effort level was increased there is an increase in the maximum angle of klap opening in the straights. It was also found that the maximum opening angle was highest in the straights and lowest in the outer corner, as also seen in Fig. 1. These finding were consistent among all subjects.

Figure 2 shows the change in stride frequency for each condition tested.

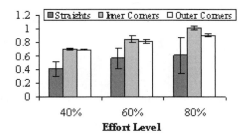

Fig. 2. Stride Frequency for changing effort level and track position.

The stride frequency was found to be substantially higher for corners than for straights and was slightly higher for inner versus outer corners.

Figure 3 is a comparison of the klap opening times for the considered conditions.

Fig. 3. Comparison of klap opening times for considered effort level and track position.

The klap opening time is the time taken to extend the klap to its maximum angle. For all data sets, as illustrated in Fig. 3, it was found that the klap opening time was relatively consistent with condition or track position.

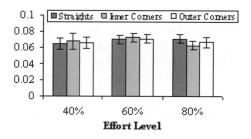

5 Conclusions and Future Work

The pilot study we undertook had several goals. We have achieved the primary objective of developing a system that will allow the characterization of the klapskate dynamics and motion for a long track skater *in situ*. Data was successfully processed for three test subjects and patterns related to effort level and track position were examined. Further work to refine the clamps and data analysis is need before this system could be widely used. The inclusion of a force sensor to measure the net force transmitted into the ice is also being investigated. Studies specifically addressing the biomechanics of the skates could then be undertaken.

Acknowledgements

The authors would like to thank the Alberta Ministry of Innovation and Science (INNSCI), the Canadian Sport Center Calgary and the Calgary Olympic Oval for supporting this work. We also thank the Oval Program Skaters for volunteering for this study.

References

Versluis, C. (2005) Innovations on thin ice. Technovation. 25 (10), 1183-1192.

Van Horne, S . and Stefanyshyn, D.J. (2005) Potential method of optimizing the klapskate hinge position in speed skating. J. App. Biomech. 21, 211 – 222.

Houdjik, H., Bobbert, M.F., De Koning, J.J., and De Groot, G. (2003) The effects of klapskate hinge position on push-off performance: A simulation study. Med. Sci. Sports Exerc. 35, 2077-2084.

Allinger, T.L. and Motl, R.W. (2000) Experimental vertical jump model to evaluate pivot location in klap speed skates. J. App. Biomech. 16, 142-156.

Houdjik, H., de Koning, J.J., de groot, G., bobbert, M.F. and Schenau, G.J.V. (2000) Push-off mechanics in speed skating with conventional skates and klapskates. . Med. Sci. Sports Exerc. 32, 635-641.

8 Golf Club Development

Synopsis of Current Developments: Golf Club Development

Ing. MMag. Dr. Anton Sabo

University of Applied Sciences Technikum Wien, Austria, anton.sabo@technikum-wien.at

In recent years, an increasing number of people have started playing golf. As in almost every sport the knowledge about the influence of the equipment on the human body and on the execution of motion is almost not available and the reaearch is still in its early stages. Regarding the huge amount of different sport equipment for each field of sport, it is very difficult to find the best for oneself. Researchers run tests and projects in different fields of sport because it is crucial to find out differences in equipment. Concerning golf, the projects deal with studies to find these differences in the range of golf clubs' properties, materials and player related parameters.

There are studies, which show differences in shaft material such as graphite and composites. Every material has its specific characteristics concerning flexural and torsional stiffness, kick-point and natural frequency. These characteristics additionally have discriminative influence on the execution, like backspin, the speed of the ball, stability of the direction of the ball, reliability of the impact at the sweet spot and the feeling of the club during the swing.

The used materials for the shaft also have an influence on the properties of a golf club. Besides shaft material also shaft properties like length also have an important influence on the game.

Most of the presented studies use video analysis of golf motion. Visual recording during measurement is very important and common and is the base to receive high quality results. However video analysis alone is not sufficient for complex measurements. Analysis of the movement using different mechanical models (e.g. 4 link model) is only one way to improve technique, playing and understanding of golf motion.

It can be observed that there are two major directions in research. On the one hand, material is tested with highly sophisticated measurement systems, trying to ensure highest reproducibility by using golf robots or testing material alone. On the other hand, researchers tend to evaluate the player's subjective impressions of different golf clubs using databases and statistics.

The range in every field of sports is wide which will result in more future studies to be done. There are new materials with new material combinations and additional innovative products every day.

In the end, their objective is similar: finding the ideal golf club for every single player. Considering the diversity of the products and the diversity of players there is still a long way to go and its final goal is still in the dark. However there is one thing we might be able to agree on: the future development in golf - whether it considers material, biomechanics or player's impressions - it will always be based on recent scientific research and will always have to focus on the human, on the player – on us.

Optimum Design and Validation of a Graphite Golf Shaft Based on Dynamics of Swing

Manryung Lee[1] and Cheol Kim[2]

[1] Kyungin Women's College, Korea, manryung@kic.ac.kr
[2] Kyungpook National University, Korea

Abstract. A player's swing motion is analyzed and the head speed is assumed as a cubic function. Then, swing accelerations and forces acting on a sweet spot of a club head are calculated. Under these forces, the deflections and twists of a shaft with optimum stacking sequences are analyzed right before the impact and at the instant of impact. An efficient optimization algorithm is introduced to optimize the stacking sequence of the shaft and wall thickness that satisfy the strength, weight, flex ratings, a kick point and dynamic torsional flexion using a genetic algorithm and finite element analyses. Unlike steel, the shaft made of graphite/epoxy materials shows many different static and dynamic behaviors during swing, depending on its ply orientations. The cost function is a minimum twist of a composite club shaft and seeks the stacking sequence satisfying flex, kick point and weight requirements. After iterative 28 generations, the optimal stacking sequence for the minimum angle of twist is obtained as $[30/45/-30/0/-45]_s$ and results in the twisting angle of 1.1°. Before the impact, the maximum deflection is 22.6 mm and the largest twist is 0.121°. At the impact, the maximum deflection is -168.991 mm and the largest twist is 0.217°.

1 Introduction

Unlike isotropic materials, anisotropic materials such as composites have a different structural feature according to ply orientations and finding the optimal stacking sequence to satisfy the design requirements is very important. Optimization researches related to composite materials have been progressed continuously (Gürdal *et al.*, 1999). Many structures composed of composite materials use limited ply orientations due to manufacturing difficulties. Many optimization methods have difficulties in handling such discrete variables as ply angles. Genetic algorithms (GA), however, are suitable for representing discrete variables and reliable for convergence to a global minimum. Due to these reasons, GA is increasingly applied to the optimization problems of the stacking sequence of composite structures. Genetic algorithms seek to mimic the biological processes of reproduction and natural selection. Natural selection determines which members of a population survive to reproduce, and reproduction ensures that the species will continue.

Riche and Haftka (1993) performed the optimization to maximize the critical buckling load of a composite panel under compression loads using genetic algorithms. Todoroki and Sasai (1999) used the recessive gene like the repair method to accomplish the optimization to maximize the critical buckling load of a composite cylinder. Liu *et al.*(2000) applied the permutation genetic algorithm to composite laminates. Soremenkun *et al.* (2001) applied the generalized elitist selection to the genetic algorithm. Soremenkun *et al.* (2001) also performed the optimization of multiple composite laminates using DARWIN.

The effective methodology to optimize the stacking sequence of composite golf shafts is introduced considering the weight, flex, kick point, and torque requirements. The design optimization combining the finite element method and genetic algorithms have been performed using ABAQUS and DARWIN that is specialized for optimizing stacking sequence of composite materials based on genetic algorithms. In this optimization procedure the thickness of the shaft can vary and the ply orientations are limited to the angles of $0°$, $±30°$, $±45°$, $±60°$, $90°$.

2 Stacking Sequence Optimization with a Genetic Algorithm

The stacking sequence optimization based on GA has been performed with the finite element analysis. In the procedure, the interface code first updates an ABAQUS input file using the stacking sequence results obtained from a DARWIN run and then reruns ABAQUS analysis. Then, the twisting angles obtained from finite element analysis are used for calculating the fitness function. This process is repeated $n \times (p+1)$ times, where n is the number of members in a GA population and p is the number of generations. The design parameters of GA used in this paper are summarized in Table 1.

Table 1. Design parameters of GA

Parameters	Value
Population size	10
Probability of crossover	0.8
Probability of mutation	0.05
Probability of gene swap	0.9

The goal of the optimization is to find the stacking sequence of laminated composite club shafts, which minimize the angle of twist, θ under the torque of 13.9 kgf-cm. The angle of twist is expressed as θ = TL/GJ (= torque×shaft length / shear modulus×polar moment) and is calculated from finite element analyses. The effective G varies depending on stacking sequence and then, θ changes. The mass of the composite shaft can be expressed as

$$W = \pi \times L \times \rho \times [(R_1 + R_2) \times N \times t - (N \times t)^2] \times g \tag{1}$$

where R_1 and R_2 are the butt outer radius and the tip outer radius respectively, N is the number of layers, L is the length of the shaft, t is one layer thickness, ρ is the density, and g is gravitation.

The fitness function f is calculated as follows,

$$f = - |\theta| \tag{2}$$

The optimization is performed to find the stacking sequence having the minimum angle of twist of the driver shaft subjected to the torque of 13.9 kgf-cm while satisfy-

Minimize θ

$$\textit{Subject to}\ \ p\ (\text{kick point}) = \begin{cases} \leq 41\% \\ 41 \square\ 44\% \\ \geq 44\% \end{cases} \tag{3}$$

$$60g \leq W \leq 80g$$

flex (natural frequence) = specified

ing the flex (1st natural frequency) and kick point requirements. Finite element analysis and optimization were performed for a composite shaft with the dimensions of 1,168 mm in length, 8.5 mm in tip diameter, and 15 mm in butt diameter. The target mass was set to be 77g with 10 sheets (wall thickness = 1.25mm). Optimum stacking sequence and wall thickness was determined from optimization. The kick point is determined from the first buckling mode under the critical buckling load and the 1st natural frequency (flex) is calculated by finite element analysis.

The finite element type S4R5 is used for modeling the composite shaft. Total of 2,000 elements and 2,210 nodes are used. After iterative 28 generations, the optimal stacking sequence for the minimum angle of twist is obtained as [30/45/−30/0/−45]ₛ and generates the twisting angle of 1.1° (Lee and Kim, 2004). The kick point of this shaft is 43% and the first natural frequency is 1.74 Hz from finite element analysis. The second best stacking sequence is [0/30/±45/−30]ₛ with twisting angle of 1.5° and the third best stacking sequence is [90/30/60/−30/−60]ₛ with twisting angle of 1.8°. Interestingly enough, the optimum design and the second best design having the same angles but different stacking sequence show the difference in twist.

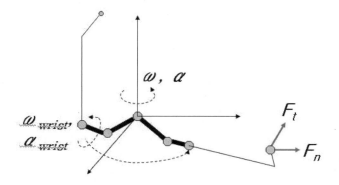

Fig. 1. A swing motion of a club shaft and force components

3 A Swing Motion and Deformation of a Shaft

An average amateur golfer's swing speed of a club head can be assumed as a cubic function of swing time *t* by analyzing golfers' motions as follows,

$$v_{head}(t) = c_1 \, t^3 + c_2 \, t^2 + c_3 \, t + c_4 \qquad (4)$$

As an example, if a person finishes the swing of a driver for 0.5 second, then four coefficients can be readily determined using the four conditions: initial and final head speeds are zero, a gradient of the cubic equation at impact (t = 0.2 s) is zero and the speed at 0.2 s is 46 m/s that is a normal head velocity right before the impact. As a result, the swing speed of the golfer specific is expressed as

$$v_{head}(t) = 1277.78 \, t^3 - 1661.11 \, t^2 + 511.111 \, t \qquad (5)$$

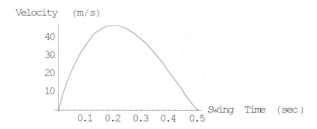

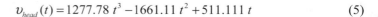

Fig. 2. The head velocity during a swing of a club shaft

The centrifugal and tangential accelerations are calculated as follows,

$$a_n = \frac{v^2}{r}, \quad a_t = \frac{dv}{dt} \qquad (6)$$

where a_n, a_t, and r are a centrifugal acceleration, a tangential acceleration, and a radius of a club head trajectory, respectively. The force components can be obtained by multiplying the head mass by accelerations. Generally, the weight of a head is 195 gf and can be transformed to the mass of 19.898 g. If a golfer has a radius of a club head trajectory, 1.6 m, then the centrifugal acceleration at the center of gravity of the head is 26.315 N and the tangential force is 0.00357 N in the swing direction at the instant right before the impact with a ball.

The total linear impact on the head face by a ball equals the corresponding change in momentum of the ball. This relationship can be expressed as

$$\int_{t_1}^{t_2} \Sigma F dt = G_2 - G_1 \qquad (7)$$

Here the momentum at time t_2 is $G_2 = mv_2$ and the momentum at time t_1 is $G_1 = mv_1$. The momentum is the product of mass and velocity. F is an impacting force during a very short period. If the mass of a ball is 45.9 g and the ball velocity is 63.5 m/s

0.000328 second after the impact, the reacting impact impulse on the head face increases to 8,886 N (907 kgf) that is almost one Ton and calculated from Eq (7). However, this impulse acts during very short period (0.000328 second) so that most of the impact is relieved and only small force is transferred to the shaft head. The impact can be converted to the equivalent static force of 20 N (0.225%). This fact is verified by experiments.

The previous loads are applied to the graphite shafts optimized, and the deflections and twisting angles at the joins of a head and a shaft are calculated at the instant right before the impact and at the impact instant. The results are summarized in Table 2. Deflection u1 directs to the ground, u2 is in the swing direction, and u3 is in the shaft direction. It is shown from Table 2 that the shaft deflects along a swing direction and to the ground before impact, however, at the impact a large deflection occurs and moves backward. There are no big differences in the twisting angles before and after impact because the graphite shaft was optimized to minimize the twist.

4 Conclusions

The stacking sequence of composite club shafts was optimized using FEM and genetic algorithm to minimize twists under the torque of 13.9 kgf-cm satisfying other

Table 2. Deflections and twists at the tip of an optimum graphite shaft

Time	Shaft stacking seq.	Twist (deg.)	u1 (mm)	u2 (mm)	u3 (mm)
Before impact	$[30/45/-30/0/-45]_s$	0.121	22.60	0.0410	-0.219
	$[0/30/\pm45/-30]_s$	0.122	22.70	0.0412	-0.220
	$[90/30/60/-30/-60]_s$	0.187	34.72	0.0632	-0.337
At impact	$[30/45/-30/0/-45]_s$	-0.217	-17.51	-168.99	15.851
	$[0/30/\pm45/-30]_s$	-0.219	-17.56	-169.37	15.924

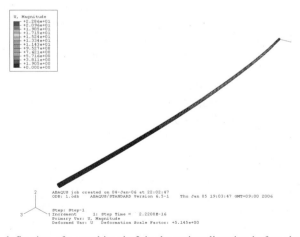

Fig. 3. The deflection of a graphite shaft in the swing direction before the impact

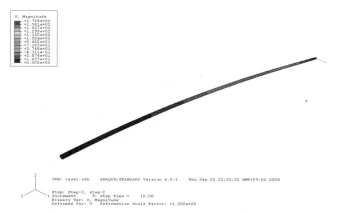

Fig. 4. The deflection of a graphite shaft backwards at the instant of the impact

specified design requirements and gets [30/45/−30/0/−45]$_s$. Individual's swing motion is analyzed and assumed as an cubic function. Deflections and twists of a shaft can be obtained and a swing can be adjusted considering the expected deformations. Before the impact, the maximum deflection is 22.6 mm and the largest twist is 0.121°. At the impact, the maximum deflection is -168.991 mm and the largest twist is 0.217°. The method developed may help to design easily and quickly composite club shafts to match individual swings by selecting the lay-up angles and sequences without expensive and intensive experiments.

References

Gürdal Z., Haftka R. T., and Hajela P. (1999) *Design and Optimization of Laminated Composite Material.* John Wiley & Sons Inc., New York.

Lee M. and Kim C. (2004) Design optimization of graphite golf shafts based on weight and dynamics of swing. *Proc. of Engineering of Sport, 5th Int. Conf.*

Liu B., Haftka R. T., M. Akgün A., and Todoroki A. (2000) Permutation genetic algorithm for stacking sequence design of composite laminates. *Comput. Methods Appl. Mech. Engrg.*, 186, 357-372.

Riche R. L. and Haftka R. T. (1993) Optimization of laminated stacking sequence for buckling load maximization by genetic algorithm. *AIAA Journal*, 31(5), 951-956.

Soremenkun G., Gürdal Z., Haftka R. T., and Watson L. T. (2001) Composite laminate design optimization by genetic algorithm with generalized elitist selection. *Computers and Structures*, 79, 131-143.

Soremenkun G., Gürdal Z., Kassapoglou C., and Toni D. (2001) Stacking sequence blending of multiple composite laminates using genetic algorithm. *Proc. 42nd AIAA/ASME/ASCE/AHS/ASC SDM. Conf.*, 2001-1203.

Todoroki A. and Sasai M. (1999) Improvement of design reliability for buckling load maximization of composite cylinder using genetic algorithm with recessive-gene-like repair. *JSME International Journal, Series A*, 42(4), 530-536.

Modelling Vibration Frequency and Stiffness Variations in Welded Ti-Based Alloy Golf Driver Heads

Simon Adelman[1], Steve Otto[2] and Martin Strangwood[1]

[1] Sports Materials Research Group, The University of Birmingham, Department of Metallurgy and Materials, m.strangwood@bham.ac.uk
[2] The R&A, St Andrews, Fife, steveotto@randa.org

Keywords: Golf Equipment, Driver, Construction, Welding

Abstract. The Characteristic Time (CT) and peak vibration frequency of a set of 65 driver heads were measured and correlated ($R^2=0.83$). The Coefficient of Restitution (CoR) of 27 of the heads was determined and the heads were sectioned in order to assess the quality of the weld between the face and crown components by measuring its size as well as inspecting it for signs of oxidation and insufficient heat treatment. In the heads with good welds the CoR was found to depend on the flexibility of both the face and the crown. However, in those with poor welds the CoR was found to have a low dependence on crown stiffness. The ratio of crown vibration/face vibration has been determined and correlated with a visual assessment of the quality of the welds. In order to further understand the effect of the weld geometry and properties on the vibration of the head the first steps towards a predictive vibration model have been taken. The natural frequencies and mode shapes of circular, elliptical, rectangular and square plates have been calculated for a variety of boundary conditions and the frequencies have been experimentally verified for a completely free circular plate. Further work is needed to develop the model to include a weld of variable geometry and material properties.

1 Introduction

Increasing the flexibility of the driver head has been associated with increases the kinetic energy of the golf ball post-impact by reducing its deformation and associated viscoelastic energy loss. One way of quantifying the flexibility of the head is to measure its coefficient of restitution (CoR) for impact with a single ball type under specified conditions. Previous research (Strangwood 2003) has identified relationships between the CoR of the head and the flexibility of its components. By modelling each component as a simply supported vibrating disc Strangwood found that the CoR of the head should be related to the geometry and material properties of its components by Eqn 1:

$$\ln(CoR.E/A) = const. - 3\ln(t) \tag{1}$$

where E is the Young's modulus and A and t are the area and thickness of the component. By plotting Eqn. 1 for each component he found the CoR to be independent of the face and sole properties, but to depend on the flexibility of the crown. The sole was found to be unimportant as it is much thicker and stiffer than the crown and so is not an effective store of elastic strain energy. Strangwood highlighted the importance of the joint between the face (where the impact occurs) and the crown (where the elastic energy can be stored). The construction of the driver head and the role of the joint between the face and crown is investigated further in this work.

Hollow all-metal driver heads cannot be made in one piece and so they are made by casting or forging 2 or 3 separate pieces and then welding the pieces together. To effectively transfer energy between the face and crown the joint between them should be identical to the base metal on either side of it in its geometry and material properties. In practice this is very difficult to achieve. Access to the finished weld is limited as the completed head is a closed structure, so while the external weld bead can be inspected and removed after welding the internal underbead cannot. The underbead protrudes into the head's interior making the weld region thicker and stiffer than the base metal. Furthermore, the uptake of oxygen during welding of titanium-based alloys increases the amount of α-phase, raising the Young's modulus of the weld. Inert gas shielding reduces oxygen uptake but often only the exterior of the weld is shielded. Oxygen from the interior of the head can still be absorbed into the liquid weld pool, stiffening the joint. One way in which the variation in material properties in the weld region can be reduced is to heat treat the head after welding. Most heads are solution treated to homogenise the microstructure of the weld and surrounding base metal by recrystallisation.

In order to understand the role of the weld, the CoR of a set of modern driver heads has been measured and their construction, materials and welds have been characterised. A model to predict how the characteristics of the weld affect the natural vibration frequency and mode shapes of a simple structure is also being developed for experimental verification.

2 Driver Head Testing

2.1 Experimental

The Characteristic Time (CT) of 65 heads was measured using the pendulum test (R & A Rules Ltd and the USGA 2003a) and their frequency spectra were measured using a Brüel & Kjær (B & K) vibration analyser. Accelerometers were attached to the face, crown and sole of each head using bees' wax and a mini-shaker was used to strike the centre of the face with a rectangular 1 ms pulse. The accelerometer signals were collected using a B & K multi channel data acquisition unit and fast-Fourier transformed to produce frequency spectra. From these data, 27 heads were selected for further testing. The area of each component was measured using Axiovision 4.0 image analysis software. The heads were sectioned using an abrasive cut off wheel and the welds were visually inspected. The thickness of each component was measured using calipers and the results used to calibrate an ultrasonic thickness gauge for speed of sound measurement and thickness measurement at locations inaccessible to calipers. Sections of each component were removed and paint-stripped before density measurement in order to calculate Young's modulus values.

2.2 Results and Discussion

The CT and CoR are both measures of the heads' flexibility and are highly correlated, with high CT and CoR values indicating high flexibility. The correlation has been quantified empirically in Eqn. 2 (R & A Rules Ltd and the USGA 2003b), which is based upon data obtained from approximately 600 driver heads:

$$CoR = 0.718 + 0.000436 * CT \tag{2}$$

The peak vibration frequency (measured on the face) also provides a measure of the heads' flexibility. A low frequency indicates high flexibility. The heads' frequency is correlated with the CoR (Strangwood 2003) and is expected to correlate with the CT. Fig. 1 plots these data for the 65 heads in this study and shows that the two quantities are well correlated ($R^2 = 0.83$).

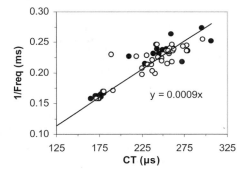

Fig 1. CT (μs) vs 1/frequency (ms) measured on the face

Twenty-seven heads (filled circles) were selected for destructive testing. The CoR of each head was calculated from Eqn. 2. and the Young's modulus of each component was calculated from Eqn. 3 (Lemaitre & Chaboche 1990):

$$E = \rho c^2 * (1 + \upsilon)(1 - 2\upsilon)/(1 - \upsilon) \tag{3}$$

where ρ is the density, c is the speed of sound and υ is the Poisson's ratio (assumed to be 0.34 for all Ti-based alloys). Figures 2(a) and (b) present plots of Eqn. 1 for the face and crown. The crown data have been categorised by the quality of the weld. The thicknesses of the underbead and the base metal were measured using calipers. The ratio of these two quantities is a guide to the geometric constraint of the weld with a high value indicating a stiffer joint. 'Poor' joints were chosen to be those with bead/base thickness ratios above the average (1.86) as well as those with unusually wide beads, signs of oxidation (gold, blue, purple and black colouration) and those which had not been heat treated to homogenise the microstructure of the weld and surrounding base metal.

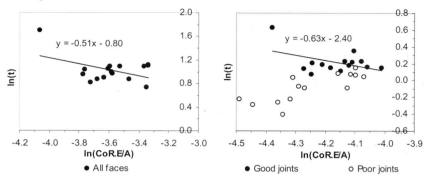

Fig 2. Plot of Eqn. 1 for: (a) the face; and (b) the crown

The face component of all the heads shows considerable scatter, but are also clustered except for one head. Inclusion of this one head gives the expected relationship, although with a gradient of 1/2 rather than 1/3. If the outlying point is not included then the data give no strong relationship. In the heads containing 'good' joints the crown component also follows the expected relationship, with a gradient of 1/1.6. The difference between the expected gradient of 1/3 and those observed in the face and crown data is likely to be due to the assumption that the components are isolated and simply supported. In the head they are actually clamped by the joints to each other and the sole. The crowns of the poor joints are also widely scattered and show no strong relationship, although they appear to give a positive slope, which would suggest increased stiffness with reduced thickness and is intuitively incorrect. Hence, the stiffness of these heads does not show a strong dependence on the crown deformation. For heads with good welds, however, there is a much greater dependence of head stiffness on the crown properties so that a more significant amount of elastic energy would be expected to be stored in both components. Double integration of the output of accelerometers attached to the face and crown during impact allowed the maximum displacement during vibration of the crown, d_c, and face, d_f, to be determined and compared. The ratio d_c/d_f is an indication of the amount of deformation transferred across the joint. The average ratio for the good joints was found to be 2.6 compared to 1.2 for the poor joints. This highlights the effect of a poor face/crown weld, which is to restrict deformation and energy transfer between the face and the crown and reduce the fraction of elastic energy stored in the crown, in line with the trends shown in Fig. 2 and discussed above.

3 Plate Modelling

3.1 Modelling

Matlab was used to model the vibration of plates of specific shape, thickness and material properties. The natural frequencies and mode shapes of the first five modes of circular, elliptical, rectangular and square plates were calculated by solving the wave equation under completely free boundary conditions (the bending moment and the Kelvin-Kirchoff edge reactions at the edge of the plate are both zero).

3.2 Experimental

Ti-15V-3Cr-3Al-3Sn (Ti-15-3) sheet of 1 mm thickness was received from Timetal. The speed of sound and density of this alloy were measured as described in Section 2.1 and the Young's modulus was calculated from Eqn. 3. The sheet was cut into circular, elliptical, rectangular and square plates using electro-discharge machining. The plates were freely supported on a foam cushion and an accelerometer was attached to the plate using bees' wax. The frequency spectrum of each plate was measured using a B & K vibration frequency analyser as described in Section 2.1.

3.3 Results and Discussion

Table 1 gives the first five predicted and measured frequencies of a 1 mm thick Ti-15-3 plate with a radius of 100 mm. Similar results have been generated for plates of different thickness, material properties and shape. The measured and predicted frequencies tend to agree to within 10%. However, some modes show less of an agreement, such as mode 3 in Table 1. The ratio, F_{Model}/F_{Exp}, does not vary systematically between modes so it is suggested that these variations are due to stiffening of the plate from the added mass of the accelerometer (~2%). The degree of stiffening varies between modes depending on the displacement at the accelerometer's location, with the maximum stiffening effect being seen when the accelerometer is located an a modal antinode. The same model has been used to predict the frequencies and mode shapes of clamped and simply supported plates. The calculated frequencies agree closely with the values predicted by Leissa (1969).

Table 1. The first five vibration modes for a 1 mm thick circular Ti-15-3 plate of 100 mm radius

Mode	F_{Model} (kHz)	F_{Exp} (kHz)	F_{Model}/F_{Exp}
1	140	131	0.9
2	243	259	1.1
3	326	421	1.3
4	549	512	0.9
5	573	584	1.0

4 Overall Behaviour

The peak vibration frequency of a driver head was shown to be a valid and useful measure of its performance. The head's frequency was correlated with its CT and the relationship was quantified empirically. The quality of the face/crown weld was assessed in each head. In heads containing good face/crown welds the CoR was shown to depend on the flexibility of both the face and the crown. However, in the heads containing poor face/crown welds (those with excessive underbeads, signs of oxidation or a lack of heat treatment) the CoR was shown to be independent of the flexibility of the crown and to depend on the flexibility of the face alone. These heads also showed a reduced ratio of crown vibration/face vibration when compared with the heads containing good welds. This shows that the presence of a poor weld between the face and the crown restricts energy transfer between the two components and significantly reduces the fraction of elastic energy stored in the crown. The presence of poor joints prevents deformation transfer from face to crown with consequent increases in overall head stiffness.

5 Further Work

In order to further understand the effect of the weld geometry and properties on the vibration of the head the first steps towards a predictive vibration model have been taken. The natural frequencies and mode shapes of circular, elliptical, rectangular and square plates have been calculated for a variety of boundary conditions and the frequencies have been experimentally verified for a completely free circular plate. The frequencies predicted for clamped and simply supported plates agree closely literature values.

Further work is needed to fully quantify the material properties of the face/crown welds in these driver heads as well as to examine more driver heads in order to build this data set. The plate model must also be developed to include a weld of variable geometry and material properties so that the effect of a variety of welds upon the mode shapes and natural frequencies of the plates may be predicted and related to the effects seen in the driver heads.

References

Leissa, A. W. (1969) Vibrations of Plates, Office of Technological Utilization, National Aeronautics and Space Administration, Washington, D.C.

Lemaitre, J. and Chaboche, J-L. (1990) Mechanics of Solid Materials. Cambridge University Press, UK

R & A Rules Ltd and the USGA (2003a) Procedure for measuring the flexibility of a golf club, Revision 1.0

R & A Rules Ltd and the USGA (2003b) Technical description of the pendulum test

Strangwood, M. (2003) Materials in Sports Equipment, Chapter 6: Materials in Golf. Jenkins, M. (Ed). Woodhead Publishing Ltd, Cambridge, UK

Optimal Adjustment of Composite-Material Club Shaft Characteristics

Moo Sun KIM[1], Sun Jin KIM[2], Dong Chul HAN[1] and Woo Il LEE[1]

[1] Department of Mechanical and Aerospace Engineering, Seoul National University, moosun@hotmail.com
[2] Department of Physical Education, Seoul National University

Abstract. In this study, we present a methodology for optimization of golf club shaft design. Shaft characteristics are considered for different player's preference, such as flexural and torsional stiffness, kick-point and natural frequency. For the case of composite-material shaft manufactured by sheet lamination process, the optimal layer orientation, stacking sequence and shaft-taper angle are sought to meet the various range of required characteristics. As an optimization tool, genetic algorithm is applied with the stacking sequence and taper angle as design variables.

1 Introduction

Golf club shaft is a primary part to determine the distance and accuracy of the trajectory of a ball. Its capabilities are estimated and affected by the flexural and torsional stiffness of a shaft. These stiffness may present various performances by composed with different player's swing feature. Therefore, depending on the preference and the capability of the player, different shaft with optimal features fitted to each player should be considered.

In these days, composite material shaft is popular due to its excellent performances and easiness in realization of the demanded features. In case of sheet lamination process to manufacture club shafts, the required characteristics can be achieved by manipulating the stacking sequence of prepreg sheets. Prepreg is a mixture of unidirectional reinforcing fibers and the resin matrix made into a sheet form. Along with the prepreg stacking sequence, the sectional diameter variation can affect the characteristics of shaft such as the kick-point and the natural frequency.

Genetic algorithm (GA) is an optimization tool that is usually used for the stacking sequence problem of composite structures (Park, Lee, Han and Vautrin 2003; Lee and Kim 2004). In this study, we suggest the methodology for optimization of golf club shaft characteristics by considering prepreg sheet stacking sequence and taper angle, which is represented by the ratio of both end diameters of a shaft, as the design variables in GA. Optimization is performed to satisfy various levels of players, from professional to weekend golfers.

2 Optimization of stiffness of shaft

2.1 Parameters for analysis of composite shaft

As the main factors to define the characteristics of club shaft, flexural and torsional stiffness should be considered, and also kick-point and natural frequency. In order to adjust these characteristics of composite material shaft, prepreg stacking sequence as well as the geometrical parameters of the shaft, are to be regarded as control parameters.

Fig. 1. Parameters for analysis of composite shaft

2.2 Static and dynamic analysis of composite shaft

To estimate the flexural and torsional stiffnesses of a shaft, classical laminate theory is applied. Cylindrical shaft is modeled as a sandwich plate with core and then analyzed when both of the bending and twisting moments acting on it. Kick-point of a shaft is defined as the maximum deflection point when the compressive forces act on both ends of the shaft. To obtain the position of the kick-point, the first buckling mode of a shaft is numerically computed by using finite element method(FEM). First natural frequency of shaft is calculated by performing modal analysis.

For the numerical analysis, shaft is divided into elements with uniform diameter that varies linearly from the butt to the tip to account for the taper (Fig. 2).

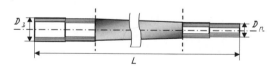

Fig. 2. Modeling of club shaft

2.3 Optimization with genetic algorithm

The aim of this study is to obtain the optimum design conditions for the club shaft whose mechanical properties satisfy the requirements of each player and his or her swing features. As mentioned previously, the parameters are prepreg stacking sequence and geometrical parameter (e.g. taper angle) that can be considered to affect the mechanical properties of club shaft. GA is an optimization method that mimics the natural phenomenon in which the superior members survive in the end of the evolution. GA searches for the superior individual through operations such as repro-duction, crossover and mutation. To implement the algorithm, sample orientations are established as design variables and stacking sequence is directly linked to the sequence of variables. In this study, 8 representative orientation are selected as 0°, ±30°, ±45°, ±60°, and 90°. Integers 0 through 7 are assigned for each orientation value.

Taper-angle as geometry condition may be represented by the diameter ratio between the tip and the butt. With this, decrease ratio in diameter can readily be obtained. Decrease ratio is represented in octal number system and linked to the strings representative of orientation to make operation easier. The octal numbers are converted later into decimal numbers to obtain decrease ratio. Figure 3 illustrates how to convert real design values and link them for the GA operation.

$$real \quad value \qquad : \left[\; 5\,0 : 0°, 45°, 90°, ..., 60°\;\right]$$
$$\Downarrow$$
$$converted \quad octal \quad number \quad : \qquad 7\;1: 0\;3\;7\;...\;5$$
$$[\text{decrease ratio : orientation}]$$

Fig. 3. Design variables and encoding into octal number system

In this study, the decrease ratio ranged from 0% to 63% and its operation such as crossover is performed separately from the orientation strings.

2.4 Optimization scheme

Depending on golf player's preferences and swing performance, appropriate characteristics of shaft fitted for a player can vary. For instance, in view of swing speed, professional demands a shaft with little bending and twisting deflection to have more precise trajectory. On the contrary, weekend golfers may prefer a rela-tively flexible shaft to increase the flying distance with as little twisting deflection as possible. Therefore, based on the swing speed or preference of a player, optimization of shaft characteristics is required. In this study, definition of a problem is estab-lished for two cases, each corresponding to professionals and weekend golfers. In case of professionals, the objective function that satisfies the possible maximum flexural and torsional stiffness is defined and in the other case for ordinary golfer, the maximum torsional stiffness and the flexural stiffness closest to any established standard value is defined as the objective function.

There are two approaches for optimization of multi-objective problems. One is to define a new objective function that includes both torsional and flexural stiffnesses at the same time and the other is to define flexural stiffness as a single objective prob-lem with the torsional stiffness as a constraint and vice versa. The other objectives as kick-point and natural frequency is defined as other constraints.

2.5 definition of objective function

Objective functions of simultaneous optimization and single optimization with constraint for both cases of professional and weekend golfers are defined as follows

Simultaneous optimization

professional : $\dfrac{1}{a_1+d_t}\left(1+\dfrac{d_{bc}}{d_b}\right)$

weekend golfer : $\dfrac{1}{(a_1+d_t)(a_2+|d_{bc}-d_b|)}$

$\left.\rule{0pt}{48pt}\right]$ *Subject to $0.40 \le kick\text{-}point \le 0.45$, 1st natural frequency$\le freq_{crit}$*

Single optimization with constraint

professional : $\dfrac{1}{a_1+d_t}$ *Subject to $d_b \le d_{bc}$*

weekend golfer : $\dfrac{1}{a_2+|d_{bc}-d_b|}$ *Subject to $d_t \le d_{tc}$*

$\left.\rule{0pt}{40pt}\right]$ *and $0.40 \le kick\text{-}point \le 0.45$, 1st natural frequency$\le freq_{crit}$*

where d_b, d_t, d_{bc} and d_{tc} represent flexural, torsional, critical flexural and critical torsional deflection respectively. a_1 and a_2 are scaling constants.

3 Results of optimization

The conditions of shaft for analyses and mechanical properties of prepreg sheet are given in Tables 1 and 2, respectively.

In GA, population size is assigned as 30. For verification of GA program, optimum angle set for minimum flexural deflection due to the pure bending moment and for minimum torsional deflection due to the pure torsional moment are searched. As is expected, within 20~30 generations, sequence of 0° for minimum flexural deflection and composition of 45° and -45° for minimum torsional deflection as ideal results are obtained.

Length	Butt diameter	Tip diameter	No of prepreg sheet
1,165 mm	15 mm	8.5 mm	10 sheets

Table.1. Geometry conditions of shaft and number of prepreg sheet

Mechanical properties of prepreg sheet	Q_{xx}	Q_{yy}	Q_{xy}	Q_{zz}
	134Gpa	7Gpa	1.75Gpa	4.2Gpa

Table.2. Mechanical properties of prepreg sheet

When bending and torsional moment are coupled and act on a shaft, the possible minimum flexural and torsional deflections are obtained for both cases with fixed and varying tip diameter. The results are used as basic values for normalization of the optimization results. (Fig. 4)

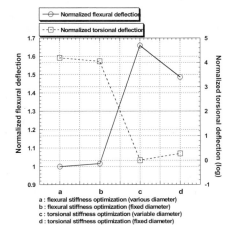

	Orientation and stacking sequence
Case a	90 90 0 90 60 90 0 0 0
Case b	90 -60 60 0 90 60 0 0 0
Case c	-60 45 0 -30 -60 60 30 -45 0 45
Case d	0 30 -60 60 -45 60 60 60 30 -45
Case a1	60 90 30 -30 0 30 -60 90 0 0
Case b1	0 45 60 0 -45 0 0 30 -60 60
Case c1	90 -60 45 0 90 -45 0 0 30 0
Case d1	0 -45 45 60 0 90 90 0 45 0
Case a2	-60 60 0 45 -60 30 60 0 0 -45
Case b2	-30 45 -45 -45 45 45 30 90 0 0
Case c2	90 –30 –60 0 0 30 –45 90 30 60
Case d2	90 60 30 -30 0 0 45 60 90 0

a : flexural stiffness optimization (various diameter)
b : flexural stiffness optimization (fixed diameter)
c : torsional stiffness optimization (variable diameter)
d : torsional stiffness optimization (fixed diameter)

Fig. 4. Figure of individual optimization results and table of orientation and stacking sequence for optimization

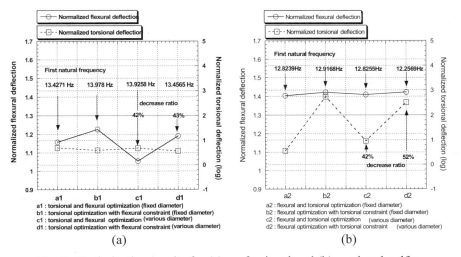

a1 : torsional and flexural optimization (fixed diameter)
b1 : torsional optimization with flexural constraint (fixed diameter)
c1 : torsional and flexural optimization (various diameter)
d1 : torsional optimization with flexural constraint (various diameter)

a2 : flexural and torsional optimization (fixed diameter)
b2 : flexural optimization with torsional constraint (fixed diameter)
c2 : flexural and torsional optimization (various diameter)
d2 : flexural optimization with torsional constraint (various diameter)

(a)   (b)

Fig. 5. Optimization results for (a) professional and (b) weekend golfer.

Optimization results of torsional and flexural stiffnesses in both cases of professional and weekend golfers are presented in Fig. 5. Every result is also normalized with the individual optimization result. Table in Fig. 4 gives the orientation and optimized stacking sequence for each case.

As shown in the results, for professional golfers, flexural stiffness obtained by simultaneous optimization is closer to the desired condition than one obtained from single optimization with flexural stiffness constraint. This may be due to the relaxation of the constraint conditions. For weekend golfers, the result of simultaneous optimization shows more variety of choices than single optimization. In this case, due to more flexible shaft characteristics, it exhibits lower natural frequency than professionals' shaft. In conclusion, single optimization result shows the inverse proportion between the flexural and torsional stiffness when only one stiffness is maximized. On the other hand, with simultaneous optimization, the stacking sequence and geometry condition obtained can maximize both stiffnesses within the appropriate ranges or other required characteristics.

4 Conclusion

In this study, we suggested the methodology to optimize stacking sequence of pre-preg sheets and geometry condition of the club shaft. For various levels of players, from professional to weekend golfers, we defined objective functions of optimization to consider their levels. We found out the proposed methodology was effective in determining the optimum design variables to satisfy different requirements.

References

Goldberg, D. E. (1989) *Genetic Algorithms in Search, Optimization and Machine Learning*, Addison Wesley.
Iwatsubo, T., Kawamura, S., Kawase, Y. and Ohnuki, M.(2002) Torsional Deflection of Golf Shafts During Golf Swing, *Proceedings of The Engineering of Sport, 4th International Conference*, pp.376-382.
Lee, M. and Kim, C.(2004) Design Optimization of Graphite Golf Shafts Based on Weight and Dynamics of Swing, *Proceedings of The Engineering of Sport, 5th International Conference*, pp. 248-255.
Park, C. H., Lee, W. I. , Han ,W. S., Vautrin, A.(2003) Weight Minimization of Composite Laminated Plates with Multiple Constraints, *Composites Science and Technology*, Vol.63, pp.1015-1026.
Reddy, J. N. (1986) *An introduction to the finite element method*, McGrawHill International edition.
Tsai, S. W. and Hahn, H. T.(1980) *Introduction to Composite Materials*, Technomic Publishing Company.

Factors Determining Backspin from Golf Wedges

James Cornish[1], Stuart Monk[2], Steve Otto[2] and Martin Strangwood[1]

[1] Sports Materials Research Group, The University of Birmingham, Department of Metallurgy and Materials, m.strangwood@bham.ac.uk
[2] The R&A, St Andrews, Fife, steveotto@randa.org

Abstract. Impact studies under dry conditions have been carried out for a range of golf balls (multi-piece solid construction balls representing a variety of compression ratings with both polyurethane (PU) and ionomer covers) using a gas cannon. The balls have been fired, at a speed of 30 m/s, at grooved and un-grooved plates at effective lofts between 20 and 70° with surface roughness, R_a = 1 μm. The ball speed, launch angle and backspin magnitude post-mpact were measured using a stereoscopic launch monitor system. Analysis of backspin was successfully carried out using effective coefficient of friction and spin ratio measures. Within the range of lofts investigated the backspin rate for each of the balls showed a maximum, which was dependent on the ball type and presence of grooves. This behaviour has been rationalised as the deformation of the cover into surface features (particularly grooves at higher loft angles) leading to a higher effective friction coefficient and greater rolling. The behaviour of two-piece balls is largely characterised by the cover hardness, but the mantle needs to be incorporated for three-piece balls.

1 Introduction

The oblique impact between a wedge and a golf ball involves a number of constituents, which are predominantly sliding, rolling and ball deformation (Maw et al. 1976 and 1981; Monk 2006) all of which contribute positively or negatively to backspin generation. The combined effect of these parameters can be treated using an effective coefficient of friction (μ_{Eff}) generated between the golf ball and the wedge. This can be derived from the launch conditions and related to the average normal and tangential forces (F_N and F_T respectively) acting throughout the impact, using Eq. 1. Previously (Monk 2006), a higher μ_{Eff} has been correlated with greater backspin rates.

$$\mu_{Eff} = \frac{F_T}{F_N} = \tan^{-1}\left(loft\ angle - launch\ angle\right) \tag{1}$$

A ball with a hard cover, such as an ionomer, spins and compresses less than a ball with a softer cover (Sullivan and Melvin 1994) and exhibits a lower μ_{Eff} (Monk 2006). A ball with a soft cover, such as polyurethane (PU), undergoes larger localised viscoelastic deformation at the impacting surface, which leads to an increase in the interaction between the face and the ball, which has been related to a larger value of μ_{Eff} and increased backspin rates (Monk 2006).

Separately, experimental data (Chou et al. 1994) have indicated that, as loft angle increases, spin rate increases up to a maximum point of 50°, before decreasing in magnitude. This trend has not been related to ball or clubface characteristics. Thus, this paper will report the effects of ball type and clubface condition on backspin generation during oblique impact.

2 Experimental

Five balls each of three types were used for the study (Table 1). Two orthogonal circumferential black lines were marked on each ball to provide lines from which to measure spin rate, and the sliding to rolling transition during impact.

Table 1. Ball type, materials and properties (Shore D hardness) for those investigated in this study; core data from Johnson (2006).

Ball	Ball Type	Construction	Core Material	Core Shore D	Cover Material	Cover Shore D
A	High Spin	2-Piece	Polybutadiene	47.2±1	PU	44±5
C	Distance	2-Piece	Polybutadiene	41.2±0.5	Ionomer	64±5
D	Mixed	3-Piece	Polybutadiene	44.0±1	PU	57±2

Hardness was measured using a Mitutoyo Shore D Durometer with a 30° steel cone, and a tip radius of 0.1 mm. Balls were fired, using an ADC Supercannon 2000, horizontally at a speed of 30 m/s, at steel plates with effective loft angles of 20, 30, 40, 50, 60 and 70°. The plates had roughness $R_a = 1$ μm and were grooved (90° V-shaped grooves, width 0.8 mm, depth of 0.4 mm and pitch of 3.5 mm) or un-grooved. The ball speed, launch angle and backspin rate post-impact were measured using a stereoscopic launch monitor system.

3 Results and Discussion

Figure 1 indicates that, initially, backspin rates increase with increasing loft angle and that, within experimental scatter, there is no significant difference in backspin with ball type at this impact speed for loft angles up to 50°. In addition, comparison of Figures 1 (a) and (b) for lofts up to 50° shows no significant effect of grooves. Over this loft range, the ball / plate interactions are characterised by increasing μ_{Eff} values, in line with previous results (Monk 2006). As loft angle increases, F_N would be expected to decrease, which will affect the degree of ball deformation (both cover and core), but the change in F_N is not matched by that in F_T so that the ratio (μ_{Eff}) increases.

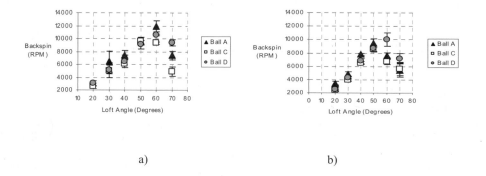

Fig. 1. Backspin as a function of loft angle for all ball types at an inbound speed of 30 m/s for a) grooved and b) un-grooved plates

However, the trend in backspin with loft angle is not monotonic; backspin reaches a peak before decreasing with increasing loft angle. Unlike the low loft angle range, the maxima in backspin and subsequent decrease with increasing loft angle are dependent on ball type and the presence of grooves. In the grooved condition, Fig. 1 (a), the backspin maximum is at a loft of around 50° for ball C with a small decrease noted as loft angle increases from 50° to 60°, but at higher loft angles (around 60°) for both balls A and D. In the un-grooved condition, ball C again shows a backspin maximum at around 50°, although the decrease in backspin at higher loft angles is greater; ball D still shows a maximum at around 60°, although at a lower backspin rate compared to the grooved condition. Ball A, however, shows a marked decrease in the loft angle to around 50° without grooves. In general, above the maximum backspin loft angle, backspin rates are greater in the grooved rather than un-grooved conditions.

For a rigid ball obliquely impacting a smooth surface, then the ball would be expected to slide up the face during impact. In this study a viscoelastic ball is impacting a roughened face and so it can be envisaged that the cover will deform around the surface features of the plate to increase friction and lead to a transition from sliding to rolling. The degree of sliding and rolling in oblique impacts can be analysed on the basis of spin ratios (Eq. 2, derived from Goodwill and Haake 2004), which have been calculated, this equation is based on rigid body analysis and therefore may be limited to lower impact speeds and harder balls, due to a lower deformation;

$$\omega_{ROLL(f)} = V_x / R, \tag{2}$$

where, $\omega_{ROLL(f)}$, is the rolling spin for the ball leaving the surface, R, is the radius of the golf ball, and V_x, is the rebound velocity parallel to the plate.

If both sides of Eq. 2 are numerically equal, then the rolling spin of the ball is equal to the measured spin and it can be deduced that the ball is in a state of pure rolling. If the rolling spin is greater (ratio > 1), then excessive rolling occurs during the impact. On the other hand if rolling spin is lower (ratio < 1) then partial sliding occurs during the impact. These ratios do not define the amount of sliding or rolling that occurs during the impact, however they are a good reference for determining a relationship between rolling and spin for each ball type. Whilst μ_{Eff} characterises the overall impact and implicitly incorporates the viscoelastic nature of the ball deformation, this is not strictly the case for spin ratio. However, spin ratio does involve speed of the ball off the inclined face and, for constant inbound speed, this will be affected by viscoelastic energy losses in the ball so that this may account for some of the differences between balls. Plotting spin ratio as a function of loft angle for the impacts against grooved and un-grooved plates, Fig. 2 (a) and (b), gives plots that are very similar to Fig. 1 (a) and (b). Comparison of Figs 1 and 2 reveals that, in the low loft angle range ball D gave higher spin ratios with the grooved plates than balls A and C, whilst, with un-grooved plates, ball A had a higher spin ratio than the other ball types. Ball D also appeared to give an abnormally high spin ratio around the backspin maximum loft angle in the grooved plate case, Fig. 2 (a), compared with its backspin rate, Fig. 1 (a).

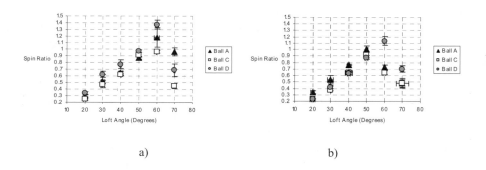

a)   b)

Fig. 2. Spin ratio as a function of loft angle for all ball types at an inbound speed of 30m/s for both a) grooved and b) un-grooved plates.

The variations in spin ratios shown by Fig. 2 for the different ball types would suggest that ball D spins more than slides in the grooved case, whilst the same is true for ball A in the un-grooved situation. Spin ratio, thus provides an analysis of the oblique impact situation and correlates well (with some scatter) with the previously-used μ_{Eff}, as shown by the relationships in Fig. 3. However, neither of these simple analyses deals with the detailed characteristics of the ball and face.

Considering the details of a golf ball's interaction with an inclined rough face, then, as the loft angle increases and F_N decreases, the degree of deformation of the cover decreases, which would be expected to reduce the interaction with the face features. This should be related to increased sliding with a consequent decrease in backspin; this is the case for loft angles above the backspin maximum.

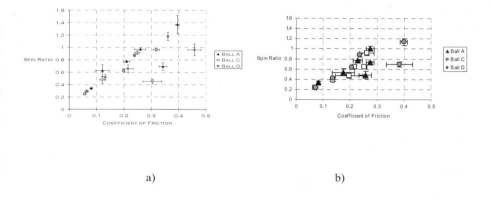

a)   b)

Fig. 3. Spin ratio against effective co-efficient of friction for all ball types at an inbound speed of 30 m/s impacting both a) grooved and b) un-grooved plates.

In terms of ball and plate characteristics then a ball with a softer cover should deform more readily and so the backspin maximum should be delayed to a higher loft angle than for a harder-covered ball. In addition, the higher local stresses associated with cover interaction with the edges of grooves would suggest that the backspin maximum occurred at higher loft angles in the presence of grooves than for an ungrooved face. These trends are approximately demonstrated, Fig. 1, in that, for grooved plates, ball C (hardest cover) gave a backspin maximum at the lowest loft angle, whilst ball A (softest cover) gave the highest backspin rate for the balls at a loft angle of 60° (around backspin maximum for balls A and D). For impacts with un-grooved plates then the differences between balls A and C are reduced with the backspin maximum occurring around the same loft angle. Thus, for the two-piece balls then the cover hardness seems to relate well to the degree of sliding and rolling in terms of backspin generation.

The analysis above holds less well for the three-piece ball D, which rolls more than balls A and C for impact with grooved plates (except at the highest loft angle), Fig. 2 (a), and has a backspin maximum at a higher loft angle than either ball A or C for impacts with un-grooved plates, Fig. 1 (b). Although ball D has an intermediate hardness cover hardness, this is thinner (0.9 mm) than those for the two-piece balls (1.4 and 1.8 mm) and overlies a very hard (67 Shore D) mantle. Hence, on impact, the cover is compressed between two hard bodies and so strain concentration occurs causing the cover to interact with the face features to a greater extent than for the covers of balls A and C resulting in higher spin ratios and μ_{Eff} values.

4 Conclusions and Further Work

Spin ratios and effective coefficient of friction values can be used to characterise backspin generation during oblique impact of multi-piece solid golf ball. Over the range of loft angles studies at a constant impact speed, all three ball types showed a backspin maximum at a loft angle, which depended on ball type and on the presence of grooves. The backspin maximum correlated with deformation of the cover into face features increasing effective coefficient of friction, which, for two-piece balls related to cover hardness, but the hardness of the underlying mantle was needed to rationalise the behaviour of the three-piece ball studied.

These studies need to be expanded to other impact speeds; groove types and finite element modelling to quantitatively account for backspin generation.

References

Chou, P. C., Gobush, W., Liang, D., & Yang, J. (1994). Contact forces, co-efficient of restitution, and spin rate of golf ball impact. Science and Golf II - Proceedings of the World Scientific Congress of Golf, St Andrews, Scotland, E & FN Spon, 296-301.

Goodwill, S. R. & Haake, S.J. (2003). "Ball Spin Generation for Oblique Impacts with a Tennis Racket." Experimental Mechanics **44**(2): 195-206.

Johnson A.D.G. (2006). The Effect of Golf Ball Construction on Normal Impact Behaviour. Ph.D Thesis, The University of Birmingham.

Maw, N., Barber, J.R., & Fawcett, J.N. (1975). "The Oblique Impact of Elastic Spheres." Wear. 101(7), 101-114.

Maw, N., Barber, J.R., & Fawcett, J.N. (1981). "The Role of Elastic Tangential Compliance in Oblique Impact." Journal of Lubrication Technology. 103, 74-80.

Monk, S.A. (2006). The Role of Friction Coefficient on Launch Conditions in High-Loft Angle Golf Clubs. Ph.D thesis, The University of Birmingham.

Sullivan, M. J., & Melvin, T. (1994). The Relationship Between Golf Ball Construction and Performance. Science and Golf II - The Proceedings of the World Scientific Congress of Golf, St Andrews, Scotland, E & FN Spon, 334-339.

Player Fitting of Golf Equipment Using a Calibration Club

Tom Mase[1], Mark Timms[2] and Cory West[2]

[1] Michigan State University, tmase@egr.msu.edu
[2] Hot Stix Technologies

Abstract. As launch and flight monitors become more prevalent in retail and research locations the need for fitting golf equipment has increased. In this paper, the fitting of golf clubs is discussed using a calibration club from which performance of other clubs is predicted. In order to do this, extensive testing, both robot and player, was completed building a database of club responses as a function of ball speed, backspin, and launch angle. From the trajectory data the best fit club for a player was determined. For drivers, the fitting simply consists of maximizing the length of the ball's flight and roll.

1 Introduction

In the early 1990s, golf equipment was going through a renaissance with material advances, increased original equipment manufacturers (OEM) research, and development efforts. Much of the R\&D was focused on the equipment itself with the thinking that better equipment benefits all golfers. Before metals arrived on the scene, there was scientific study of equipment fitting, although it may have not been recognized as fitting. After working on the perfect model of the golf swing, Cochran and Stobbs (1968) explored the possibilities for design of equipment. Even though it was contrary to their search, the research team found individual golfers exhibit patterns characteristic of themselves and different from those of players having very similar swings. Back when metals were making the transition from steel to titanium, there was a beginning of fitting. Pelz (1990) conducted testing on different shafts hit by U.S. PGA Tour professionals and was able to conclude what shafts were better for short, medium, and long hitters.

Currently, there are so many outstanding golf club products it is worthwhile to pick and choose amongst these to match a golfer to a club and/or ball. Making the best choice and customizing the choice has been limited to the big OEMs until recently. Before the early 2000s, the only launch monitors were in companies like Acushnet (Gobush, Pelletier and Days, 1994). These were research devices utilizing photogrammetry made for the development of product rather than fitting the right equipment for a specific player. Without launch monitors, non-OEM research on fitting required testing by taking the distance a ball traveled. For instance, high handicap golfers benefit in 3-wood and 5-iron distance when using 2-piece balls over other types of balls (Hale, Bunyan and Sewell 1994). However, it was somewhat of an arduous task to come to this conclusion. Winfield (1999) took advantage of Acushnet's in-house launch monitor \cite{launch:gobush94} to conduct real ball and club fitting that measured launch conditions and computed flight based on trajectory models (Aoyama 1990, Smits and Smith 1994).

Because only a few companies had the ability to rapidly evaluate a golf shot's launch conditions, fitting of golf equipment was restricted to R\&D departments. But starting in 2002 commercial launch monitors became available commercially launching a golf equipment fitting frenzy.

2 Launch Monitors

The reason golf club fitting is becoming widespread in recent years is due to affordable, commercial launch monitors. These instruments utilize photogrammetry and Doppler radar and range in price from several hundred to tens of thousands of US dollars. Economies of scale and competition are reducing the price allowing for wider availability at point-of-sale, professional fitters, and professionals.

With the exception of the radar unit, initial conditions are captured: ball speed, launch angle, and backspin. Radar based instruments allow for trajectory measurement (velocity as a function of time) as well as spin. While initial conditions suffice for most fitting, it is preferable to have data from the entire trajectory.

As discussed in Winfield 1999, if one knows the launch conditions along with the lift and drag coefficients as a function of Reynolds number and spin ratio a reasonable path may be computed. Most launch monitors operate in this fashion: measure the initial conditions and calculate the path.

3 Optimum Launch

By taking an aerodynamic model for a ball (Smits and Smith 1994), it is easy to compute the carry distance for a variety of launch conditions. The carry distance computed was used as the objective function in an optimization algorithm to determine what backspin and launch angle a player should have for a given initial ball velocity. A simplex method for minimizing the carry distance as a function of backspin and launch angle converged quickly (<1,000 iterations) determining the Smits and Smith (1994) ball optimum launch (Fig. 1).

4 Driver Fitting

Driver fitting has the most obvious objective function for fitting: maximize distance. The choice that most all consumers make is to hit the ball further. Some high level

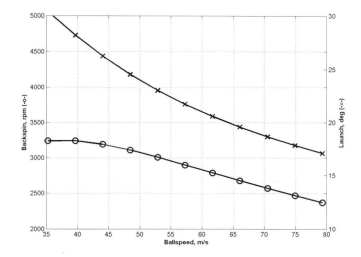

Fig. 1. Optimized launch Conditions for carry and distance over a range of ball velocities.

players will forfeit distance for a specific flight characteristic such as a lower flight. Fitting a player to the best driver may involve trying many products to obtain the best distance. Without a math based fitting algorithm the process may be a random "walk" through products hoping to find the right one before the customer fatigues.

Thus, the need for calibration club fitting which reduces the number of products tried. In this method, a player hits a calibration club and the launch conditions are measured with a launch monitor. Key to fitting with the calibration club method is having tested a broad variety of clubs for comparison. A database of many driver's launch conditions, tested on a robot, have been hit and analyzed. Included in the database are the launch conditions of the calibration driver. Each club model and loft has been tested over a wide range of club head speeds allowing the fitting of most all players. Based on the relative performance of the drivers searched in the database, a driver can be recommended to the customer.

Each driver model in the database was tested at nominal club head speeds of 35.7, 40.2, 44.7, 49.2, and 54.7 m/s. Ball speed was used as the independent variable since the launch monitor records ball speed. At all club head speeds the appropriate shaft was used for testing. Some club speeds may have more than one shaft flex to be tested. In general, two similar model driver heads tested at the same speed with different flex shafts produced similar launch conditions for transition club speeds. Figure 2 shows backspin data for a single club head model, multiple lofts, tested on a robot as a function of ball velocity. Ball speeds with double data points are the transition speeds, and the two data points represent a shot with two different shaft flexes. It is seen that the data for a given loft head blends well into a single data set.

Figure 2 show the backspin for a single model, 400 cc, driver over a range of club head speeds and club head lofts. In spite of the scatter of the data points, it is easy to see from linear fits to each loft that this driver creates a set of characteristic launch conditions. Linear regressions of the data as a function of velocity generate well defined characteristic lines offset as the loft changes. Note that the 8.5 degree loft driver data looks different indicating that it should be re-tested. There was most likely a setup problem in testing that particular driver.

Having characteristic curves like this is the starting point for the fitting. When a player hits these drivers the spin and launch conditions may not match the absolute numbers shown, but the relative differences remain in tact. It is assumed that the data shown in Fig. 2 for robot testing will remain linear and simply shift the club model's ordinate values. The player's launch conditions can be scaled onto the database or the database parameters can be scaled to the player.

Figure 3 shows player data superposed on the linear fits for the model considered. The error bars for the players shots indicate a standard deviation span. For each driver, 5 to 8 shots were recorded. If more shots were taken it is anticipated that the standard deviation would be significantly reduced. In addition, backspin is one of the most difficult launch parameters to measure with accuracy. It is seen that backspin from the drivers hit by the player reasonably match that of the robot. Some of the differences between spins is not quite the same as that of the robot, however, Table 1 shows what may cause this difference. For the robot data, several clubs were hit and averaged into a single line. The different clubs all had the same manufacturers loft, but individually the lofts vary. On average, the nominal lofts are slightly different than the published lofts due to manufacturing variation. However, when a player grabs a single club it can be different from the nominal. In Fig. 3 it appears that the best loft is between the 8.5 and 9.5 degree heads, however, the heads tested were 9.0 and 10.5 degrees even though they were labeled as 8.5 and 9.5 degree products. Thus, the player should most likely play with a 9.5 degree driver in this model.

Table 1. Driver properties for lofts tested.

Published Loft, deg	Nominal Loft, deg	Measured Loft, deg	CG Height mm	CG Depth mm	CG MOI g-cm^2
7.5	8	8.5	23.8	17.6	4030
8.5	9	9	22.8	16.3	4030
9.5	10	10.5	23.7	16.1	4101
10.5	11	11	23.5	17.2	4101
11.5	12	12	23.6	16.1	3960

Once the database parameters have been scaled for player hitting the calibration club, predicted launch conditions for each club in the database were computed. From this the database clubs were sorted to determine the best performing clubs for the player. Two sorting methods have been used to do this based on initial backspin or carry distance based on a trajectory code. If the aerodynamics parameters of the

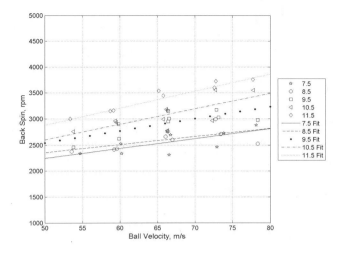

Fig. 2. Single model driver backspin as a function of ball velocity and loft.

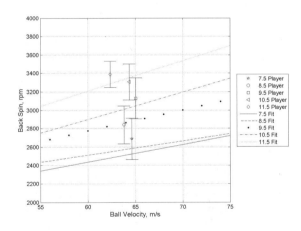

Fig. 3. Player test data superposed on driver spin lines.

ball are well known, the latter method is preferred. However, for wide scale fitting the ball used may vary. In this case the initial spin criterion works well.

Figure 4 shows a ranked comparison of 40 different loft and brand drivers compared to the 9.5 degree driver hit by the player (65.3 m/s ball speed, 2800 rpm backspin, 14 degree launch angle). There are 23 drivers that will give the player a slight improvement in distance over the calibration club. Low ranking drivers are those imparting too much spin on the ball. The distance for each driver was calculated using a simple trajectory code (Smits and Smith 1994) along with the USGA's ITR roll model.

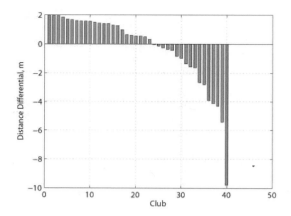

Fig. 4. Comparison of 40 driver models referred to a single 9.5 degree calibration driver.

5 Acknowledgments

The lead author would like to acknowledge the support of Hot Stix Technologies for parts of this work.

References

Aoyama, S.: 1990, A modern method for the measurement of aerodynamic lift and drag on golf balls, *in* A. Cochran (ed.), *Science and Golf: Proceedings of the First World Scientific Congress of Golf*, E. & F.N. Spon, London, UK, pp. 199--204.

Cochran, A. and Stobbs, J.: 1968, *The search for the perfect swing*, Triumph Books, Chicago, IL.

Gobush, W., Pelletier, D. and Days, C.: 1994, Video monitoring system to measure initial launch characteristics of golf ball, *in* A. Cochran and M. Farrally (eds), *Science and Golf II: Proceedings of the 1994 World Scientific Congress of Golf*, E & FN Spon, London UK, pp. 325--333.

Hale, T., Bunyan, P. and Sewell, I.: 1994, Does it matter what ball you play?, *in* A. Cochran and M. Farrally (eds), *Science and Golf II: Proceedings of the 1994 World Scientific Congress of Golf*, E & FN Spon, London UK, pp. 362--368.

Pelz, D.: 1990, A simple, scientific, shaft test: steel versus graphite, *in* A. Cochran (ed.), *Science and Golf: Proceedings of the First World Scientific Congress of Golf*, E. & F.N. Spon, London, UK, pp. 264--269.

Smits, A. and Smith, D.: 1994, A new aerodynamic model of the ball in flight, *in* A. Cochran and M.Farrally (eds), *Science and Golf II: Proceedings of the 1994 World Scientific Congress of Golf*, E & FN Spon, London UK, pp. 340--347.

Winfield, D.C.: 1999, Golf club and ball fitting using launch condition measurements, *in* M. Farrally and A. Cochran (eds), *Science and Golf III: Proceedings of the 1998 World Scientific Congress of Golf*, Human Kinetics, Chicago, IL, USA, pp. 548--553.

Evaluation of Long and Short Shafts of Golf Club by Real Swing

Takuzo Iwatsubo and Daiki Nakajima

Faculty of Engineering, Kansai University, Japan, iwatsubo@ipcku.kansai-u.ac.jp

Abstract. This paper proposes an evaluation method of long and short shaft of golf club by real swing. Head speed of the club, stability of direction of a ball, reliability of impact at the sweet spot and feeling of the club under the swing are used for the evaluation. First the swing is analysed by four link model and head speed and torque under the swing are calculated by Newton-Euler's method. Then difference of swing by short and long shaft club is discussed for high and middle level golfer. Next stability of direction and reliability of impact at the sweet spot are measured and these mean values and distributions are obtained. Questionnaire of feeling of the club of swing are performed and these data are statistically analyzed and obtained principal component and coefficient of correlation of each item. From these data, the long and short shafts of golf club are evaluated by evaluation function which would be decided by players requirement.

1 Introduction

It is a great glamorous for the golf players to drive a ball long distance and accurate direction. Research on the high performance golf club is divided into two subjects one is on club head which have high restitution coefficient and stable fly direction of ball in spite of miss shot, and the other is on club shaft which have good feeling for swing, length of shaft, light weight shaft, torsion and stiffness of shaft and so on.

With regard to the club shaft, a relation between shaft length and swing of player on stability of golf club impact at the face center and the feeling of golf club and so on is studied[1][5][6].

However these research are limited to the special items and these items may have trade off relation even if it is desired to satisfy the optimum condition of these items at the same time. And as the golf player is very sensitive for his club, so feeling of the club under the swing and impact is very important. From this reason synthetic evaluation of golf club is needed in order to design an optimum golf club.

This paper proposes a synthetic evaluation method of shaft length of a golf club by evaluating a driving length, reliability of direction of hit ball and feeling of swing and impact, in order to apply a development of golf club and to find optimum golf club for individual players.

2 Analysis of down swing motion by using 4 link model

2.1 Measurement of swing motion

4 link model is used in order to analyze a down swing motion as shown in Fig.1., where angle α is the angle between the swing plane and horizontal plane.

In order to measure the down swing motion, image of movement of joint is taken by two high speed cameras, which flame speed is 250 (flame/s). The movements of joint are digitalized and direct liner transformation (DLT) method is used to obtain the real swing motion. Tables1 shows data of players and golf club, respectively. The weight of club head and shaft stiffness are same and shaft length of the club is changed.

With regard to force and moment of each joint are obtained by using Newton- Euler method and the measured data of each joint.

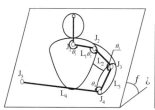

Fig.1. Four link model

Table1 Data of clubs

Club length (in)	Club weight (kg)	Shaft Stiffness	Location of C.G.(m)	Inertia moment at C.G (kg·m²)* 10⁻²
42	0.2837		0.821	4.56
45	0.2853		0.884	5.25
48	0.2839		0.898	6.55
51	0.2842		0.946	7.72

2.2 Results and discussion of swing motion

As a relation between driving range and head speed is close, driving range can be evaluated by head speed. Fig.2 shows mean of the head speed of the test players for 4 different drive shafts. It is known from this figure that the driving range increases for increase of shaft length and the inclinations for all players are almost same. This means head speed increase for the increase of the shaft length. Fig.3 shows joint torque of wrist, as an example, which is obtained by New-ton – Euler method, where the origin of time and the beginning of the swing are collision to the ball. It is known from this figure that the joint torque increases gradually from the begin-ning of the swing motion but it decreases and becomes negative just before the impact. So it can be used for the evaluation of head speed.

Next in order to evaluate the integral of positive and negative joint torques of each joint are calculated and shown from the beginning of the down swing to the impact in Figs. 4 and 5. These values are average of the torques for 4 different shaft length. The integral value indi-cates the amount of the joint torque. These figures show integral of positive torque of neck, shoulder, elbow and wrist, where the amount of torque of elbow is small because the elbow is usually kept straight. There is not so good relation between the integral values and the head speed of the club for the integration of positive torque. But head speed becomes large for the large negative integrated torque value. This is because rotational speed of the top link becomes large for increase of negative joint angle of the wrist as known from a knowledge of double pendulum motion. It is known with regard to the shaft length that the amount of integral value of the positive torque becomes large for increase of shaft length and that the large torque is needed for swing of the long shaft. The amount of the integral value of negative torque be-comes maximum for 45 inch club. This is because 45 inch club is popular for test players and they can swing very efficiently. But head speed of the 51 inch is the highest in four clubs in real data. This data shows head speed depend on both the negative torque and shaft length.

3.Evaluation of dispersion of hit ball direction

Fig.6. shows a target area of a driving range. Test players shown in Table 1 shot a ball to the target three times for 4 different length clubs and probability of success through the target is measured. Fig.7. shows the probability of success for shaft length. It is known from this figure that the probability of success become worse for increase of shaft length. This is because test player uses usually 45 inch club and he can not adapt 51 inch club.

4.Evaluation of meet rate to the center of face

The meet rate to the center of face is defined as the mean and variance of length between center of club face and center of ball trace obtained by shot mark sensor as shown in Fig.8. Table 2 shows the mean and variance of the length. The meet rate of 45in shaft is the best for

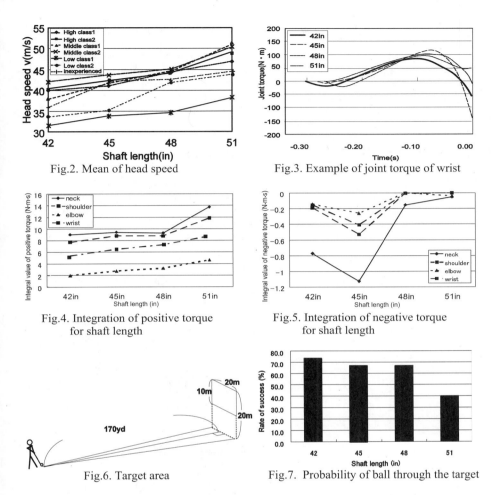

Fig.2. Mean of head speed

Fig.3. Example of joint torque of wrist

Fig.4. Integration of positive torque
for shaft length

Fig.5. Integration of negative torque
for shaft length

Fig.6. Target area

Fig.7. Probability of ball through the target

mean and variance and that of 51in is the worst. This result is the same as that of disperse of
hit ball direction.

5. Evaluation of feeling of shaft length and club

In order to evaluate the feeling of shaft length and club, questionnaire is performed for test
players. From this data, mean of each item, coefficient matrix of correlation and principal
component are calculated. Items of the questionnaire are written in Tables 4 and 6 and answer
is obtained by 5 grade from -2 to +2point followed by SD method.
Table 3 shows mean of each item of questionnaire and Table 4 shows coefficient matrix of
correlation in order to investigate a relation between principal component an other items.
Table 5 shows eigen value, contribution rate and cumulative contribution rate of the principal
component. Condition of significant principal component is decided from the statistics, thus
the first two eigen values can be adopted, that is, the first principal component is stiffness of a
shaft and the second one is the sound of club head at the impact.
Next, eigenvector can be obtained from the eigenvalue. Then point of principal component for
each shaft length can be obtained by using eigenvector and point of questionnaire, which is

shown in Table 6. It is known from the table that good evaluation for the shaft stiffness is 45in>48in>42in>51in and for the sound of club head at the impact is 48in>51in>45in>42in.The components to contribute to the shaft stiffness have strong relation with stability of ball direction, feeling of swing, feeling of follow through, feeling of wrist control and feeling for charge. Therefore shaft stiffness is effected by the swing. On the other hand, the components to contribute to the impact sound have strong relation with the driving range, feeling of head speed for swing. Therefore the impact sound is effected by the head speed and driving range.

6. Synthetic evaluation of shaft length of driver
6.1 Evaluation method

Head speed(x_1), stability of ball direction(x_2), meet rate to the sweet spot (x_3), shaft stiffness(x_4) and impact sound(x_5) with regard to the questionnaire of feeling of shaft length are obtained in the above chapters. Point for 4 kind of club shaft is calculated for each evaluation items x_1 - x_5 and weighting functions of each evaluation item are decided for a proper player. Then synthetic evaluation function S is calculated as follows;

$$S = \sum_{i=1}^{5} a_i x_i \qquad (1)$$

where a_i is a weighting function of each item. Best fitted shaft length for the proper player is decided from the maximum value of the synthetic evaluation function.

Table 2 Mean and variance of impact point

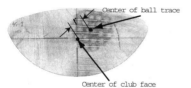

Center of ball trace

Center of club face

Fig.8. Distribution of impact points

(in)	mean	varianc
42	8.73	54.13
45	2.91	30.32
48	3.19	41.8
51	13.49	150.7

Table3   Mean of each item

Anerage data	1.driving range	2.stability of ball direction at	3.raising of ball	4.feeling for swing	5.feeling for follow through	6.feeling of wrist control	7.feeling of heavy club
42in	-1.25	0.67	-0.08	0.83	0.42	0.67	-1.00
45in	0.42	1.25	0.33	1.17	0.92	0.58	0.50
48in	1.25	-0.25	0.50	-0.17	0.33	-0.08	0.92
51in	-0.33	-1.67	-0.25	-1.67	-1.25	-1.00	-0.67

	8.feeling for change	9.shaft stiffness	10.good feeling at the meet	11.feeling of head speed for	12.feeling at the impact	13.impact sound 1	14.impact sound 2
42in	0.42	0.17	-0.50	-1.50	-0.67	-0.92	-0.42
45in	1.50	0.58	0.92	-0.08	0.75	0.17	0.00
48in	-0.17	0.17	0.50	1.17	0.75	0.25	0.75
51in	-1.33	-1.08	-0.50	0.67	-0.58	0.25	0.17

	1	2	3	4	5	6	7	8	9	10	11	12	13	14
1	1	-0.02	0.84	-0.05	0.23	-0.15	0.97	0.09	0.26	0.78	0.83	0.89	0.79	0.90
2	-0.02	1	0.49	1.00	0.96	0.97	0.24	0.98	0.95	0.56	-0.57	0.42	-0.41	-0.40
3	0.84	0.49	1	0.47	0.71	0.40	0.94	0.53	0.73	0.89	0.40	0.95	0.38	0.60
4	-0.05	1.00	0.47	1	0.96	0.99	0.21	0.96	0.95	0.51	-0.60	0.38	-0.47	-0.42
5	0.23	0.96	0.71	0.96	1	0.93	0.47	0.94	1.00	0.68	-0.36	0.61	-0.27	-0.14
6	-0.15	0.97	0.40	0.99	0.93	1	0.11	0.91	0.91	0.39	-0.68	0.27	-0.59	-0.47
7	0.97	0.24	0.94	0.21	0.47	0.11	1	0.34	0.50	0.90	0.66	0.98	0.67	0.77
8	0.09	0.98	0.53	0.96	0.94	0.91	0.34	1	0.93	0.67	-0.45	0.52	-0.24	-0.33
9	0.26	0.95	0.73	0.95	1.00	0.91	0.50	0.93	1	0.70	-0.32	0.64	-0.24	-0.10
10	0.78	0.56	0.89	0.51	0.68	0.39	0.90	0.67	0.70	1	0.36	0.97	0.52	0.43
11	0.83	-0.57	0.40	-0.60	-0.36	-0.68	0.66	-0.45	-0.32	0.36	1	0.51	0.92	0.94
12	0.89	0.42	0.95	0.38	0.61	0.27	0.98	0.52	0.64	0.97	0.51	1	0.58	0.62
13	0.79	-0.41	0.38	-0.47	-0.27	-0.59	0.67	-0.24	-0.24	0.52	0.92	0.58	1	0.78
14	0.90	-0.40	0.60	-0.42	-0.14	-0.47	0.77	-0.33	-0.10	0.43	0.94	0.62	0.78	1

Table4   Coefficient matrix of correlation

		Eigen value	Contribution rate	Cumulative contribution rate (%)
1	shaft stiffness	7.59	0.54	54
2	impact sound 2	5.96	0.43	97
3	impact sound 1	0.44	0.03	100
4	feeling at the impact	1.95E-10	1.39E-11	100
5	good feeling at the meet	1.49E-10	1.06E-11	100
6	feeling for follow through	1.60E-11	1.14E-12	100
7	feeling of head speed for swing	1.49E-14	1.07E-15	100
8	driving range	1.36E-15	9.71E-17	100
9	stability of ball direction at the driving range	3.33E-16	2.38E-17	100
10	feeling of heavy club	3.05E-16	2.18E-17	100
11	feeling for swing	2.78E-16	1.98E-17	100
12	feeling for change	-1.67E-16	-1.19E-17	100
13	raising of ball	-5.27E-16	-3.77E-17	100
14	feeling of wrist control	-1.55E-15	-1.11E-16	100

Table5 Principal component

	42in	45in	48in	51in
Shaft stiffness	-0.02	2.58	0.80	-3.36
Impact sound	-2.79	-0.39	1.96	1.22

Table6 Point of questionnaire

6.2 Calculation method of a point

The point is decided from +2 to –2 for each evaluation item. Data which is used for the evaluation is the results obtained from higher than middle class players whose hand cap is less than 25. The point of the head speed is calculated as follows; As minimum and maximum head speed are 27.93m/s and 51.31m/s, respectively. These values are decided as minimum point(-2) and maximum point (+2), respectively. Then other data is decided linearly and means are obtained for each shaft length. The result is shown in Table 7. The point of stability of ball direction at the driving is calculated as follows; as the test player hits a ball 3 times for each golf club and minimum and maximum probability passing the target are decided to 0 to 100%, the point is –2 to +2 from this data by taking the mean. The result is shown in Table 8. The point of meet rate to the center of face is calculated as follows; as the meet rate has mean length from the center of the face to the center of ball trace and its variance, so the point is divided to these two items each 50%. As maximum and minimum lengths are 33.25mm and 0mm so the point is calculated by counting the measured data and these values are taken mean for each shaft length and taken point from –1 to +1. With regard to the variance, as maximum and minimum values are 163.47 and 28.34, respectively, these values are pointed linearly from –1 to +1. The results are obtained by adding these two factors as shown in Table 9.

	42in	45in	48in	51in
Head speed (m/s)	40.22	42.25	44.46	49.38
Point	-1.31	-0.72	-0.03	1.43

Table7 Point of head speed

	42in	45in	48in	51in
Probability (%	73.33	66.67	66.67	40.00
Point	1.46	1.33	1.33	-0.80

Table8 Point of flying direction of ball

	42in	45in	48in	51in
Length of mean (mm)	8.72	2.91	3.20	13.49
Point of mean	0.48	0.82	0.81	0.19
Variance	52.12	30.25	41.43	150.12
Point of variance	0.46	0.68	0.57	-0.57
Point of meat rate	0.94	1.5	1.38	-0.38

Table9 Point of nice meet at the center

6.3 Example of synthetic evaluation

The synthetic evaluation function is shown in Eq.1 and x_i (i=1-5) is calculated in the previous section. If a_i is decided for arbitrary player, S can be calculated for each club of shaft length. If the player decides weighting functions as follows; Feeling of impact sound $a_1 = 1.1$, Shaft stiffness $a_2 = 1.2$, Meet rate $a_3 = 1.4$, Stability of direction $a_4 = 1.4$, Head speed $a_5 = 1.0$, then the value of synthetic evaluation function can be calculated as; S_{42in}=0.5, S_{45in}=4.46, S_{48in}=5.32, S_{51in}=-1.57 From the result it is known under these weighting function, 48in club shaft is best for the player from the synthetic evaluation of the shaft length.

Finally, the evaluation function is selected for shaft length in this research, but if shaft stiffness, center of gravity of the club or other parameters are selected as an object, optimum club for a player can be selected by using similar method.

7. Conclusion

Evaluation method of shaft length which satisfy a player's request is proposed by using synthetic evaluation which consist of a driving range, stability of ball direction at the impact, feeling of club and shaft and meet rate to the center of face.

Driving range is evaluated by using head speed and joint torque, where DLT method and Newton – Euler method are used in order to obtain a player's real data. It is known that head speed is proportional to the shaft length. But meet rate to the center of face and dispersion of hit ball direction becomes worse for longer shaft. Evaluation of feeling of shaft length is performed by using questionnaire of 14 items, principal component analysis and correlation analysis. It is known from this analysis that the predominant and significant principal components are shaft stiffness and impact sound. Shaft stiffness is closely related with stability of ball direction, feeling of swing, feeling for follow through, feeling of wrist control and feeling for charge. The impact sound is closely related with driving range and feeling of head speed for swing.

Finally, synthetic evaluation function is defined as a sum of each weighted evaluation function and synthetic evaluation is done. As an example, weighting function is decided and the most adequate club length is given for player who gives weighting functions.

References

Hanjyo, M. Science of Golf- living body information science teach us, *Kodansha*.

Iwatsubo, T., et. al., (1990) Optimum design of golf club. 56-524,C, pp1053-1059, *JSME*.

Iwatsubo, T., et. al., (1993) Analysis of ball trace due to golf swing, 3rd conference of sports Engineering '93, Hachioji, pp27-34

Iwatsubo, T., et. al., (1999) Investigation of sweet area of golf club. Dynamics and Design Conference '99. No99-7

Jorgensen, P. T., Science of Golf. *Maruzen books*

Mita, T. and Osuka k., Introduction to a Robot Control Engineering. *Koronasya*

Mizoguchi, M., Hashiba T. and Yonezawa T. ,(2001) Effects of shaft length on golf swing motion. Symposium on Sports Engineering '01, Tokyo, pp. 102-105.

Muro j. and Ishimura S., Mean, variance, standard deviation. *Tokyo tosho*

Iron Golf Club Striking Characteristics for Male Elite Golfers

Alex J. McCloy[1], Eric S. Wallace[1] and Steve R. Otto[2]

[1] University of Ulster, UK, alexmccloy@hotmail.com
[2] R&A Rules Limited, UK

Abstract. Ball launch condition data and clubhead data were measured in order to gain an understanding of the striking characteristics associated with a group of elite golfers using different iron golf clubs. Ten right-handed male golfers were used as subjects (handicap -0.5 ± 1.7). The testing was carried out in a dedicated indoor golf facility. Each golfer hit eight shots with each of four of his own iron clubs (3-iron, 5-iron, 7-iron and pitching wedge). Launch conditions and clubhead data were measured using a stereoscopic high-speed camera system. The mean club head speeds for the group decreased and the mean clubhead angle of attack increased as the club became more lofted. The more lofted clubs produced a higher mean spin rate and higher mean launch angle. Whilst these findings are as expected, the study is novel in providing a scientific database for the competences with irons associated with this elite skill level. It also validates the choice of a single subject that may be used in the future construction of a simulation model designed to investigate iron club striking properties.

Introduction

The advent of state-of-the-art golf launch monitors permits the routine capture of detailed ball launch conditions, widely used for shots struck by a golf robot. Whilst the use of a robot provides high levels of reliability relating to the swing, the fact that golfers actually use the clubs and balls in the game situation requires an attempt to be made to analyse their clubhead kinematics and ball launch data, in an attempt to discriminate between different clubs, balls, and golfers' swings, as well as combinations of these factors. Launch conditions are defined as the golf ball's initial direction, spin and velocity, which together determine the impact phase of the subsequent flight of the ball. The majority of research to date, with golfers as subjects, has focused on ball launch conditions associated with driver golf clubs and driving performance in terms of the player's ability to generate clubhead speed (Egret, Vincent, Weber, Dujardin & Chollet 2003; Fradkin, Sherman & Finch 2004, Werner & Greig 2000). Whilst drivers are used predominantly off the tee with the player's main aim of propelling the golf ball a large distance with a degree of accuracy, iron clubs are predominantly used for approach shots to the green where the player's priority is in achieving accuracy, determined by optimum displacement of the golf ball from its starting position. There is a dearth of research data pertaining to iron golf clubs, yet an understanding of the ball launch conditions for these clubs is necessary if player swing characteristics and club properties are to be evaluated.

Furthermore, such evaluations may be carried out experimentally or through the application of simulation studies, with the latter requiring input from experimental data. The main purpose of the present study was to investigate the launch characteristics associated with a range of iron golf clubs for elite golfers. The specific objectives of the study were to provide baseline club head and launch data for elite golfers using 3-irons, 5-irons, 7-irons and pitching wedges and to select a representative golfer from this elite group whose data may be later used in simulation studies.

Methods

Ten male, right-hand dominant subjects (22.8 ± 3.1 yr, 1.83 ± 0.42 m, 80.6 ± 9.1 kg, and -0.5 ± 1.7 handicap) volunteered for the study. All were considered elite players, with nine subjects representing highly skilled handicap amateur golfers who had competed regularly in senior golf for their clubs and universities. The other subject was a former senior international amateur with a playing handicap of plus two when he recently turned professional. This skill level for the group was chosen in an attempt to minimise the effects associated with wide ranging ball striking abilities and to attenuate any possible effects associated with learning during the testing sessions. All subjects were injury-free at the time of testing and they all agreed to participate in the test by providing written informed consent as per the University's ethical requirements for human subject testing. The tests took place in an indoor biomechanics laboratory at the University, while the subjects stood on an artificial practice mat with shots struck into a purpose-erected indoor net. This environment minimised the interference effects associated with a number of external factors (for example, wind speed/direction, temperature and sun glare, ground conditions), thus ensuring test consistency, (Fradkin 2004 et al). The indoor testing also prevented the subjects from obtaining feedback in relation to shot outcome.

All subjects were instructed to bring along their own three-iron, five-iron, seven-iron and pitching wedge, with standard loft values of approximately 22 degrees, 28 degrees, 36 degrees, and 48 degrees, respectively. The subjects also brought their own golf shoes and golf glove, if required. A dozen premium golf balls were provided for each subject to do the test. Balls were marked with a black line around the circumference, which allowed spin rates to be obtained. Prior to testing, subjects were allowed to follow their own preferred warm-up routines, as they would for a competitive round of golf. This also allowed the subjects to familiarise themselves with the playing surface, the short (0.015m) rubber tees being used, and the lighting in the facility. Using the known frequency of the cameras and the distance between successive ball images it was possible to calculate the club head and ball launch conditions. Two reflective markers, one on the toe and the other on the heel, were placed on each clubhead. These markers enabled the clubhead to be identified by the software and their displacements recorded with a known time base to yield velocity prior to and during impact with the golf ball. Four images of the clubhead were captured prior to impact. The first image was taken when the club head triggered a laser beam behind the ball/tee position. Clubhead velocity was reported as the average of the values for the two markers over the distance of 0.08m prior to impact.

The order of club usage was three-iron, five-iron, seven-iron and pitching wedge for all subjects and each subject performed eight shots consecutively with the same club. Subjects were requested to execute each shot as close to actual playing conditions as possible. Data were stored for subsequent analysis if both the subject and the experimenter agreed that the shot was deemed to be representative of the player's ability with any given club.

Results

The key variables selected for analysis were the group means and standard deviations (SD's) for pre-impact velocity and angle of attack for the clubhead (defined as the angle between the club path and the horizontal), and the immediate post-impact ball launch variables of velocity, angle and spin rate (Table 1).

As the clubs become more lofted, that is from 3-iron – pitching wedge, the angle of attack increased, with the negative sign indicating the downward movement of the clubhead. The clubhead velocity decreased with increasing loft, as did the ball velocity. The Velocity Ratio, given as ball velocity divided by clubhead velocity, as expected follows a similar trend. Launch angle and spin rate both increased with increasing loft. The lowest standard deviations – relative to the mean values –were noted for club head velocity and ball velocity. Greater relative variances were noted for launch angle and spin rate, which were similar in magnitude, with the largest relative variance noted for the clubhead angle of attack.

Table 1. Clubhead and ball launch condition variables for the different iron golf clubs.

	3 iron	5 iron	7 iron	Pitching Wedge
Clubhead Angle of Attack (deg)	-5.9 ± 1.7	-6.7 ± 1.5	-8.1 ± 1.3	-9.4 ± 1.4
Clubhead Velocity (m/s)	41.6 ± 1.7	40.4 ± 1.6	38.7 ± 1.7	36.2 ± 2.2
Ball Velocity (m/s)	59.1 ± 2.8	56.0 ± 2.4	51.4 ± 2.3	42.3 ± 2.7
Velocity Ratio	1.42 ± 0.05	1.39 ± 0.04	1.33 ± 0.04	1.17 ± 0.06
Launch Angle (deg)	13.7 ± 1.9	16.7 ± 1.9	20.4 ± 1.7	25.9 ± 2.9
Spin Rate (rpm)	4214 ± 566	5243 ± 511	6844 ± 680	9536 ± 1048

Scatterplots representing a total of 320 data points for all shots and all subjects were produced to illustrate the relationships between pairs of variables, and to locate the position of the selected single subject's data in relation to the group data in order to determine the validity of this subject as a representative of this elite golfer group in future planned modelling studies (Figs. 1 & 2). In each plot, the single subject's data are identified by the iron number and the white data point. The single subject's data points are considered to be adequately contained in the groupings for each iron for all variable comparisons.

Figure 1. Scatterplot of all data points for Backspin v Launch Angle (r=0.93)

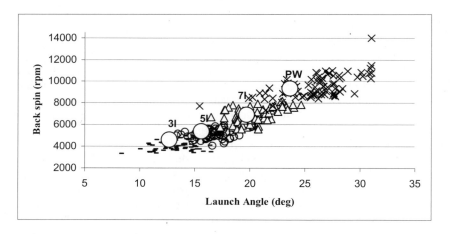

Figure 2. Scatterplot of all data points for Backspin v Ball Velocity (r=-0.85)

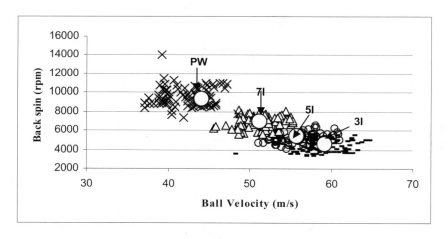

Discussion

The objectives of the present study were to provide baseline clubhead and launch data for elite golfers using 3-irons, 5-irons, 7-irons and pitching wedges and to select a representative golfer from this elite group whose data may be later used in simulation studies. Little previous published scientific research has focused on launch conditions associated with irons, possibly because there is a wide range of lofts offered by the range of irons available to the golfer and thus there is no perceived need to study this aspect, and furthermore there is generally no desire by golfers to maximise either clubhead velocity or ball velocity when using irons. Instead, the golfer can select an iron club from a range of separate clubs in an attempt to achieve the desired shot outcome. This is in stark contrast to the driver, which has been the subject of many launch conditions studies, where generally only one club is carried by the golfer and the intention is to achieve optimum launch conditions to maximise distance. Yet it is widely accepted in the golfing world that there is a wide range of ball striking ability levels by golfers using irons across different skill levels. Furthermore, with the growing practice of clubfitting where the objective is to match player characteristics and equipment properties, irons are often used – particularly the 5-iron and 7-iron. Maltby (1995) identifies 11 variables that may be applied to the clubfitting process. Of these variables, lie angle, shaft flex, club head design, club length, swingweight and total weight may be considered to be the most important in relation to iron clubs. In today's clubfitting processes, these variables may be readily altered, yet there is no known scientific database to provide normative data. In the present study, no attempt was made to impose standardised clubs with known properties on these players – instead the intention was to provide a database for elite golfers for these 4 iron clubs, when using their own clubs with which they were very familiar. As such, many expected findings were realised, whilst some interesting findings were observed.

The noted negative angles of clubhead attack for all irons clearly shows that elite players strike the ball with a descending club head up to the moment of impact, associated with a 'clean strike' of the ball. The increase in angle of attack associated with the higher lofted clubs is as expected, and is due to the steeper swing plane with these shorter clubs. The results for clubhead velocity are also as expected, in that while no attempt is generally made by elite players to consciously swing faster, the longer clubs nonetheless provide greater club head velocity due to, at least in part, the physics of their length advantage for a given angular velocity (swing speed). The comparison of the club head velocities for the 5-iron and pitching wedge show the 5-iron to be 1.115 times faster which is in broad agreement with Egret et al (2003) who reported the five iron speed as 1.1 times faster than the pitching wedge speed for a similar skill level (mean (SD) handicap 0.4 ± 1.1) of golfers. The noted decrease in ball speed with increasing loft angle, in turn associated with concomitant increases in launch angle and spin rate, is accounted for by the more glancing nature of the impact with the higher lofted clubs. The impact conditions associated with greater loft reduce the component of the force delivered by the club head along the ball launch direction. As the loft of the club increases the force acting in the tangential direction increases, while the normal force decreases. The net effect of these dynamics is manifested in the relatively larger gains in ball velocity compared to clubhead velocity, with an increase noted in ball velocity for the 5-iron compared to the pitching wedge representing a 32.3 % gain, compared to a gain of 11.5 % in clubhead velocity for these clubs. This is further illustrated by the Velocity Ratio, shown to decrease from a value of 1.39 for the 5-iron to a value of 1.17 for the pitching wedge. The Golftek[TM] system (www.golftek.com, Accessed 12th January, 2006), which refers to the Velocity Ratio as the 'Solid Hit Factor', suggests a typical value of 1.3 for 5-irons, with a value of 1.5 widely reported for drivers. The somewhat higher 5-iron mean value of 1.39 recorded for this subject group may be due to their better ball striking abilities. The mean (SD) launch angle of 16.7 (1.9) deg for the 5-iron in the present study falls within the range of 16 ° to 20 ° suggested by Werner & Greig (2000) as the optimum launch angle for a 5-iron. Chou, Gilbert & Olsavsky (1994) using a mechanical golfer, reported that as the loft of an iron golf club increased there was a linear increase in both launch angle and back spin (a 4 degree loft increase resulted in a 3 degree increase in launch angle and a 1100 rpm increase in spin). The 4 degree increase in loft also produced a 3.6 mph (1.6m/s) drop in ball speed. Whilst these results do not compare directly with those in the present study, the trends and orders of magnitude are similar.

In conclusion, the present study provides an original database of clubhead and ball launch conditions data for expert golfers using a range of iron clubs. Such data may be used for future empirical studies on comparisons with different golfer skill levels or variations in iron club properties. In addition, an examination of individual data has permitted the identification of a single subject considered to be representative of this elite skill level. Thus, the application of this subject's data in future modelling studies overcomes some of the selection bias assumptions, as discussed in the literature (for example, Reboussin & Morgan 1996), associated with single-subject design.

Acknowledgements

We would like to express our thanks to the R & A Rules Limited for their valuable contributions to the study.

References

Chou A., Gilbert P. & Olsavsky T. (1994) Club head designs: How they affect ball flight. In: A. Cochran (Ed), *Golf the Scientific Way* pp. 15 – 25.

Egret C.I., Vincent O., Weber J., Dujardin F.H. & Chollet D. (2003) Analysis of 3D kinematics concerning three different clubs in golf swing. International Journal of Sports Medicine, 24, 465 – 469.

Fradkin A.J., Sherman C.A. & Finch C.F. (2004) How well does club head speed correlate with golf handicaps? J. Sci. Med. Sport 2004; 7:4:465 – 472

Golftek (www.golftek.com, Accessed 12th January, 2006)

Maltby R.D. (1995) *Golf club design, fitting, alteration and repair: the principles and procedures*, Maltby Enterprises; 4th Rev. Edn.

Reboussin D.M. & Morgan T.M. (1996) Statistical considerations in the use and analysis of single – subject designs. Med. Sci. Sports Exerc. Vol. 28, No. 5, pp. 639 – 644.

Werner F.D. & Greig R. (2000), *How golf clubs really work and how to optimize their performance*, Jackson Hole, Wyo: Origin Inc.

A Golf Swing Robot Emulating Golfers Considering Dynamic Interactions Between Arms and Clubs

Chaochao Chen[1], Yoshio Inoue[2] and Kyoko Shibata[3]

[1] Kochi University of Technology, Intelligent mechnical systems engineering,
086403w@gs.kochi-tech.ac.jp
[2] Kochi University of Technology, Intelligent mechnical systems engineering,
inoue.yoshio@kochi-tech.ac.jp
[3] Kochi University of Technology, Intelligent mechnical systems engineering,
shibata.kyoko@kochi-tech.ac.jp

Abstract. An impedance control method, in view of the dynamic interactions between human arms and golf clubs, has been investigated for a golf swing robot. The control method has been separately implemented to a digital controller simulating a golf swing robot and to a prototype of golf swing robot. Both the results show that the swing motions of the golf swing robot with the impedance control method agree with those of different golfers.

1 Introduction

A large amount of research has been devoted to improve golfer's swing skills and golf club performance for decades. Among these studies, golf swing robots have also formed a large body of literature (Suzuki and Inooka 1999; Ming and Kajitani 2003). In these works, professional golfer's swing motion was expected to be emulated by robots and the evaluation of golf club performance was replaced by robots instead of golfers.

Though much progress has been achieved in this area, there still remains a long-standing challenge for a golf swing robot to accurately emulate the fast swing motions of professional golfers. It is noted that conventional golf swing robots on the market are usually controlled by the swing trajectories directly measured from professional golfers' swings. The swing motions of these robots, unfortunately, are not completely the same as those of advanced golfers, in that they do not involve the dynamic interactions featured by different characteristics of human arms and golf clubs. Suzuki et al. proposed a new golf swing robot model in 1999. In their model, the robot like advanced golfers, was able to utilize the interference forces resulting from the individual dynamic features of different golf clubs on the arms. However, the difference between golfer's arm and the robot's arm in mass (actually, moment of inertia of arm) was not considered in their research. This would result in distinct dynamic interactions between arms and clubs, and consequently different swing motions of golfers and the robot.

In our study, an impedance control method based on velocity instruction is proposed, in which the dynamic interactions between human arms and golf clubs are considered. A model of golf swing considering the shaft bending is given. A digital controller using the method is established to simulate the swing motion of a golf swing robot. A prototype of golf swing robot with the method is developed. The comparison of swing motions is carried out among golfers, the digital controller and golf swing robot.

2 Dynamic Modeling of Golf Swing

Dynamic model of golf swing is shown as Fig. 1. The rotations of the arm and golf club are assumed to occur in one plane inclined with an angle θ to the horizontal plane. The shaft is treated as an Euler-Bernoulli beam, in which the elastic modulus, inertia and cross sectional area are constant along the beam length. The damping of the shaft is neglected. Since the center of gravity of the club head is regarded as on the central axis of the shaft, only the bending flexibility of the shaft is considered.

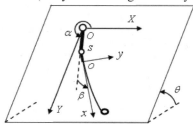

Fig. 1. Dynamic model of golf swing

Equation of motion of golf swing is derived by the Lagrange's method and assumed mode method, which can be written as Eq. 1.

$$B(\theta)\ddot{\theta} + h(\theta,\dot{\theta}) + g(\theta) + D\dot{\theta} = \tau \tag{1}$$

Where

$$B = \begin{bmatrix} B_{11} & B_{12} & B_{13} & B_{14} \\ B_{12} & B_{22} & B_{23} & B_{24} \\ B_{13} & B_{23} & B_{33} & B_{34} \\ B_{14} & B_{24} & B_{34} & B_{44} \end{bmatrix}, \theta = \begin{bmatrix} \alpha \\ \beta \\ q_1 \\ q_2 \end{bmatrix}, h = \begin{bmatrix} h_1 \\ h_2 \\ h_3 \\ h_4 \end{bmatrix}, g = \begin{bmatrix} g_1 \\ g_2 \\ g_3 \\ g_4 \end{bmatrix}, D = \begin{bmatrix} 0 & 0 & 0 & 0 \\ 0 & c & 0 & 0 \\ 0 & 0 & 0 & 0 \\ 0 & 0 & 0 & 0 \end{bmatrix}, \tau = \begin{bmatrix} \tau_1 \\ \tau_2 \\ 0 \\ 0 \end{bmatrix}$$

B is the inertia matrix; h is the non-linear force vector; g is the gravity vector and τ is the input vector; q_1 and q_2 are the first and second time-varying mode amplitudes of shaft elastic displacement, respectively; c is the wrist damping coefficient.

3 Impedance Control Design

The dynamical equation of a mechanical system is always expressed as Eq. 2.

$$M\ddot{x} + C\dot{x} + Kx = F \tag{2}$$

Where M, C and K are respectively denoted as inertia, viscosity and stiffness, which are called mechanical impedance (Hogan 1984). Here, the golfer's arm as a mechanical system is investigated. The virtual system representing the dynamic model of a golfer's arm and the robot system expressing the dynamic model of a robot's arm are shown in Fig.2 and Fig.3.

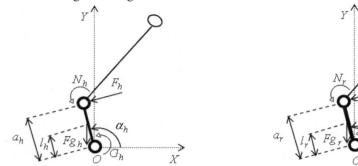

Fig. 2. The virtual system   **Fig. 3.** The robot system

Equations of motion of the virtual and robot systems are given by Eq. 3 and Eq. 4.

$$J_h\, \ddot{\alpha}_h = G_h + F_h\, a_h + fg_h(\alpha_h) + N_h \tag{3}$$

$$J_r\, \ddot{\alpha}_r = G_r + F_r\, a_r + fg_r(\alpha_r) + N_r \tag{4}$$

where the subscripts h and r denote the golfer and robot, respectively; $fg(\alpha) = Fgl\cos\alpha$, Fg is the gravity force of the arm; l is the distance between the center of gravity of the arm and joint O; G is the shoulder input torque; F is the reaction force from the club to arm, and the direction of this force is perpendicular to arm; N is the reaction torque from the club to arm; J is the moment of inertia of the arm about the joint O; a is the arm length, and $a_h = a_r$.

The dynamic parameters J_h and J_r in Eq. 3 and Eq. 4 are defined as the impedance. With the various arm masses for the golfer and robot, the impedance J_h and J_r are varied. Consequently, the swing motion of the robot is not the same as that of the golfer, even if the input torques of the shoulder joints are equal. According to the Euler method, accelerations $\ddot{\alpha}_h$ and $\ddot{\alpha}_r$ are approximated to the following expressions.

$$\ddot{\alpha}_h^n = \frac{\dot{\alpha}_h^n - \dot{\alpha}_h^{n-1}}{\varDelta t}\ .\ \ n = 1,\ldots N. \tag{5}$$

$$\ddot{\alpha}_r^n = \frac{\dot{\alpha}_r^n - \dot{\alpha}_r^{n-1}}{\varDelta t}\ .\ \ n = 1,\ldots N. \tag{6}$$

where $\varDelta t$ is the sampling time; N is an integer; $\dot{\alpha}^n, \ddot{\alpha}^n$ are the angular velocity and acceleration in the nth sampling period.

Substituting Eq. 5 into Eq. 3, and Eq. 6 into Eq. 4, and after some manipulations, Eq. 7 and Eq. 8 are obtained.

$$\dot{\alpha}_h^n = \dot{\alpha}_h^{n-1} + \frac{\varDelta t}{J_h}\left(G_h^{n-1} + F_h^{n-1}\, a_h + fg_h\left(\alpha_h^{n-1}\right) + N_h^{n-1}\right). \tag{7}$$

$$\dot{\alpha}_r^n = \dot{\alpha}_r^{n-1} + \frac{\varDelta t}{J_r}\left(G_r^{n-1} + F_r^{n-1}\, a_r + fg_r\left(\alpha_r^{n-1}\right) + N_r^{n-1}\right). \tag{8}$$

In the impedance control method, the arm angular velocity of the golfer is regarded as the control input reference for the robot. As is shown in Eq. 7, the reaction force F_h^{n-1} and the reaction torque N_h^{n-1} should be known in advance, if we expect to acquire the control input reference $\dot{\alpha}_h^n$. Since a PI controller is used to assure that the arm angular velocity of the robot is equivalent to that of the golfer, the motions of the same club used by the golfer and the robot are the same. Therefore, the reaction force and reaction torque from the same club to the arms of the golfer and robot are equivalent, that is, $F_h^{n-1} = F_r^{n-1}$ and $N_h^{n-1} = N_r^{n-1}$. Then the velocity reference $\dot{\alpha}_h^n$ for the golf swing robot can be calculated from the following equation.

$$\dot{\alpha}_h^n = \dot{\alpha}_h^{n-1} + \frac{\Delta t}{J_h}\left(G_h^{n-1} + F_r^{n-1}\,a_h + fg_h\left(\alpha_r^{n-1}\right) + N_r^{n-1}\right). \tag{9}$$

The whole configuration of the control system is described as Fig. 4.

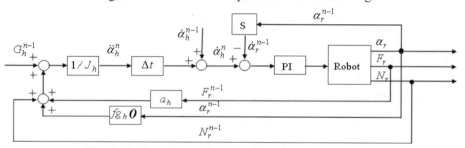

Fig. 4. Configuration of the control system

4 Simulation and Experiment

A fourth-order-Runge-Kutta integration method at intervals of 1.0×10^{-4} s was used to drive the golfer swing model and digital controller. The control input references for the digital controller were changed every 4ms. The simulation flowchart of digital controller is shown as Fig. 5.

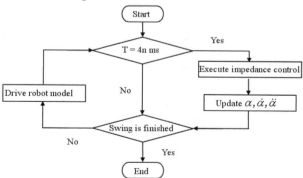

Fig. 5. Flowchart of digital controller simulation

The golf swing robot consists of an actuated joint driven by a direct drive motor and a passive joint with a mechanical stopper. The stopper achieves the wrist cocking and uncocking action. Two force sensors were used to measure the reaction force and torque from the club to arm. The sampling time for the robot feedback control was 4ms. A flexible solid beam made of aluminium was used to replace the shaft, and it was clamped at the grip. The natural frequencies and vibration mode shapes of the club were obtained by numerical calculation and experimental modal analysis. The photograph of the robot is shown in Fig. 6.

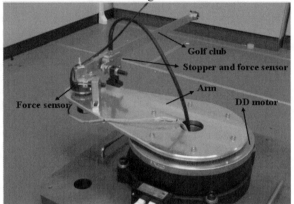

Fig. 6. Prototype of golf swing robot

It was assumed that the swing commenced with $\alpha = 120°$, $\beta = -90°$ and $\dot{\alpha} = 0$, $\dot{\beta} = 0$. Since a pause usually occurs at the top of the downswing, the bending of the shaft at the initial of the downswing was ignored. Two golfers with different arm masses were noted as H1 and H2, and the arm masses were 5kg and 6kg, respectively; The digital controller and golf swing robot were defined as DC and R, and the arm mass was 3.779kg. It is noted that there are many kinds of torque functions of the shoulder joint such as constant, ramp and trapezoid (Suzuki and Inooka 1999) applied in the previous research work. Here, a trapezoid-shaped torque (Fig. 7) was employed at the shoulder joint for the golfers. The wrist damping coefficients of the golfers and robot were considered as the same.

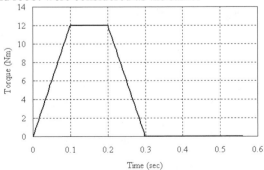

Fig. 7. Torque of shoulder joint

Fig. 8 shows the arm and club rotating angular velocities for H1, DC, R and H2, DC, R. From these figures, it is clearly indicated that the swing motions of the golf swing robot with the impedance control method are consistent with those of golfers

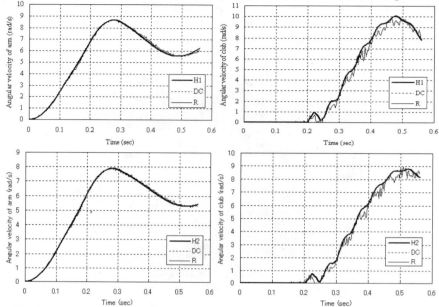

Fig. 8. Comparison of swing motions for golfers, digital controller and robot

5 Conclusions

An impedance control method based on the velocity instruction was proposed to control a golf swing robot. By this method, not only are the interference forces from the club on the arm considered, but also the influence of different human arms on the golf swing motion is involved.

A digital controller simulating the swing of a golf swing robot and a prototype of golf swing robot were developed to verify this control method. The simulation and experiment results show that the robot with the impedance control method can emulate the swing motions of different golfers.

References

Hogan, N. (1984) Adaptive control of mechanical impedance by coactivation of antagonist-muscles. IEEE Transaction on Automatic Control, Vol. AC-29, No. 8, 681-690.

Ming, A. Kajitani, M. (2003) A new golf swing robot to simulate human skill-accuracy improvement of swing motion by learning control. Mechatronics, 13, 809-823.

Suzuki, S & Inooka, H. (1999) A new golf-swing robot model emulating golfer's skill. Sports Engineering, 2, 13-22.

9 Outdoor Sports

Synopsis of Current Developments: Outdoor Sports

Ulrich Hartmann
Technische Universität München, hartmann@sp.tum.de

To understand the perspectives taken in the assembly of this synopsis, first I will say a few words about my own field of research – which is focused in the area of exercise physiology issues in connection with practical performance diagnosis, computer-aided interpretations as well as the occurring relevant questions concerning this subject. The main work has been done in the area of top elite sports (energy supply mechanisms, muscular adaptation, high altitude physiology), but also many studies have taken place in the field of basic working / exercise physiology and advanced sport science, for the industry and also for the use of different facilities in the practical field of sport and the army.

Summary of Topics Addressed

I was requested to help with the evaluation of some of the submitted and pre-selected papers, and to summarize the contents and trends in this synopsis. Most of the reviewed papers dealt with different issues in outdoor equipment. Aspects of research were the kinematics and dynamics of walking with different backpack designs (Foissac et al.). A further contribution concentrated on the thermal comfort (thermal isolation, water vapour resistance, water absorption, microclimate inside the boots) of special footwear under high altitude climate conditions (Rosa et al.). Finally, Millet gave an interesting overview over a large variety of perspectives in fatigue research that should be useful as a general reference for further research in this field.

Conclusions and Future Perspectives

In general all papers focussed on clear questions, clear study designs and mainly clear results. Due to small inadequate methodological proceedings in few cases where the results were contradictory mainly clear recommendations could be given to the reader. In some of the studies and under the given aspects of research the authors tended to focus the given results primarily onto technical but less onto physiological / biological aspects; a deeper co-operation between different scientists could be helpful to share the questions and to jointly interpret the results in this case! The majority of the studies can be admitted to be of high relevance to practice and a useful area of research. None of the studies was extraordinary or with signposting investigations in its field, some of the given topics were already matter of research in the industry or the military. Nevertheless, the presented papers are useful and according to the results and summaries there are still some open and some new questions for further research under the given topics.

Influence of a New Backpack Design on Kinematics and Dynamics of Walking

Christophe Gillet[1], Matthieu Foissac[2], Sébastien Leteneur[1], Philippe Freychat[2] and Franck Barbier[1]

[1] Laboratoire d'Automatique, de Mécanique et d'Informatique Industrielles et Humaines, Université de Valenciennes, France, christophe.gillet@univ-valenciennes.fr
[2] Decathlon Research Center, Villeneuve d'Ascq, France

Abstract. The backpacks influence the posture of the walkers. The object of this study is to quantify the modifications of the posture but also the modifications during gait, in differs configurations load: (a) without backpack, (b) with a commercial backpack, and with a proto-type of which it is possible to make move part of the load, (c) behind of the trunk and (d) in front of the trunk. To evaluate this new design, 13 subjects were asked to walk, in straight line at their natural walking speed. Kinematics and dynamics data were recorded by a Vicon optoe-lectronic system and 2 force platforms (Kistler). The results show that the forward load condi-tion (d) allows the raising of the walker trunk. Consequently, the postural attitude of the wal-ker is closer to normal walk attitude (without load). For the dynamic data, no significant difference was noted between the 3 conditions of walk with backpack.

1 Introduction

Many authors studied the modifications of walk while carrying a backpack, and many fields were investigated : physiological parameters (Holewiijn 1990, Bobet and Norman 1984), temporal parameters (Kinoshita 1985), kinematics parameters (Mar-tin and Nelson 1986, Barbier et al. 1998, Lafiandra et al. 2003), kinetics gait parame-ters (Kinoshita 1985, Quesada et al. 2000, Lafiandra and Harman 2004) and subjec-tive perceptual method (Legg and Mahanty 1985, Legg et al.1997).

The basic characteristic of a backpack is the addition of a mass relying on the back part of the trunk. This positioning of the load induces an imbalance of the trunk backwards that the hikers must compensate by having their trunk inclined forwards, of about 6 to 11 ° (Kinoshita 1985). This load position prevents also the carriers from reaching easily the content of their pack in order to catch food, drink or a clothe while walking for example. Very few backpacks offer the possibility of moving rapidly and efficiently a part of their content to the front.

A new backpack (cf. Fig. 1.) has been designed to bring easily three pockets with a total capacity of about 10 l from the back to the front of the walker.

The objective of this study was to determine whether this backpack improved or not the kinematics and dynamics of walking.

Fig. 1. New backpack studied.

2 Methods

13 male subjects (26 +/- 3 years old, 79 +/- 12 kg) were asked to walk at their natural walking speed in 4 conditions : (a) without backpack (WB), (b) with a typical trekking backpack (BP) and with the new backpack having its 3 pockets at the rear of the walker (BPR) (c) and at the front (BPF) (d).

The backpacks were loaded in an homogeneous way with the same load of 15 kg. The mass difference of the backpacks is thus induced by the difference in mass of the backpacks with vacuum (0.5kg). For the new backpack, the mobile top pocket had a loading of 3 kg and the sliding side pockets weighted 1 kg each one.

To calculate the position of the backpacks' center of gravity, it is possible to suspend them by a point. The vertical line associated to this point passes through the centre of gravity of the backpack. In a frame attached to it, the equation of the gravity line was calculated. Repeating the operation, a second line equation is obtained. Theoretically, these two lines cut themselves at the centre of gravity of the backpack. However, these lines are not convergent because of measurement errors due to the 3D acquisition and unknown factors of the backpack manipulation. Nevertheless it is possible to approximate this point by minimizing the distance which separates two points belonging to each line (Guelton et al. 2002).

The subjects walked along a 6 m path and 3D data were recorded with an optoe-lectronic system (Vicon 612) with eight cameras, recording at 120 Hz, while they walked on two force platforms (Kistler, type 9286A).

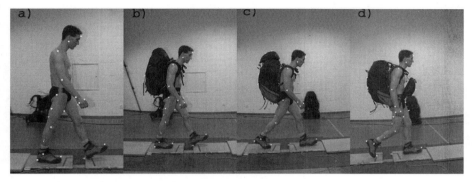

Fig. 2. Four walking conditions. a) without backpack, b) with a traditional backpack, c) with the new backpack with its pockets at the back and d) with the new backpack with its pockets at the front.

All the data were analyzed using repeated-measures analysis of variance (ANOVA). To compare each condition to others, a post-hoc Tukey test was perfor-med.

3. Results

As shown in Fig. 3, trunk inclination was found to be of 4.9 +/- 2.7 ° without back-pack, 14.2 +/- 3.2 ° while carrying backpack with the total load in the back, and 10.6 +/- 2.9 ° with the three pockets at the front (p<0.001 for all differences).

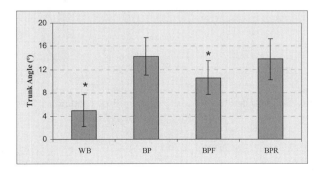

Fig. 3. Trunk angle for the four conditions : WB, BP, BPF and BPR. * = p<0.001 between both WB and PPF vs. other conditions.

The distances between backpack's and subject's centers of gravity were significantly different between the load conditions: 40 cm for BP, 37 cm for BPR and 22 cm for BPF (Fig. 4).

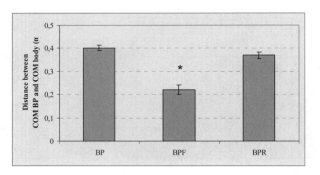

Fig. 4. Distance between the center of mass of the backpack and the center of mass of the body for the three load conditions : BP, BPF and BPR. * = p<0.001 between BPF vs BP and BPR.

Ankle, knee (cf. Fig. 5) and hip joint moments showed statistical differences between WB and all backpack conditions, but no difference was found between the backpack conditions. The same results were found when regarding peak ground reaction forces (both vertical and antero-posterior reception and propulsion).

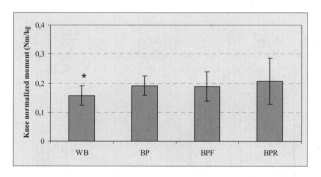

Fig. 5. Knee normalized moment for WB, BP, BPF and BPR conditions. * = p<0,001 between WB and both BP, BPF and BPR.

Hip oscillations in the horizontal plane are statistically greater without backpack than with any other backpack condition (25° of amplitude vs. 16°, p<0.001) but here again no statistically significant difference was found between the backpacks conditions.

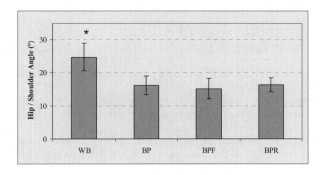

Fig. 6. Angle between hip and shoulder for WB, BP, BPF and BPR conditions. * = p<0,001 between WB and both BP, BPF and BPR.

The dynamic data (forces and moments) did not show any difference between the load conditions. However, the BPF condition was found to be the preferred load condition in terms of comfort expressed by the subjects for 62 % of them (23 % ranked it second and 15 % third).

Lloyd and Cooke 2000 showed that for uphill walking, the oxygen consumption was lower by 5 % when part of the load was placed in front of the trunk. This could explain the positive feelings of the subjects found in our study. However these authors studied loads heavier than in this study and a slower walking speed (25.6 kg vs 18 kg and 3 km/h vs 4.5km/h)

4 Conclusion

The new backpack design allows a more upward position, by moving dramatically the center of gravity of the load forward. This posture modification resulted in a better comfort expressed by the majority of the subjects. The better comfort could be explained by less work of the trunk muscles to compensate the load located in the back of the carrier. This could have an indirect impact on energetic cost of walking but it wasn't investigated in our experiment. Finally, the new backpack has no significant influence on forces and moments distribution in joints.

Acknowledgment

This study was funded by DECATHLON Production. The Vicon system and force platforms were financed by FEDER and the région Nord-Pas-De-Calais. The authors acknowledge Jacques Delacroix, Régis Leroux and Richard Kaczmarek, for their help and very efficient technical collaboration.

References

Barbier, F., Salom, O. & Angué, J.C. (1998) Assessing the ergonomics of 2 rucksacks. The Engineering of Sport, Editor S.J. Hake, Blackwell Science, pp 243-250, ISBN 0-632-05048-9.

Bobet, J & Norman, R.W. (1984) Effects of load placement on back muscle activity in load carriage. Eur. J. Appl. Physiol. 53, 71-75.

Cook, T.M., Neumann, D. A. (1987) The effects of load placement on the EMG activity of the low back muscles during load carrying by men and women. Ergonomics. 30 (10), 1413-1423.

Guelton, K., Gillet, C., Delporte, L., Barbier, F., Lepoutre, FX. (2002) An easy method to determine the centre of gravity of some lifeless body with a motion analysis system. Arch. of Physiology and Biochemistry. 110, 90.

Holewijn, M. (1990) Physical strain due to load carrying. Eur. J. Appl. Physiol. 61, 237-245.

Kinoshita, H. (1985) Effects of different loads and carrying systems on selected biomechanical parameters describing walking gait. Ergonomics. 28 (9), 1347-1362.

Knapik, J., Harman, E., Reynolds, K. (1996) Load carriage using packs: a review of physiological, biomechanical and medical aspects. Appl. Ergon. 27(3), 207-216.

LaFiandra, M., Wagenaar, R.C., Holt, K.G., Obusek, J.P. (2003) How do load carriage and walking speed influence trunk coordination and stride parameters? J. Biomech. 36(1) 87-95.

Lafiandra, M., Harman, E. (2004) The distribution of forces between the upper and lower back during load carriage. Med. Sci. Sports. Exerc. 36(3), 460-467.

Legg, S.J., Mahanty, A. (1985) Comparison of five modes of carrying a load close to the trunk. Ergonomics. 28 (12), 1653-1660.

Legg, S.J., Perko, L. & Campbell, P. (1997) Subjective perceptual methods for comparing backpacks. Ergonomics. 40 (8), 809-817.

Lloyd, R. and Cooke, C.B. (2000) The oxygen consumption associated with unloaded walking and load carriage using two different backpack designs. Eur. J. Appl. Physiol. 81(6), 486-492.

Martin, E. & Nelson, R.C. (1986) The effects of carried loads on the walking patterns of men and women. Ergonomics. 29 (10), 1191-1202.

Quesada, P.M., Mengelkoch, L.J., Hale R.C. & Simon S.R. (2000) Biomechanical and metabolic effects of varying backpack loading on simulated marching. Ergonomics. 43 (3), 293-309.

Study of the Loss of Thermal Properties of Mountain Boots in an Expedition to Mount Everest

David Rosa, Enrique Alcántara, Juan Carlos González, Natividad Martínez, Mario Comín, María José Such, Pedro Vera and Jaime Prat

Instituto de Biomecánica de Valencia (IBV), enrique.alcantara@ibv.upv.es

Abstract. This paper presents the results of the research carried out during the expedition of the Polytechnic University of Valencia (UPV) in the south face of the Mount Everest. The objective of the study was to evaluate the modification of the thermal properties of two mountain boots worn during the expedition. Both boots were tested before and after the expedition using the laboratory Textile Transmission Test (TTT). This test provides information about the global thermal isolation of the footwear, the global water vapour resistance of the footwear, and the water absorption capability. A breathable/waterproof sock is used to study the transport of water. Two different models of boot were worn by the subjects (One of them developed with a high thermal isolation and the other commercial). The hours of use were 175 (60 estimated kilometers) and the mean environmental conditions were, temperature 5°C and humidity 36% RH. The results of the study showed statistically significant differences in the dry thermal insulation and the wet heat exchange, but not in the global water transfer and absorption of the boots. The results of this study show that actually there is a relevant degradation of the thermal properties with use of sport equipment for high altitudes, which can increase injury risk for sportsmen. This highlights the need to advance in the control of performance during the whole life cycle of the equipments for extreme conditions in order to reduce injury risk of on high altitude climbers.

1 Introduction

Thermal comfort and protection is a critical factor in choosing clothing and footwear for high altitude mountain climbing. Frostbit is a major and rather frequent injury in climbers, especially in the feet. However, there is little information about how mountain boots deteriorate during expeditions and if they keep an adequate level of thermal protection.

On the occasion of the climbing expedition to the Mount Everest organized by the Universidad Politécnica de Valencia, a parallel scientific expedition was realised by the Institute of Biomechanics of Valencia with the main objective of advancing in the knowledge of the thermal behaviour of the human body in real situations of high altitude. A partial goal was to study the loss of thermal properties of two mountain boots during the expedition. This paper presents the results of this study.

2 Materials and Methods

The study consisted in measuring the thermal properties of two boots of high charac-teristics of thermal isolation before and after the expedition to Mount Everest. The thermal properties were measured at the alb using the Textile Transmission Test for footwear applications.

The idea of this test is to determine the thermal resistance (Rc) and the transmis-sion of water vapour (MVTR) of the complete boot. This experiment is done intro-ducing a sock inside the boot that allows two different testing conditions: dry when there is not transference of water vapour, or wet conditions using an imperme-able/breathable sock.

An electric resistance and a sensor of temperature are placed inside the sock for controlling the experiment. The water is heated up to 35°C. The energy needed to maintain the water at this temperature for two hours of study is registered. Simulta-neously, the environmental conditions are controlled to be able to establish the flow of existing heat between the interior of the set (boot, sock and equipment) and the environment. On the other hand, the loss of weight of the set is measured. This is used to evaluate the breathability and the sweat absorption of the boot. Besides, a sensor of microclimate (temperature and humidity) is placed in the toe zone to meas-ure the climatic conditions of the air that has remained enclosed between the boot and the sock.

The study parameters are showed in the following table:

Variable	Units	Description
Thermal resistance	m2*°C/W	Global insulation of the boot
Breathability	Grams	Grams of water perspired by the boot during the test.
Absorption	Grams	Grams of water absorb for the boot during the test.
Insole Absorption	Grams	Grams of water absorb for the insole during the test.
MTVR	g/m2*24h	Index of mass of vapour water prespired, normalized by the surface of the boot and considered for 24 hours. This index is normalized so that the values are compa-rable independently of the size of the footwear
Temperature Toe Capsule (°C)	°C	Average temperature obtained in the capsule placed in the toe zone (stationary phase of the test).
Humidity toe cap-sule (%HR)	%	Average humidity obtained in the capsule placed in the toe zone (stationary phase of the test)

Table 1. Parameters obtained for the Thermal Transmission Test

The two boots tested are described below.
    -BESTARD Boot (Fig.1)
    The boot consists of two differentiated parts:

Outer shell.
The exterior part is made from leather Perwanger of 3 mm. The lining consists of Cambrelle, insulating thermal "Primaloft" 400 grs. and Gore-Tex membrane.
Interior booty.
The interior part consists of a Gore-Tex lining, sole of rubber and resistant fabric (canvas) of Cambrelle with water repellent treatment. A removable anatomical Insole is included with two layers, Cambrelle and thermal material "Primaloft" of 200 grs.

-FAL Boot (Fig2)
This boot is made with leather box and "Cordura", both water repellent. It incorporates Gore Tex Duratherm lining, breathable and with wind-shield effect; Coolmax in the collar. The sole is Vibram Vertige with crampons. It also features a perimetral protection of rubber on the upper, rigid insole and double isolation of tongue.

Fig. 1. Bestard boot

Fig. 2. FAL boot

Four boots were tested. The hours of use (for the two boots) during the expedition were 175 (60 estimated kilometres) and the mean environmental conditions were, temperature 5°C (minimum -11°C) and humidity 36% RH. The altitude of use is 5.300 metres in a glacier terrain.

Data analysis consisted of an Analysis of Variance considering boot, conditions of test and condition of use as factors, as well as the interaction between them.

3 Results

The results showed significant differences in thermal properties between the tests before and after the expedition. Both boots lose properties during the expedition. The interaction between boot and condition of use was significant and thus, analysis was done separately for each boot. Descriptive statistics are showed in table 2.

Results for the BESTARD boot.

Variable	Dry Study		Wet Study	
	New	Used	New	Used
Thermal resistance (m2*ºC/W)	0,51	0,44	0,33	0,27
Breathable (gr)	0	0	3,27	3,77
Absorb (gr)	0	0	8,57	11,63
Absorb insole (gr)	0	0	0,7	1,03
MTVR (g/m2*24h)	0	0	590,30	680,55
Temperature Toe Capsule (ºC)	31,25	30,36	32,66	31,55
Humidity toe capsule (%HR)	53,96	58,48	82,69	73,01

Results for the FAL boot.

Variable	Dry Study		Wet Study	
	New	Used	New	Used
Thermal resistance (m2*ºC/W)	0,41	0,33	0,25	0,20
Breathable (gr)	0	0	2,68	3,36
Absorb (gr)	0	0	6,93	6,63
Absorb insole (gr)	0	0	0,93	0,70
MTVR (g/m2*24h)	0	0	484,8	664,6
Temperature Toe Capsule (ºC)	31,2	30,1	29,9	29,9
Humidity toe capsule (%HR)	49,9	54,3	93,0	80,8

Table 2. Results for the average variables.

Results for the BESTARD boot.

The results (Fig.3) showed a significant decrease only for thermal resistance in both testing conditions: dry and wet.

Thermal resistance showed a significant difference of 35.3 % between dry and wet testing conditions. Besides, a significant decrease in the thermal resistance after the expedition has been detected for both types of study: study in dry (13,7 %) and study in wet (18,2 %).

Results for the FAL boot.

Figure 4 shows the results for thermal resistance in two conditions, dry and wet and before and after the expedition.

Significant differences were found for thermal resistance between dry and wet conditions before the expedition (40%). Besides, a significant decrease in the thermal resistance has been detected for both types of study: study in dry (20 %) and study in wet (20 %).

Also, significant differences were found in the breathability: when the boots are used, it increases (25,4% more). In the study about the absorption of the boot and the absorption of the insole, no significant differences were detected.

When analyze the microclimate conditions at the toe area, the same thing happens that with the previous boot.

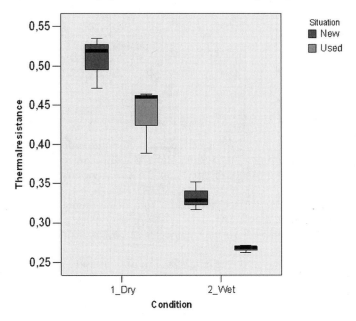

Fig.3. Box&Whisker Diagram for
Thermal Resistance (Bestard boot)

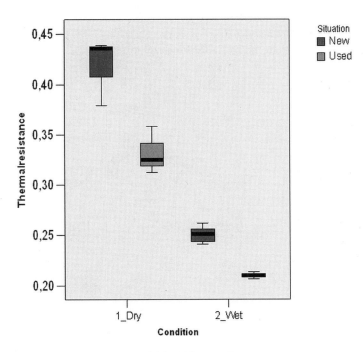

Fig.4. Box&Whisker Diagram for
Thermal Resistance (FAL boot)

4 Discussion and Conclusions.

This study presents very novel information about the thermal properties of equipment for extreme conditions. The results show a significant reduction of thermal properties due to the use during an Expedition. This reduction is rather high in some cases:
    - The thermal resistance of the boot had a significant reduction up to 18,2% for one boot and 20% for the other.
    - The breathability for the FAL boot increases (25,4% more).

These results show that as the climbers stay in high altitude, the risk of injury increases as their loss of physical capacity due to altitude effects is accompanied by a reduction in the protection provided by clothing, in this case footwear. This should be strongly considered for further research about clothing and footwear for these activities. Differences found due to the use suggest the paramount importance of going on further investigations about the product degradation.

Acknowledgments

The authors wish to thank to Calzados FAL, Calzados BESTARD, LORPEN and Universidad Politécnica de Valencia for their support.

The Role of Engineering in Fatigue Reduction

Guillaume Millet[1], Stéphane Perrey[2] and Matthieu Foissac[1]

[1] Université Jean Monnet Saint-Etienne, France, guillaume.millet@univ-st-etienne.fr
[2] Université de Montpellier, France
[3] Decathlon, France

Abstract. The purpose of this short review is to discuss how fatigue of an athlete can be limited by using recent innovations in sport engineering. Through a general definition and illustrations in some locomotions, several aspects of fatigue limitation will be presented : shocks reduction, influence of weight, distribution of muscular work and efficiency. In the second and third parts, recent textile innovations aiming at decreasing fatigue by different ways (elastic compression stocking, drag, temperature regulation) and new equipments dedicated to improve recovery will be illustrated.

Introduction

It is generally accepted that fatigue depends on the nature, duration and intensity of exercise. This task dependency is obviously conditioned by the athlete's level so that training is the best way to reduce fatigue. However, for a given athlete and exercise, equipment can play an important role to reduce fatigue amplitude. Fatigue is often quantified as a strength loss and/or an increase of energy cost of locomotion when the exercise duration or intensity are high enough. In that context, the following review will try to illustrate how engineering can reduce muscular disruption and/or lowered the energy cost drift. For that purpose, several main concepts will be discussed : (i) reduction of muscle damages, (ii) effects of amplitude and localization of the equipment's mass, (iii) distribution of muscular work among different muscular groups and (iv) gross efficiency. Fatigue can also be prevented by wearing new clothes, in particular by enhancing heat balance and muscle oxygenation. Finally, engineering can not only reduce fatigue but also helps to improve recovery. Here, the effects of electromyostimulation and elastic compression stocking will be discussed.

1 Locomotions

1.1 Shocks Reduction

Muscular damages are known to impair muscular function. During running, while the eccentric phase of the stretch-shortening cycle is probably the main cause of muscle damages, these ones can also probably be explained by the repetition of shocks when running on roads or hard tracks. While "in vitro" experiments have shown that soft soles reduce peak force, no differences in the magnitudes of vertical impact peak force were found between soft and hard shoes during running on a force plate (Clarke, Frederick and Cooper 1983, Komi, Gollhofer, Schmidtbleicher and Frick 1987). It has even been reported a greater force impact with cushioned shoes to compare with hard shoes (Nigg, Luethi, Denoth and Stacoff 1981).

Increasing cushioning characteristics is generally well-perceived by runners but can have by-effects in terms of energy cost. Bosco and Rusko (1983) have indicated that running with soft shoes required greater energy consumption than running with normal shoes at high velocity. Opposite results have nonetheless been reported (Frederick, Howley and Powers 1986). Thus, both mechanically and energetically speaking, the optimal shoe stiffness and its influence on performance and fatigue is still under debate.

In mountain biking, shocks reduction may be obtained by suspension. It has been recently suggested that full suspension bicycle provides a physiological and psychological advantage over the hard tail bicycle during simulated sub-maximal exercise on bumps (Titlestad, Fairlie-Clarke, Whittaker, Davie, Watt and Grant 2006). When comparing front suspension (FS) and front and rear (or dual, DS) suspension, elite bikers rode significantly faster on DS than FS bikes during 30 min riding tests (Nishii, Umemura and Kitagawa 2004). Interestingly, serum creatin kinase (CK), considered as a indirect index of muscle damages, increased significantly 24 h after the trial when cyclists exercised with the FS bike. However, as for running, cushioning may have by-effect in terms of efficiency (MacRae, Hise and Allen 2000).

1.2 Effects of Weight

The control of the equipment's weight is essential when considering the energy expenditure and thus the potential metabolic consequences in terms of fatigue. When mass is carried at the extremities, its influence is dramatic. For example, the increases in VO_2 due to foot loading (7.2% per kg of load) were nearly twice as great as those due to thigh loading (Martin 1985). As a result, running shoes' weight is considered by manufacturers as a key factor (Divert 2005). Probably because lower angular acceleration of limbs, adding mass and changing mass distribution of poles do not generate any extra energy cost of uphill walking (Foissac, Seux, Berthollet and Millet 2006b).

Regarding the mechanical effects of weight, it has been found that loads up to 1.1 kg at each ankle produced no significant change in stride frequency or stride length (Cavanagh and Kram 1989), but an other study has demonstrated that that 0.5 kg added to each foot produced small but significant increases in stride length, swing time and flight time and a decrease in peak ankle velocity (Martin 1985).

1.3 Distribution of Muscular Work

Poles can reduce fatigue because of the work distribution between upper and lower limbs muscles. Although imposing no metabolic consequence, subjects using poles reduced their activity in several lower limbs muscles and reported lower ratings of perceived exertion (Knight and Caldwell 2000, Foissac et al. 2006b). Many ways of carrying a load exist: Traditional backpacks, double packs (with a part of the load distributed at the front of the walker), jackets loaded or weight-carrying in hands were the most studied. Shoulder backpack was 10% more expensive than double pack (Datta and Ramanathan 1971). In spite of these results, backpack is the most used load-carrying system for a hiking purpose because of its multiple ergonomic advantages. It gets also a good thermal comfort and no vision or respiration impairments on the contrary to double packs, which may be essential to limit fatigue. With recent improvements in normal backpack, differences in $\dot{V}O_2$ between double and normal backpacks have been reduced to about 5% (Lloyd and Cooke 2000).

Finally, the characteristics of the connection between the pack and the wearier are important because it may redistribute work between eccentric and concentric phases of walking. When modifying the stiffness of the connection, noticeable modifications of the mechanics of backpack and of energetics of walking can be observed (Foissac, Belli and Millet 2006a).

1.4 Efficiency

Any equipment allowing a better efficiency can potentially reduce fatigue of an athlete compared with the same intensity without this equipment. In cycling, this is typically the case when aerodynamic and rolling resistances are reduced. Several experiments have also be done to test the effects of new chainrings but the results have been mostly found controversial. For example, delta efficiency measured with the Rotor - a new pedaling system that makes each pedal independent from the other so that cranks are no longer fixed at 180 degrees - was reported to be higher than a traditional chainring between 60 and 90% of VO_2max (Santalla, Manzano, Perez and Lucia 2002) but these results were not confirmed by an other study from the same laboratory (Lucia, Balmer, Davison, Perez, Santalla and Smith 2004) in trained cyclists. Different types of non-circular chainrings have also been tested in the literature, e.g. "Shimano Biopace", "Eng10", "Harmonic", "Pro Race eccentric" etc. Most but not all of them were found to be not more efficient than traditional circular ones. These conflicting results may be due to the intensity of the tests Zamparo et al. (2002) However, to the best of our knowledge, none of these experimental chainrings is currently utilized in competition by elite cyclists.

In speed skating, a relatively new device allowing better efficiency is used by elite ice skaters. In the "Klapskate", the rigid connection between the shoe and the blade has been replaced by a hinge mechanism beneath the ball of the foot. The hinge of the klapskate allows the foot to plantar flex at the end of the push-off while the full blade continues to glide on the ice. This enhances the effectiveness of plantar flexion during the final 50 ms of the push off and increases work per stroke and mean power output (Houdijk, de Koning, de Groot, Bobbert, Van Ingen Schenau 2000). Gross efficiency was found to be higher with klapskates compared with conventional skates (16.3% vs 14.8%) (Houdijk, Heijnsdijk, de Koning, de Groot and Bobbert 2000).

2 Textile

2.1 Temperature Regulation

Many fabrics and clothing "systems" have been designed to enhance heat balance and provide greater thermal comfort. Exercise in the heat may cause individuals to suffer from heat stress that can increase fatigue. A few studies do exist on the role of several thermal parameters in the determination of exercise performance ; all suggest that precooling improves distance run performance. Arngrïmsson *et al.* (Arngrimsson, Petitt, Stueck, Jorgensen and Cureton 2004) concluded that wearing a cooling vest during warm-up is effective in improving subsequent simulated 5-km run performance. In other studies, precooling has improved performance in tasks requiring all-out efforts as short as 70 s (Marsh and Sleivert 1999) and as long as 60 min (Hessemer, Langusch, Bruck, Bodeker and Breidenbach 1984). These data suggest that performance in track events between 600 and 10,000 m might be improved through precooling.

Light-weight cooling vests designed for sporting activities have been shown to provide a significant thermoregulatory advantage in hot and humid environmental conditions, as evidenced by lower core and skin temperature, lower sweat rate and improved ratings of thermal and moisture perception (Webster, Holland, Sleivert, Laing and Niven 2005) : endurance times for running at 95% of VO_2max were increased by up to 49 s and perceptions of the thermal state and skin wetness showed greater levels of satisfaction.

2.2 Effect of Compressive Garments

Effect of compressive garments has been noted in prevention of injury and effects of blood stasis, as occurs in cardiovascular insufficiency. These garments are known as gradient compression stockings. The first exercise-related research on compressive garments (Berry and McMurray, 1987) found lower lactate concentrations after an exhaustive exercise bout when the stockings were worn during the exercise. In a series of investigations of Lycra-type compression shorts, it has been noted (Kraemer, Bush, Bauer, Triplett McBride, Paxton, Clemson, Koziris, Mangino, Fry and Newton 1996, Kraemer, Bush and Newton 1998) enhanced athletic performances (e.g. repetitive jump power). Among possible mechanisms proposed to explain this result is an increased resistance to fatigue.

In running wearing the elastic tights with a control pressure of 30 mmHg at the ankle alters the performance by 2.3% on a 5000 run test; the lower performance was related to a shorter stride length (Chatard 1988). It has been shown a significant lower energy cost in trained runners by wearing compression and elastic tights compared to shorts at 12 km/h but not at higher running speeds (Bringard, Perrey and Belluye 2006). These authors further showed a lower (- 36 %) VO_2 slow component at ~80% of VO_2max for 15 min wearing compression tights compared with conventional shorts. This latter finding shows that muscle fatigue relating to the slow component phenomenon is reduced when wearing compression tights. Overall these studies show a distinct performance advantage of wearing compression tights during exercise. These results support an hypothetic mechanism of improved oxygen avail-ability, probably by improving blood delivery or reduced interstitial spaces.

2.3 Hydrodynamics

It is widely accepted that reducing active drag and increasing buoyancy decrease the physiological cost of any given swimming speed (Capelli, Zamparo, Cigalotto, Francescato, Soule, Termin, Pendergast and Di Prampero 1995) and may lag fatigue occurrence. This has been shown for a neoprene wet suit (Toussaint, Bruinink, Coster, De Looze, Van Rossem, Van Veenen and De Groot 1989). With the availability of a new generation of suits that cover larger parts of the body and are made in different materials than the traditional suits, there is a potential for drag reduction.

A Lycra specially designed suit covering the torso of male swimmers reduced the energy demand of swimming, compared with a standard racing suit, presumably due to the drag-reducing characteristic of the suit (Starling, Costill, Trappe, Jozsi, Trappe and Goodpaster 1995). However, when comparing swimmers propelling through a pool wearing training suits with swimmers wearing sharkskin suits, Toussaint et al. (Toussaint, Truijens, Elzinga, van de Ven, de Best, Snabel and de Groot 2002) found no significant reduction in drag from the sharkskin suits. Note that it is admitted that loose fitting competition or drag-reducing suits can increase drag when compared with tight-fitting suits (Toussaint et al. 2002, Walsh 1998) and may act in favour to much higher muscle fatigue over time.

3 Recovery

3.1 Electromyostimulation

Mainly because temporal pressure that sometimes exists during training or racing periods and because the type of recruited motor units differs between voluntary and electrically induced contractions, electromyostimulation (EMS) can help athletes to improve their recovery. EMS is a mean of increasing muscle blood flow, especially when stimulating at low frequencies. For high-intensity exercise, active recovery period usually results in accelerated blood lactate and H^+ clearance. For exercises inducing muscular damages (e.g. long distance running, eccentric strength training), the increased blood flow to an injured area may help remove cellular debris, increase nutrient and neutrophil or macrophage delivery arrival, thus may speed repair. However, two recent studies have failed to show any superiority of EMS recovery compared with passive or traditional active recovery (Lattier, Millet, Martin and Martin 2004, Martin, Millet, Lattier and Perrod 2004) with the exception of a tendency toward a better performance during an all-out test after the EMS intervention.

TENS currents, i.e. currents that do not induce muscle contraction, are also widely used in clinical settings as an effective non-pharmacological modality for the treatment of pain of various aetiologies. Denegar and Perrin (1989) demonstrated a temporary reduced perception of pain and increased range of motion when applying low-frequency EMS to the upper arm of subjects experiencing delayed onset muscle soreness (DOMS). As there was no control group, time effects could also have been responsible for the results. Conversely, under blinded placebo-controlled conditions, Craig et al. (Craig, Cunningham, Walsh, Baxter and Allen 1996) did not report any measurable effect of EMS on DOMS and range of motion. In summary, despite its large use by athletes, that there is no strong scientific evidence regarding the EMS' ability to enhance the recovery process from any type of fatigue. However, it should be recognized that only short-term studies have been performed. It is likely that, as for the strength training purpose of EMS, this technique should not replace but complete the traditional active recovery.

3.2. Elastic Compression

Although a variety of compressive tights are available for sportsmen for practising purpose, these tights are considered by many to be beneficial for recovery treatment and related exercise symptom relief. One of the most recognized action of compression tights occurs during recovery from exercise and relate to prevention of DOMS. The increased microcirculation provided by compressive garments may prevent post-exercise damage and pain. By using near infrared spectroscopy, Bringard *et al.* (Bringard, Denis, Belluye and Perrey 2006) showed in sportsmen a significantly lower venous pooling and a higher calf oxygen saturation level with compression tight in comparison to elastic tight and shorts in lying supine and upright standing positions. It appears that compression compared to elastic tights have positive effects on muscle oxygenation and venous function at rest, and could be useful to oxygenate fatigued muscles after exercise. In this context, wearing elastic compression stockings in 63 year old sportspeople during an 80 min recovery period between two maximal exercises lead to a significant 2.1% increase in 5 min performance, involving reductions in hematocrit and lactate concentrations (Chatard, Atlaoui, Farjanel, Louisy, Rastel and Guezennec 2004).

Conclusion

This review shows that equipment is a non-negligible aspect of the fatigue limitation, specially during prolonged exercise when energy cost is among the determining factors of performance. Indeed, a better efficiency, i.e. lower metabolic fatigue at a same velocity or power, is often implicated in fatigue reduction with engineering advances. It is however not the only way as illustrated by the parts of the review that focused on work redistribution, temperature regulation, shocks reduction or elastic compression. It should nevertheless be noted that all equipments that allow a performance improvement do not necessarily help to limit fatigue (e.g. light shoes). Finally, this review also shows that (i) scientific evidences exist about the interest of using some equipments that are not actually used in competition (e.g. new chain-rings) and (ii) on the contrary, there is a lack of scientific data about some equipment widely used on the field (e.g. electromyostimulation).

References

Due to the pages restriction, the references were not listed. To receive them, please contact : guillaume.millet@univ-st-etienne.fr

10 Shoes & Apparel

Synopsis of Current Developments: Shoes and Apparel

Thomas L. Milani

Chemnitz University of Technology, Germany, thomas.milani@phil.tu-chemnitz.de

In the last 30 years intensive research was realized in the area of sport shoes and apparel technologies. During this period fundamental research resulted in basic functional criteria, for example for the performance of sports, especially of running shoes. The main criteria were the prevention of sports injuries and the athletic performance. Sport shoes should avoid high impact loads and reduce the rearfoot movement, especially overpronation and the pronation velocity. In the area of soccer shoes main research was performed in the weight aspect of shoes and in the field of stability and traction. Furthermore, sport shoes and sport apparel should fulfil certain aspects of comfort that was realized by innovative materials. The intensive research resulted in improvements of the used material and the design and construction features of sport equipment.

Yet, the improvement in midsole materials does not necessarily improve the biomechanically measured cushioning properties of running shoes. Furthermore, the recent tendency to design running shoes with better stability properties seems to result in increased impact loads. The main injury related information of variables like GRF- and tibial acceleration values is based on studies that have shown that repetitive impact loads may result in bionegative effects onto bone and cartilage tissue. Otherwise, in biomechanical research this relationship and the relation of impact loads to running injuries are in discussion. Recent studies have shown that repetitive impact loads in a physiological range result in biopositive effects like improved stability of bones [Kersting et al.].

Nigg concluded that impact loads are not related to running injuries. Furthermore, Nigg hypothesized that impact loads may be necessary to act like a feedback mechanism to tune muscles and reduce the loads in joints and tendons. The range and the magnitude of biopositive and bionegative signals in running are still unknown and have to be analysed in further studies. The combination of biomechanical and neurophysiological measurements and innovative mechanical test procedures may result in new ideas and essential criterias for running shoes.

Traditionally, material test procedures are used to analyse mechanical properties of running shoes. The Dynatup-Impacter INSTRON (Corporate Headquarters, Canton, Massachusetts USA) and the impacter of EXETER Research (Brentwood, NH 03833 USA) are used to analyse cushioning properties of running shoes. Various variables are used to describe the mechanical properties: peak acceleration at impact, material deformation, stiffness and the hysteresis. Obviously, there are advantages in mechanical test procedures like reliability and standardization of the test protocol. Nevertheless, scientific studies show no relationship when comparing the peak acceleration at mechanical impact with the tibial acceleration in running. Therefore, mechanical tests and test procedures have to be used carefully and the results have to be interpreted with respect to these results.

In spite of intensive research the complex relationship of injuries, comfort and functional aspects of sport technologies are discussed controversially. For example, literature review of scientific articles of the last years revealed that improvement in impact aspects of running shoes that are related to knee problems and ostheoarthritis came across with higher incidence of injuries that are related to rearfoot movements, e.g. pronation in the subtalar joint. The history in footwear research development indicates that the strategy of „cushion – support – guidance"may be the reason for higher incidence of problems in rearfoot motion. Since excessive rearfoot movement does not automatically result in running injuries Grau et al. suggest considering rearfoot movement individually. New concepts in running footwear development suggest supporting the foot by low profile designs. Low profile designs reduce lever mechanics at touch down and relieve the possibility of foot sensory information.

Research in the field of sport shoes resulted in fundamental knowledge of the biomechanics of the foot and lower extremities. New manufacturing procedures and innovative test strategies were developed and resulted in new materials with functional and high comfort properties. Nevertheless, further research has to be performed to obtain more knowledge in the relation of injuries and sports technologies.

Literature

Kersting, U.G., Knöcherne Adaptation des Calcaneus an laufinduzierte mechanische Belastung. 1997, Deutsche Sporthochschule: Köln.

Nigg, B.M., The role of impact forces and foor pronation: A new Paradigm. Clin. J. Sports Med., 2001. 11: p. 2-9.

Grau, S., H. Baur, and T. Horstmann, *Pronation in sport shoe research.* D. Z. Sportmed, 2003. 54(1): p. 17-24.

Intermittent Pneumatic Compression Technology for Sports Recovery

Tom Waller[1], Mike Caine[1] and Rhys Morris[2]

[1] Sports Technology Research Group, Wolfson School of Mechanical and Manufacturing
Engineering, Loughborough University, UK, t.m.waller@lboro.ac.uk
[2] Medical Physics and Bioengineerng, University of Wales College of Medicine, Cardiff, UK

Abstract. Intermittent pneumatic compression (IPC) technologies are widely used in clinical populations to aid the reduction of limb oedema and for the prophylaxis of deep vein thromboses (DVT). IPC application within athletic populations is not however widespread. The main mechanism for the effectiveness of IPC is that it augments venous and arterial blood flow via the periodic inflation of external cuffs. We believe that this may be beneficial to the warm-down activities of athletes. The removal of waste products may help to reduce injury risk and the phenomenon of delayed onset muscle soreness (DOMS). A new implementation of the technology has been developed to test the extent of any potential warm-down effects induced by IPC treatment in athletes. This paper presents a pilot study in which male participants were exposed to IPC following intensive exercise. The specific treatment comprised 60sec inflation and 60sec deflation of a calf-thigh three compartment sequential compression garment (ratio 70:65:60mmHg) on each leg. This cycle was implemented by an electric pump with the participants in the partially supine position. The recovery protocol was designed to assess the ability of IPC to reduce the symptoms of delayed onset muscle soreness (DOMS) elicited by a high intensity repeated shuttle run. A 1 hour IPC treatment was implemented in this case. Vertical jump was used to identify any change in performance pre and post trial. Visual analogue scales were used +1, +24 and +48 hours after the tests to assess the presence of DOMS. During these tests, heart rate and blood pressure measurements were recorded.

1 Introduction

Intermittent Pneumatic Compression (IPC) has been used as a mechanical method of deep vein thrombosis prophylaxis for a number of years. These systems comprise the pumped inflation and deflation of air bladders within cuffs that can cover the foot, calf or whole leg. The inflation can be applied uniformly or sequentially with a variety of pressures at rapid or moderate rates. The core mechanism for the efficacy of these systems is the prevention of stasis by augmenting venous and arterial blood flow (Morris and Woodcock 2002 and 2004). Similarly this technology has been used for the reduction of lymphedema (Pappas and O'Donnell 1992) and to enhance localised muscle recovery (Wiener, Mizrahi and Verbitsky 2001).

Given these anti-inflammatory and muscle recovery mechanisms it is believed that IPC may have application in the treatment of delayed onset muscle soreness (DOMS). DOMS is a phenomena often associated with eccentric muscle contractions whereby damage to muscle fibres from periods of exercise causes muscle pain that can persist for several days. This has disruptive effects upon the training schedules of athletes and therefore solutions to reduce the duration of soreness are of great benefit.

One such condition that is often associated with the occurrence of DOMS is high intensity shuttle running (Thompson, Nicholas and Williams 1999). This paper therefore describes a pilot study to investigate the influence of IPC upon the occurrence of DOMS after a period shuttle running.

2 Methods

Nine healthy male participants mean SD (age 25.2±1.72yrs, height 184.8±9.94m, body mass 87±10.46kg, body fat 14.9±3.61% and VO_{2max} of 53.1±2.69ml·kg^{-1}·min^{-1}) attended the High Performance Athletics Centre at Loughborough University on four separate occasions. It was requested that participants did not consume any food for 2 hours prior to arriving and refrained from alcohol, caffeine and rigorous exercise for at least 24 hours prior to the tests.

The first of these sessions was a test to estimate maximal oxygen uptake (VO_{2max}). This was done using a progressive shuttle run (Ramsbottom, Brewer and Williams 1988) and permitted the grouping of participants during a modified Loughborough Intermittent Shuttle Test (LIST, Nicholas, Nuttall and Williams 2000) tailored to exacerbate delayed onset muscle soreness. All participants were also habituated with the segmented bladder, Intermittent Pneumatic Compression (IPC) and vertical jump equipment (Jump Meter, Just Jump). Body mass and composition were measured using a set of electronic scales (BF Scales, Tanita) and a 4 site skinfold calliper method (Jackson and Pollock 1985) respectively.

Each participant was then assigned to a group (n>1) comprising others with similar progressive shuttle run scores for completion of the following tailored LIST test:

- 12 x 20m at jogging speed (4 x 3 sequential bouts interspersed with the sprint)
- 12 x 20m at maximal running speed (sprint) (4 x 3 sequential bouts interspersed with the jog)
- 18 x 20m at walking pace
- Repeat (approximately 12 times, equivalent 1hr total duration)

The shuttle runs were repeated on three separate occasions by each participant (at least 3 days apart) and preceded a treatment session that comprised either: a rest for one hour; a one hour low-pressure IPC treatment (20:15:10mmHg) or; a one hour high-pressure IPC treatment (70:65:60mmHg). All treatments were carried out in the partially supine position and the order of application was randomized.

Body mass, vertical jump and calf / thigh circumference were recorded prior to and immediately following the completed treatment session. Heart rate was recorded during the tests (Polar Team System, Polar Electro). A vital signs monitor (Smart-signs Assist, Huntleigh Healthcare) was used to record heart rate and blood pressure during the treatment sessions. Following completion of the sessions, participants completed a soreness diagram (Fig. 1) with the aid of a 10 point scale with anchors ranging from 1 (Not sore) to 10 (Very, very sore). Participants were prompted to rate their perceived level of soreness and identify the location by marking the diagram accordingly. This was repeated without supervision +24 and +48 hours post-testing.

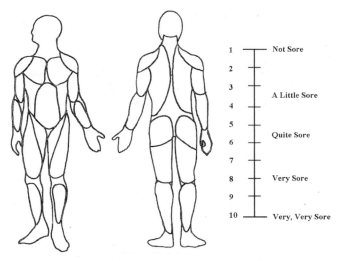

Fig. 1. Soreness diagram and accompanying visual analogue scale

Repeated measures t-tests were used to investigate any differences in data between the high pressure, low pressure and no pressure treatments. Significance was taken at $p<0.05$. All results are presented as mean values ± 1 standard deviation.

3 Results

3.1 Shuttle Session

Consistent heart rate plots were recorded by all subjects across each of the shuttle run sessions. The heart rates observed during all trials ranged from a minimum of 118.9±12.3bpm to a maximum of 181.8±9.1bpm. These heart traces were also used to check consistency of the shuttle run sets and sprint component durations. The plots show that both the set and sprints were consistent for all subjects with 6.8±0.7 and 3.0±0.3min for the set and sprint times respectively.

3.2 Treatment Session

Over the hour of treatment, hart rate gradually approached resting values from elevated rates immediately post-shuttle run. During this time both the high and low pressure treatments produced significantly lower mean heart rate values than the no treatment condition (83±6.9 and 82.4±9.1 vs. 87.3±8.1bpm respectively). Incidentally the high and low pressure treatments were not significantly different. Diastolic blood pressure was also significantly higher without any leg compression (76±2.4 vs. 71.7±2.0 and 71.5±1.9mmHg for the high and low pressure treatments respectively). As per the heart rate plots, there was no significant difference between high and low pressures.

3.3 Performance Measures

There was no significant or consistent change in the measured calf and thigh circumferences. Vertical jump performance was reduced on all occasions however the magnitude of the reduction was smaller following high and low pressure IPC treatments. The high pressure treatment produced a significantly smaller mean reduction than both the low pressure treatment and no treatment (1.9±1.4 vs. 4.4±3.8 and 5.9±3.4cm respectively).

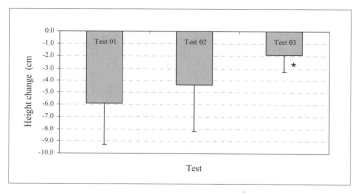

Fig. 2. Reduction in vertical jump performance (*significant, $p < 0.05$)

3.4 Soreness Measures

Mean perceived soreness measured at the shins, calves, quadriceps and hamstrings was significantly reduced by high and low pressure IPC treatment +1 hour (2.1±1.2, 3±1.1 c.f. 4±1.5), +24 hours (1.3±0.2, 2±0.8 c.f. 3±1) and +48 hours (0.6±0.2, 1.1±0.3 c.f. 1.9±0.7) compared to no treatment after the shuttle runs. In each case the low pressure treatment produced significantly lower ratings than after no treatment, and the high pressure treatment produced significantly lower ratings than both the low pressure treatment and no treatment.

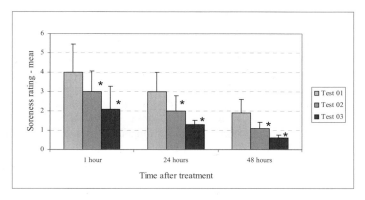

Fig. 3. Mean soreness rating for shin, calf, quadriceps and hamstring (*significant, $p<0.05$)

4 Discussion

Analysis of the shuttle run heart rate data shows that good consistency was maintained throughout the trials in terms of duration and effort of each shuttle component. The relative changes in the remaining data can therefore be considered reliable.

During the IPC treatment session, heart rate and blood pressure were both lower compared to the no treatment condition. It is interesting to note however that both the low and high pressure values are similar. This could suggest that only a low pressure is necessary to reduce cardiac work, however, further inspection of the equipment highlights a more likely explanation. The low pressure IPC pump applied pressure to both legs simultaneously whereas the high pressure pump compressed each leg in turn. This may have resulted in a lower net effect with the leg under high pressure and the leg receiving no pressure forming a lower equilibrium comparable to the lower pressure treatment. This factor does not seem to have had an effect on the performance and soreness data.

The vertical jump results clearly show a benefit of IPC given the reduction in the performance deficit. This improvement is greatest and reaches significance with the high pressure treatment suggesting that IPC is capable of maintaining performance levels even after high intensity work. The mechanism for this benefit may be explained by the soreness data given that the lowest soreness values were recorded following the high pressure IPC treatment. This reduced soreness may have permitted a more pain free muscle activation in the final vertical jump manoeuvre. Interestingly there is also a reduction in soreness with the lower pressure treatment. This suggests that there may be other mechanisms contributing to the reduced soreness such as the thermal insulation or supportive benefits. In addition to this it is clear that this treatment could be optimised. The pressures applied were in accordance with the manufacturer's instructions and chosen to produce a magnitude of effect in blood dynamics. A program of optimisation may be required to better understand the benefits of increasing pressure and perhaps the application of heat or cold.

We had expected to see some increase in calf or thigh circumference brought on by fluid collecting in the limbs during treatment. We hypothesized that the IPC treatment would abate this swelling. The data suggests however that there is no change. This could be due to insufficient muscle damage taking place, however, it is more likely that the results are due to postural differences. The first measurement was taken on arrival at the test session on a cold muscle whereas the second measurement was taken immediately following 1 hour in the supine position. It may therefore have been more appropriate to take a third measurement between the completion of the shuttle run and prior to the treatment session.

5 Conclusion

This study has shown that IPC provides benefits in terms of abating performance reduction following high intensity exercise and reduces soreness shortly after exercise as well as 48 hours later. This favourably suggests that athletes undertaking IPC as part of their training regime should be able to increase their training volume with a reduced risk of discomfort and injury. Further investigations will be useful in optimising the technology thus increasing the magnitude of these effects.

Acknowledgements

The authors would like to thank Huntleigh Healthcare for supporting the project and supplying the IPC equipment and garments.

References

Jackson, A.S. and Pollock, M.L. (1985) Practical assessment of body composition. Phys. Sports Med. 13, 76-90.
Morris, R.J. and Woodcock, J.P. (2002) Effects of supine intermittent compression on arterial inflow to the lower limb. Arch. Surg. 137, 1269-1273.
Morris, R.J. and Woodcock, J.P. (2004) Evidence-based compression: Prevention of stasis and deep vein thrombosis. Ann. Surg. 239, 162-171.
Nicholas, C.W., Nuttall, F.E. and Williams, C. (2000) The Loughborough Intermittent Shuttle Test: A field test that simulates the activity pattern of soccer. J. Sports Sci. 18, 97-104.
Pappas, C.J. and O'Donnell, T.F. Jr. (1992) Long-term results of compression treatment for lymphedema. J. Vasc. Surg. 16, 555-564.
Ramsbottom, R., Brewer, J. and Williams, C. (1988) A progressive shuttle run to estimate maximal oxygen uptake. Brit. J. Sports. Med. 22, 141-144.
Thompson, D., Nicholas, C.W. and Williams, C. (1999) Muscular soreness following prolonged intermittent high-intensity shuttle running. J. Sports Sci. 17, 387-395
Wiener, A., Mizrahi, J. and Verbitsky, O. (2001) Enhancement of tibialis anterior recovery by intermittent sequential pneumatic compression of the legs. Basic Appl. Myol. 11, 87-90.

A New Protocol for Testing the Tensile Strength of Rugby Garments – A Preliminary Study

Bryan C. Roberts and Mike P. Caine

Wolfson School of Mechanical and Manufacturing Engineering, Loughborough University, Loughborough, UK, B.C.Roberts@lboro.ac.uk

Abstract. Contact sports such as rugby demand high performance apparel capable of withstanding large stresses during game-play. In order to ensure that failures are minimised the fabric requires testing. Current British standards detail methods of tensile testing fabrics from bulk stock as opposed to fully manufactured garments. It is important for the sporting goods industry that tensile strength data is tested from panels of the made-up garment instead of material from bulk stock. Generally manufacturers are interested in the minimum force required for a garment to fail. This paper describes a protocol for tensile testing of made-up rugby shirts and their critical components. Summarised are findings from one full protocol test which illustrate typical strengths, and design weaknesses for a modern rugby shirt.

1 Introduction

In today's professional game of rugby, players normally wear multi-panel, synthetic shirts which incorporate the latest in textile technology. The modern shirt is typically lightweight, with high levels of moisture management and comfort cooling properties desired. As a result the shirts have become less robust and exhibit a tendency to rip during competitive play. Designers and testers alike require a simple reproducible method of testing shirts during development so as weaknesses or faults in the design, or choice of fabric, can be identified and remedied. Currently the major publicly available means of testing fabric is by methods detailed in textile standards. International, British and American standards describe methods of testing bulk stock fabric before it is cut and stitched to make the final garment. Many authors (including Furter 1985 and Saville 1999) conclude that the material's mechanical properties are altered during the manufacturing process. Therefore in today's sporting goods industry it would be more prudent to test the garment after production.

Modern principles of tensile testing suggest that opposite parts of a fabric specimen should be gripped and pulled apart until the specimen breaks. BISFA (2000) describes the breaking force as the maximum force applied to a test specimen carried to rupture during a tensile test. The term strength at break is commonly misused to represent the average of the results of breaking force measurements. Saville (1999) explains that the tensile strength at break may not be the maximum force as elongation of the specimen may continue after the maximum tensile force has developed. This study describes a protocol for testing the tensile strength of garments using commercially available instruments and modified strip method based British standards.

2 Proposed Protocol

Literature concerning the tensile testing of fibres, yarns and fabrics from bulk stock is extensive. However, there is little research detailing the tensile testing of finished garments, indeed we are unaware of any test recommendations for rugby shirts. Previous standards are used as a benchmark in order to suggest a new tensile testing protocol. Two were chosen for modification: BS EN ISO 13934-1:1999 and BS EN ISO 13935-1:1999. These have similar principles in that the breaking force is recorded after a fabric specimen of specified dimensions (the latter having a seam in the middle) is extended at a constant rate until it ruptures. Furthermore, both standards are carried out in a controlled atmosphere, as described in BS EN ISO 139: 2005. Likewise both methods use a Constant Rate of Extension (CRE) machine in accordance with BS EN ISO 10012: 2003.

2.1 Rate of Extension

Both standards dictate a constant rate of extension of $100mm \cdot min^{-1}$ however, according to Nishikawa, Murray and Flanders (1999) a person's hand velocity, when reaching, can exceed $200 \ cm \cdot sec^{-1}$ (or $120000mm \cdot min^{-1}$). Consequentially a much higher constant rate of extension than that described in the standard would be desirable. Unfortunately however, the rate we were able to use was dictated by the maximum attainable on the CRE machine available, namely $500mm \cdot min^{-1}$. Furter (1985) notes that the maximum force can be affected by the rate of extension; the breaking force increases, with all staple yarns, with an increasing extension rate. The percentage increase is dependant on the type of fibre and blends, but typically for cotton based fabrics an increase of <5% is expected up to $500mm \cdot min^{-1}$.

	BS EN ISO 13934-1:1999	BS EN ISO 13935-1:1999
Width of specimen (mm)	50mm ± 0.5mm	50mm
Gauge length (mm)	200mm ± 1mm ≤ 75% elongation at maximum force; 100mm ± 1mm > 75% elongation at maximum force	200mm ± 1mm
Overall specimen width (mm)	50mm ± 0.5mm	100mm
Overall specimen length (mm)	"long enough to allow gauge length"	"long enough to allow gauge length"
Bulk specimen length (m)	> 1m	≥ 700mm
Bulk specimen width (m)	"full width" at least 3m from the end of a bulk piece	350mm
Test specimen distance from edge of bulk (mm)	150mm	≥100mm

Table 1. Strip specimen sizes from British standards

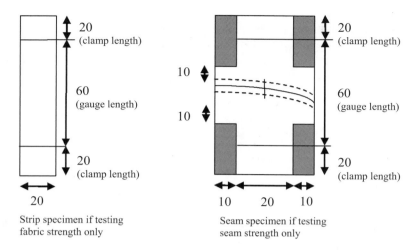

Fig. 1. New protocol specimen sizes (not to scale). Numbers denote dimensions in mm.
Shaded areas are removed prior to testing of the specimen.

2.2 Specimen Sizes

Both standards allow interested parties to agree on specific sampling procedures,
however in the absence of an appropriate procedure or material specification BS EN
ISO 13934-1:1999 and BS EN ISO 13935-1:1999 suggest the following (Table 1).

The main difference between sampling bulk stock and finished garments is that
the space envelope for testing is significantly reduced. Therefore specimen dimensions were reduced as per Fig. 1. A minimum of two specimens per area are required.

It is important that the gauge length (distance between the two effective clamping
points of a testing device, BS EN ISO 13934-1:1999) of all specimens remains constant. According to Furter (1985) and Saville (1999) the maximum tensile strength is
directly proportional to the gauge length. Tensile strength of textiles is determined by
the "weak-link effect" whereby the textile will break at its weakest point. The longer
the specimen the greater the probability that it will have more weak points, and thus
break at a lower load. In reality though, as Saville (2004) describes, faults will be
randomly distributed along the length of the material with a normal distribution of
strengths around the mean value. By reducing the gauge length we would expect our
results to mimic that of the mean tensile strength of the material and as a consequence possibly overestimate the strength of a larger fabric panel.

2.3 Bias Specimens

A tackle occurs when the ball-carrier is held by one or more opponents and is
brought to ground (IRB Law 15 2005). When tackled the fabric on the jersey may be

pulled in one of many directions. The standards describe methods of testing in the warp and weft direction only. This does not meet the demands of rugby and therefore bias specimens specifically 45°bias and 135°bias should be tested. Initial analysis of reference specimens (see 2.5) will highlight if bias specimens are required throughout the rest of the shirt.

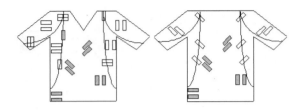

Fig. 2 An example protocol showing placement of specimens on the front and back of a shirt (not to scale). Grey depicts reference specimens.

2.4 Reference Specimens

The main panel, usually at the front and back of the shirt, should be divided into six specimens; two in the warp, weft, 45°bias and 135°bias. None of these specimens should contain the same warp or weft threads and are referred to as reference specimens see Fig. 2. The reference specimens provide a benchmark against which other specimens can be compared. This allows the effect of seams or two multiple fabric specimens to be ascertained.

3 Methodology

The tensile strength of one international prototype shirt was tested to highlight potential weaknesses. In total, 55 specimens were tested. Specimens were cut from each shirt using sharpened scissors, coded according to orientation, and dimensioned to within 0.5mm width and 1mm gauge length following the above procedures. All specimens were tensioned using an Instron 3366 CRE machine, with 25mm x 50mm jaws, mean temperature 20.03°C ± 1.12SD and mean humidity 65.1% ± 1.35SD.

Data were recorded electronically via a computer linkup to the Instron machine. Force (N) and extension (mm) were collected every 40ms. Force versus extension graphs were automatically produced and results collated for comparison.

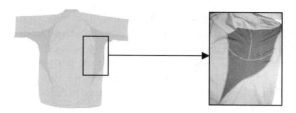

Fig. 3. Weakest part of protocol shirt.

4 Results

The maximum load that a unseamed specimen could withstand was 337.86N, whilst the maximum load that a seamed specimen could withstand was 250.73N. The minimum load that a unseamed specimen could withstand was 131.92N, whilst minimum load that a seamed specimen could withstand was 85.91N. Specimens cut from the shirt in the weft direction were weakest; warp fabric specimens were around 70% stronger than weft specimens. Only six specimens ruptured across the seam; in the others the fabric failed first. One of these seam specimens was considerably weaker than any other part of the shirt. This seam ruptured at 85.91N; 61% of the strength of the next weakest seam specimen (141.29N), the site is depicted in Fig. 3.

5 Discussion

The data obtained provided an invaluable insight into the overall strength characteristics of the shirt. Of particular value was the ability to identify one discrete weak spot as shown in Fig. 3. In this instance the vertical seam acts like a rip cord against the horizontal seam when tensioned in the warp direction along the vertical seam. Designs like these if left unchanged increase the likelihood of failure during match play.

Each rugby garment manufacturer has their own particular traditions and style of shirt, therefore it would be impossible to detail an exact placement of every specimen. Instead future tester should follow these guidelines and test the appropriate area:

1. Different fabrics sewn into the main body of the garment should be sampled including their interactions with the other fabrics around it. The seam joining two panels should be in the middle of the test specimen either in the warp or weft direction as appropriate.
2. The tester should sample seam junctions in the warp and weft direction i.e. where different seams meet.

3. Specimens should be taken were panels are connected by a new type/ design of stitching. Seam positioned in the middle of the specimen.
4. For seamless garments, the point where two different fibres or fabrics join should be tested in both the warp and weft direction if possible.
5. Specimens should not touch the edge of the garment (the hem, sleeve, or collar); 15mm distance from the edge is preferable.
6. Two fabric strip specimens from the same panel should not touch. They must be separated by a minimum of 5mm.
7. As in British standards two seam specimens can touch and lie next to each other.
8. Reference samples ought to be included.

6 Conclusions

With the ever increasing demand for high quality garments, designers and manufacturers require suitable testing methods to ensure their garments, in this case rugby shirts, meet the desired mechanical performance specifications. The tensile strength of garments is critical in the game of rugby and other contact field sports. This study aims to provide the user with a method of testing garments for tensile breaking force focusing on the strip method. It is evident from the example shown that designers can create weaknesses in garments simply by altering the style. By using this protocol designers can be more confident in their finished product; furthermore they should be able to condense the development time. Ultimately, tests of the nature described, help garment designers to produce more functional and comfortable shirts whilst ensuring that they can withstand the demands of rugby. Further development and validation of the test methodology described is justified.

References

BISFA. (2000) *Terminology of man-made fibres.* BISFA- The International Bureau for the Standardisation of man-made fibres, Belgium.
BS EN ISO 10012:2003 Measurement management systems. Requirements for measurement processes and measuring equipment.
BS EN ISO 139: 2005 Textiles – standard atmospheres for conditioning and testing.
BS EN ISO 13934-1:1999 Textiles- tensile properties of fabrics- determination of maximum force and elongation at maximum force using strip method.
BS EN ISO 13935-1:1999 Textiles- seam tensile properties of fabrics and made-up textile articles – determination of maximum force to seam rupture using the strip method.
Furter, R. (1985) *Strength & elongation testing of single & ply yarns: experience with USTER tensile testing installations (Mott-Qcas).* Textile Institute and Zellweger USTER AG, Manchester.
IRB. (2005) *The laws of the game of rugby union.* IRB, Ireland.
Nishikawa, K.C., Murray, S.T., and Flanders, M. (1999) Do arm postures vary with the speed of reaching? Journal of Neurophysiology. 81, 2582–2586.
Royds, P. (Sir) (1949) *The history of the laws of rugby football.* Walker & Co, London.
Saville, B.P. (1999) *Physical testing of textiles.* Woodhead Publishing Ltd, Cambridge.

Modelling Traction of Studded Footwear on Sports Surfaces using Neural Networks

Bob Kirk[1] , Matt Carré[2], Stephen Haake[3] and Graeme Manson[4]

[1] University of Sheffield, Sports Engineering Research Group, bob.kirk@shef.ac.uk
[2] University of Sheffield, Sports Engineering Research Group, m.j.carre@shef.ac.uk
[3] Sheffield Hallam University, Sports Engineering, CSES, s.j.haake@shu.ac.uk
[4] University of Sheffield, Dynamics Research Group, graeme.manson@shef.ac.uk

Abstract. Traditional regression techniques have shown limited use in the development of empirical models for the traction performance quantities of studded footwear on surfaces. This is due to the unknown and often non-linear relationships between performance parameters, such as traction force, and input variables, from the shoe and surface. Experimental data has been used to train artificial neural networks to model the relationship between stud parameters, namely cross-sectional area, length and two shape coefficients, with dynamic traction as the output variable. A variety of neural network structures and optimisation algorithms were evaluated. The most promising network gave an average prediction error of 10%, compared to an error of 36% when an optimised linear model is employed. This study shows that the neural network technique has powerful potential in understanding the effect of shoe and surface parameters and in the optimisation of traction forces experienced by athletes.

1 Introduction

The interaction of studded footwear on sports surfaces plays a major role in terms of athlete performance and injury risk. In order for footwear and surface manufacturers to better their respective products it is vital to understand the influence of the input parameters, such as stud length and shape, on the performance criteria, such as the traction force and penetrability.

Sports literature contains many examples of authors who have measured traction forces and moments by means of devices which exhibit simplified shoe-surface interactions. For example, McNitt, Middour and Waddington (1997) used a device capable of measuring translational traction and rotational traction, the latter often linked to the increased risk of cruciate knee ligament injuries (Torg, Theodore, Quedenfield and Landau 1974). McNitt *et al.* (1997) investigated the effect of grass cutting height showing that shorter grass provided higher traction values. Nigg (1990) has shown that traction forces do not follow a linear relationship with normal force, opposing the classical Coulomb friction law, highlighting the risk of extrapolating relationships observed at unrealistic conditions to those observed in realistic athlete situations, as attempted by Bowers and Martin (1975).

Authors have also attempted to understand shoe-surface interactions from measurements of ground reaction forces and pressures by virtue of in-sole pressure sensors and forces plates. Asai, Igarashi, Murakami, Carré and Haake (2002) carried out such studies but it is difficult to pinpoint the influence of specific aspects of different shoes as the shoes used in these studies differed greatly in terms of material, cleat size and cleat shape.

The authors have carried out traction experiments using a sled, with different types of studs attached, pulled across the surface using a constant speed motor (Kirk, Haake, Senior and Carré 2005). The methodology of such testing aims to identify, isolate and control stud and surface variables in order to achieve a clear understanding of their role on traction parameters.

This paper concentrates on the stud variables, with all traction measurements taken on a single artificial surface, XL turf. As a result, traction quantities will depend on:

$$T = f(n, A, l, \mu, N, v, S_1, S_2)$$ (1)

where T is the traction performance parameter; n is the number of studs attached to the plate; A is stud cross-sectional area (m^2); l is stud length (m); μ is the coefficient of friction between the stud and surface; N is the normal force (N); v is velocity (ms^{-1}); S_1 and S_2 are dimensionless shape parameters.

The results presented in this paper were obtained in an attempt to reduce the amount of input parameters required to train a neural network (discussed later). The studs tested were all aluminium and between 6 and 18 mm in length. Five of each stud type were used and all experiments were carried out at a constant normal force. Consequently, the parameters in Eq. 1 can be reduced so that only area, length and shape coefficients are considered. Rectangular, hexagonal and cylindrical studs were oriented to give 5 different effective stud shapes orthogonal to the direction of travel. Figure 1, shows the 2 shape coefficients defined.

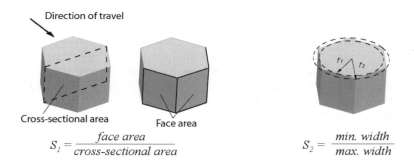

$$S_1 = \frac{face\ area}{cross\text{-}sectional\ area} \qquad S_2 = \frac{min.\ width}{max.\ width}$$

Figure 1- Shape coefficients S_1 and S_2 defined from stud geometry.

It is the authors' intent to determine an empirical relationship for the parameters in Eq. 1.

It has been shown (Kirk *et al.* 2005) that linear fitting of independent (input) variables, such as stud cross-sectional area and length, to dependent (output) variables, such as dynamic traction force, using correlation analysis in the form of Pearson correlation coefficients is not adequate. First, some of the variables are inter-related.

For example, as stud length increases, stud area will also increase for a fixed stud width. Second, it has been observed (Kirk *et al.* 2005) that the input variable versus output variable plot is often non-linear.

More complex regression analysis can be carried out by assigning input variables higher order polynomial expressions, or other functional forms. Techniques such as analysis-of-variance, as attempted by Owende and Ward (1996), can then be used to determine the required constants and evaluate the fit of the chosen relationship. The drawback with this approach is that the functional form of the relationship, for example the order of the polynomial, must be predefined. The inter-relating nature of the input parameters in traction further complicates this task.

2 The Neural Network Technique for Regression

Artificial neural networks are based on the parallel architecture of the human brain. A network consists of layers of artificial neurons which interconnect, as shown in Fig. 2.

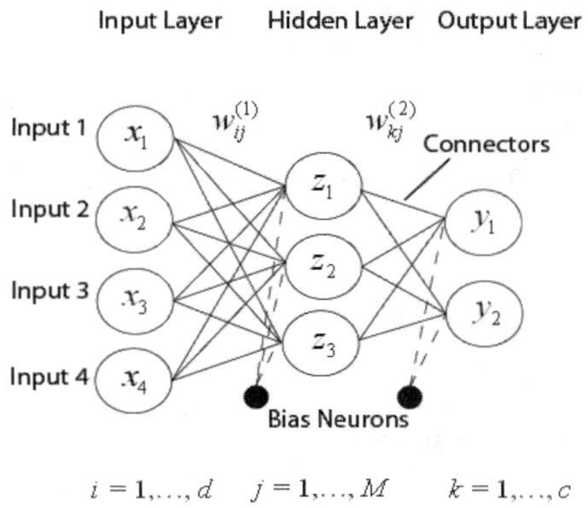

$$z_j = \tanh\left(\sum_{i=1}^{d} w_{ij}^{(1)} x_i + b_j^{(1)}\right) \quad (2)$$

$$y_k = \left(\sum_{j=1}^{M} w_{kj}^{(2)} z_j + b_k^{(2)}\right) \quad (3)$$

$$i = 1,\ldots,d \quad j = 1,\ldots,M \quad k = 1,\ldots,c$$

Fig. 2. Schematic of Artificial Neural Network.

The neurons in the first level correspond to the input variables while the output neurons represent the output variables. Each input value is multiplied by the weight, w_{ij}, of each connector. The hidden neurons pass the sum of their respective inputs through hyperbolic tangent activation functions. Finally, the sum of the products of hidden values, z_j, and weights gives the output value, y_k. Neural networks learn by example. By feeding the network known values of inputs and outputs, optimisation algorithms are used to determine the optimum neuron weights to fit the inputs to the outputs. This procedure is referred to as training the network. A network consisting of only a layer of inputs and a layer of outputs performs classical linear regression. By employing a hidden layer of neurons the complexity of the function describing the input to the output can be arbitrary (Hornik, Stinchrome and White 1989).

Neural networks have been used for regression problems where the influences of multiple parameters on an output are unclear and where the measurements of input variables are intrinsically noisy. Successful applications include: optimisation of engine efficiency and performance in engine management systems based on sensory readings of engine temperature, speed and pressure; and cooking time evaluation based on food type and weight in microwave ovens. Neural networks have also been widely used in probabilistic classification problems such as medical screening programmes.

Three sets of data are desired to implement a successful neural network. First, a set of data is required to train the network as described above. A second set of data is then required to ensure that the network does not over-learn, simply fitting the noisy data rather than the underlying function, analogous to using a polynomial curve of too high an order to fit data. Thus a validation set of data is used, which has not been used during training, to stop the training when the validation error reaches a minimum. Last, a test set of data is required as a final assessment of the network (Tarassenko 1998). Ideally the number of data patterns (known inputs and outputs) employed in each of the three sets should be equal.

There is no concrete rule as to the number of patterns a network requires for successful learning, as this depends on the complexity of the unknown network function. Information theory suggests that this should be at least equal to the number of connectors in the network (Tarassenko 1998) whereas Widrow and Lehr (1990) have argued this number should be several times larger for good learning.

Using a network with many hidden units increases the complexity of the underlying network function but also the amount of patterns required to train the network, as the number of connectors is increased.

3 Applications of Neural Networks to Traction Data

The experimental measurements were carried out with the prior intent of training a neural network. The minimum amount of parameters were varied to reduce the amount of experimental results required. Twenty-four patterns were obtained for the 4 input variables discussed in Section 1 with dynamic traction as the output. According to Tarassenko (1998) a network of 3 hidden units requires at least 19 training patterns, while 4 hidden units requires 25 patterns, to be confident that the network is learning the underlying function rather than that of the noise. Due to the small amount of data available in the current problem, 22 patterns were used as training data, while the 2 remaining patterns were used to evaluate the network. The patterns to be used in each set were chosen at random 10 times to give 10 different training sets and 10 test sets. Similar processes of cross-validation have been used where training data is scarce in order to ensure as much data as possible is used in training the network (Bishop 1995). It is, however, acknowledged that more data is needed in future work to train, validate and test a reliable network.

Two types of neural networks have been evaluated for the traction data: a Linear Model (LM) using the iterated-least squares optimisation technique, used as a benchmark to the non-linear neural network; and a Multi-Layer Perceptron (MLP) using 3 different optimisation algorithms. Each model was allowed to train for between 100 and 1000 epochs, for 1 to 20 hidden units. The neural network package Netlab was used which is a series of Matlab programs (Nabney 2002).

4 Results

Table 1 indicates the training performance of key networks. It can be seen that the training error decreases with the number of hidden units. However, the test error increases for more than 3 hidden units, showing that more complex networks tend to provide a better fit of the data they are trained on, but not of the underlying function between traction and the input variables. The best performing network consists of 3 hidden units and has a average test error of 10.1% with a standard error of ± 3.1% for the 10 networks evaluated on different data samples.

Type	Hidden Units	Optimisation Algorithm	No. of iterations	Ave. % Training Error	Ave. % Test Error
LM	-	Re-iterated Least Squares	-	14.2	35.6
MLP	2	Scaled Conjugate Gradients	100	7.9	15.2
MLP	3	Scaled Conjugate Gradients	400	2.6	10.1
MLP	4	Scaled Conjugate Gradients	500	0.7	12.3
MLP	5	Quasi-Newton	400	0.3	15.6
MLP	10	Conjugate Gradients	200	0.1	22.0

Table 1- Summary of training performances of neural networks of varying architecture

Figure 3 shows plots of traction force versus stud area and length for different values of shape coefficients. Available experimental points are also plotted, which can be seen to lie close to the networks' surfaces. Such neural network outputs can be used to understand the role of each input variable on traction force.

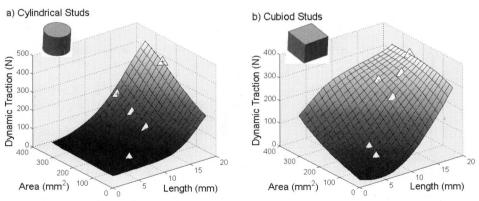

Fig. 3. Plots of dynamic traction versus stud area and length for 2 different shape coefficients: a) Cylindrical studs; b) Cuboid shaped studs. Experimental data shown as white triangles.

Fig. 3a) shows that the rise in traction force with length is greater for studs of larger cross-sectional area. The differing surface plots between Fig. 3a) and Fig. 3b) highlight the significant effect of the shape parameters on traction. For example, changing a stud of 12 mm length and 18 mm diameter from a cylinder to a cuboid causes a 20% rise in traction. This is one example of the predictive capabilities trained neural networks offer researchers and manufacturers of shoes and surfaces.

5 Conclusions and Future Work

This study has shown that artificial neural networks provide strong potential in the understanding of stud-surface interactions. It has been shown that a simple multilayer perceptron with 3 hidden units is capable of predicting traction forces to within 10% of experimental values, based on only 22 experimental measurements. More training data is required before such a network can be implemented as a reliable predictive tool. Networks which account for more variables are to be investigated. This will require many more experimental measurements to ensure accurate learning as the complexity of the network increases.

References

Asai, T., Igarashi, A., Murakami O., Carré M. J., Haake S. J. (2002) Foot loading analysis of primary movements in football. In: S. J. Haake, S. Ujihashi (Eds.) *The Engineering of Sport 4*. Blackwell Science Ltd., pp 602-608.

Bishop, C. M. (1995) *Neural Networks for Pattern Recognition.* Oxford University Press, Oxford.

Bowers, K. D. and Martin, B. (1975) Cleat-surface friction on new and old astroturf. Medicine in Science and Sports and Exercise. **7,** 132-135.

Hornik, K. M., Stinchrome, M. and White, H. (1989) Multilayer feedforward networks are universal approximators. Neural Networks, **2,** 359-366.

Kirk, R. F., Haake S. J., Senior, T. and Carré, M. J. (2005) Determining relationships associated with traction of studded footwear on artificial sports surfaces. Asia-Pacific Congress on Sports Technology, Tokyo, Japan, 336-342.

McNitt, A. S., Middour, R. O. and Waddington, D. V. (1997) Development and evaluation of a method to measure traction on turfgrass surfaces. Journal of Testing and Evaluation, **25,** 99-107.

Nabney, I. T. (2002) *NETLAB- Algorithms for Pattern Recognition.* Springer, London.

Nigg, B. M. (1990) The validity and relevance of tests used for the assessment of sports surfaces. Medicine and Science in Sports and Exercise. **22,** 131-139.

Owende, P. M. O. and Ward, S. M. (1996) Characteristic loading of light mouldboard ploughs at slow speeds. Journal of Terramechanics, **33,** 29-53.

Tarassenko, L. (1998) *A Guide to Neural Computing Applications,* Arnold, London.

Torg, J. S., Theodore, M. D., Quedenfield, C. and Landau, S. (1974) Journal of Sports Medicine. **2,** 261-269.

Widrow, B. and Lehr, M. A. (1990) 30 years of adaptive neural networks: perceptron, madaline, and backpropagation. Proceedings of the Institute of Electrical Engineers, **78,** 1415-1442.

The use of Stereoscopy and High-Speed Video for the Measurement of Quasi-Static and Dynamic Shoe Loading Scenarios

Paul J. Gibbs, Sean R. Mitchell, Andy R. Harland

Loughborough University, p.j.gibbs@lboro.ac.uk

Abstract. The need for efficient and cost effective methods of recording the structural response of running shoes under simple mechanical testing was identified in order to allow virtual prototyping. Digital video was chosen over other methods due to its availability and relatively low cost. Two complementary experiments were set up to test the performance of the data collection methods: (1) Quasi-static, uni-axial loading was applied up to 1.5 kN (+/-1N) over a period of 30 seconds and the 3D structural response recorded using two 50 Hz video cameras, (2) Dynamic, uni-axial loading to 1.5 kN (+/-200N) was performed using a drop-test rig. Shoe deformation was recorded using a high-speed video system operating at 4 kHz. Testing was performed on two shoe types with no uppers, one with a foam midsole, one with a TPU structure midsole. Analysis of the video data was done using ImageJ in conjunction with Visual Basic routines. Results showed a resolution of 0.68 mm for the quasi-static tests (in the plane of the camera) and 0.46 mm for the dynamic testing. These results were acceptable for use, although the resolution of depth using the two 50 Hz camera was only 6.81 mm, so use of another 3D method is recommended.

1 Introduction

The midsole of a running shoe exists to provide cushioning during a running strike. The emergence of structure based midsole designs requires more design iterations or the use of virtual prototyping to reduce potential structural failures. Traditional material testing (such as ASTM F1614-99) does not provide sufficient information to allow full verification of virtual prototypes. To better understand the structural response of these irregularly shaped, highly non-linear moulded materials, a set of preliminary tests was conducted to look at the spatial motion of points on the midsole coupled with force and cushioning response.

The emphasis of this paper is to report the spatial recording methods under simple repeatable loading. The loading is not necessarily intended to provide conditions representative of a running foot strike.

2 Testing Methods

2.1 Sample Preparation

To allow tracking of points on the surface of the midsole during loading, the midsoles were sprayed with a light coating of either black or white paint, to match the colour of the main midsole material. This was done to avoid any erroneous results occurring from cracks in the paint during the tests. Pins with 4 mm spherical heads were then inserted into the samples at points of interest. The pin points were cut down to around 5 mm to minimise any effect on the properties of the materials while still restraining the pin heads from vibration. The pins used for each shoe were of a contrasting colour; originally white pins on black shoes and red pins on white shoes (red was chosen as preliminary tests indicated it gave better contrast under the lighting conditions). Shoes were given a light coat of anti-shine spray immediately before testing to reduce unwanted glare. Tests were then performed against a background the same colour as the shoe, thus highlighting the pinheads.

2.2 Quasi-static

Testing was performed on an Instron 3366 series (screw-thread) machine. To remove any significant inertial effects, while still allowing testing to be completed in the available time, each test was performed over 30 seconds. Running impacts are generally accepted to give peak forces of 2 kN (Bates *et al.* 1983), but as the test samples had no upper sections (which contain some cushioning materials) samples were tested to 1.5 kN to avoid permanent damage to the limited number of samples before dynamic testing.

Four midsoles of two types: adidas Supernova (foam cushioning) and adidas Ultraride (Thermoplastic Polyurethane (TPU) structure cushioning) were tested, with each midsole being loaded over 30 seconds, unloaded over 3 seconds and rested for 30 seconds. This was repeated 3 times in succession for each midsole. Impacting was carried out using a 50 mm diameter domed indenter in the heel and a 50 x 77 mm asymmetric indenter in the forefoot region.

To record the structural response of the shoe, the test was videoed using two Canon XM1 miniDV digital video cameras. A stereoscopic set-up was used (Fig. 1) to investigate the feasibility of generating depth data in addition to the horizontal and vertical position data required. Working space around the Instron machine was limited, so the cameras were mounted next to each other with a lens separation (base

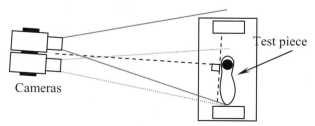

Fig. 1. Stereoscopic (static) test set-up

pair distance) of 120 mm. Correct alignment of both cameras was achieved by sighting the image on each camera through holes accurately drilled in a plate centered about the impactor axis of the Instron. By aligning the holes in the center of the screen, the centers of the CCD of the cameras could be set at 120 mm and parallel to each other.

2.2 Dynamic Testing

Testing was performed using a drop-test machine fitted with the same impactor as used in the quasi-static testing. A weight stack of mass 70 kg corresponding to the generalised 'average' runner was dropped from a set height onto each midsole by use of a winch and quick-release catch. Due to the nature of the drop test, each midsole was subjected to a repeatable initial impact, followed by a few smaller impacts, the number and energy of these impacts being dependent on how the weight stack rebounded. The midsole was then rested for at least 3 minutes before being re-tested. This was repeated 3 times for each of the 4 midsoles of each type.

An accelerometer was fitted to the impactor to allow determination of the force-time curve during the test. Spatial displacements of the shoe and the impactor were recorded using a Photron Fastcam APX high-speed digital camera at 4 kHz. Fans were used to keep the test area at room temperature (22 +/- deg. C) under the camera's spotlights. The aim of the dynamic test was to create a peak impact force of 1.5kN, to allow comparison with the quasi-static data. This value was derived from the accelerometer readings, and was repeatable within a range of +/-200N. All tests showed a loading time of 0.04 (+/-0.0025) seconds for the primary impact, a duration which is similar to that described in Knoerr and Rouiller (2002).

3 Analysis

3.1 Object Recognition

Spatial measurement results in a large amount of data being generated. Each impact produces around 1000 images, each with 30 points, so an automatic tracking system was developed.

Video was decompressed into AVI format using RADVideo (RAD Game Tools). ImageJ (Rasband 1997), was then used to create macros to automatically recognise and record positions of the midsole pins. For all the tests, the video images were first cropped, then thresholded to remove unwanted objects in the frame. ImageJ's 'Analyze Particles' function was then used to find and analyse the remaining objects

During analysis of the quasi-static results it was apparent that while using red-on-white pins it is only possible to get reasonable results using a colour camera: the contrast between the red pins and the shadows in the image was not enough to distinguish them using grayscale video. As the high-speed camera was grayscale the white shoes were re-sprayed black, and white pins added (Fig. 2). Due to the filtering system in place in the tracking program, errors caused by cracks in the paint could be significantly reduced.

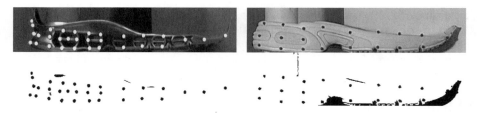

Fig. 2. Object lighting/colour results. Top row: Black shoe, white pins (left), white shoe, red pins. Bottom row: Respective recognised objects.

The resolution of the system was calculated to be 1 pixel = 0.46 mm for the high-speed camera in horizontal and vertical directions, and 1 pixel = 0.68 mm for the quasi-static camera. The resolution of depth for the 3D setup was 6.81 mm, which is too poor for use in this instance. The speed of all cameras was sufficient to capture motion without blurring for each test. Accuracy of both tests is slightly compromised due to positioning of the cameras; the stereoscopic set-up means both cameras must have the shoe fully in view, so part of the image is 'wasted' at either side on both cameras. Using one camera tightly cropped to the midsole would produce better results. Placement of a second camera at an angle, with the use of more complex calibration techniques, such as that detailed by Zhang (1999) should allow much better depth perception. The high-speed camera must be placed a sufficient distance away from the drop area to minimise vibration from the impact transferred through the floor. A longer zoom lens or better isolation of the camera would improve resolution for the dynamic test.

3.2 Particle Tracking

A MS Visual Basic program was used to organise the files and track the pin positions through each video. Objects not fitting user-inputted criteria of height, width, area, perimeter and circularity are removed by the first stage of the program. This leaves only the objects corresponding to pins, or objects that are visually similar (usually reflections). As a final stage of filtration, the user can input the positions of the pins in the first frame (by clicking on the image). The tracking algorithm then attempts to follow each selected object through to the next frame. A check is performed on each object, and the closest object within a given radius is considered the position of the given pin in the next frame (Fig. 3). As the maximum speed of a pin seen on any of the results is around 1.5 pixels per frame, this method gives good results without

Next position

Last position

Limiting radius

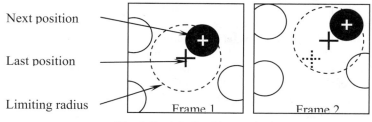

Frame 1   Frame 2

Fig. 3. Particle Tracking method

requiring more complex tracking algorithms. Should any object be obscured during a frame (often due to it merging with a reflection and being removed) the algorithm holds the last known position of the object and picks it up when it becomes visible again - provided it has not moved out of the given tracking radius (15 pixels).

3.3 Extraction of Results

The tracking algorithm outputs positions for each pin over time. This allows the extraction of various metrics from the results, by looking at the slopes and stationary points of the positional graphs. For example, the spatial position of the impactor will give the velocity and acceleration of the test, along with the maximum compression, rebound height and coefficient of restitution of the midsole. Tracks of other points allow comparison of two midsoles. In this paper the position of the tip pin is used as it is most distant from the impact location and so will be most dependant on any small differences in the midsole structure (the tip pin for forefoot testing is the furthest pin from the impactor that is reliably visible to the camera).

4. Results

The measurement system set out in this paper generates a large amount of data, which has been reduced to single value measurements for the purpose of presentation in this report. Figures 4 & 5 show values determined from the displacement of the impactor. Maximum compression is a direct measurement of the peak deflection of the upper surface of the midsole, while energy loss is calculated from the maximum height the impactor returns to on its first rebound.

Figure 6 shows the maximum displacement of the midsole tip during an impact. The spread within one type of shoe is similar to the maximum compression values, but the TPU shoes show a generally higher tip deflection. This measurement is a function of the stiffness and damping characteristics of a midsole, so it is not surprising that there are some differences between two shoe constructions.

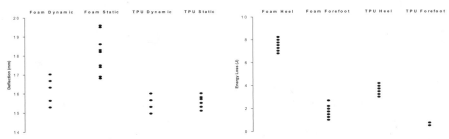

Fig. 4. Maximum compression of midsole heel region under heel impact test.   Fig. 5. Energy loss during heel and forefoot dynamic impacts.

Figure 7 illustrates the force-deflection response of a foam shoe under heel loading. The strain-rate dependant behavior of the midsole is clear as the dynamic test requires 30% less midsole compression to reach 1.5kN loading. The negative loading on the dynamic curve is due to the impactor rebounding and the shoe moving on the plate.

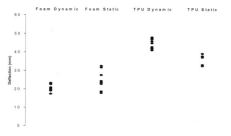

Fig. 6. Maximum deflection of Tip for quasi-static and dynamic heel impacts.

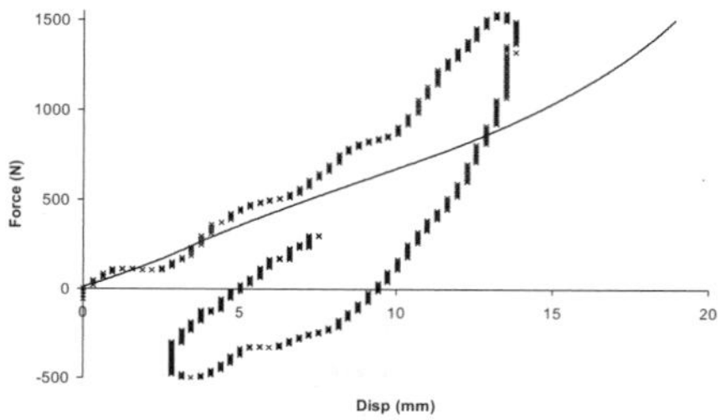

Fig. 7. Static (line) and Dynamic heel loading on a foam shoe.

5 Conclusions

The structural response of the shoe, measured by the maximum deflection of the tip, shows fairly good correlation between shoes of the same model and batch, and indicates that the TPU shoes may be stiffer in bending, especially considering that there is another foam section contained within its upper.

While there must be some debate on the worth of various structural metrics for the characterisation of shoe performance, the methods used to collect this data have shown themselves to be fast, cheap and reliable for 2-dimensional data collection.

The expandable capabilities of ImageJ make it a very powerful tool indeed, especially considering that its license requires that it be distributed free.

The stereoscopic technique used in this paper is not recommended, as the resolution is poor.

References

Bates, B.T., Osternig, L.P. Sawhill, J.A. and James, S.L., (1983) An Assessment of Subject Variability, Subject-Shoe Interaction and the Evaluation of Running Shoes Using Ground Reaction Force Data. Journal of Biomechanics, 16(3), pp.191-192.

Knoerr, K. and Rouiller, V. (2002) *Integrating High Performance PU-Materials in Sportshoe's,* Polyurethanes Expo, Oct 13-16 2002, pp533-537.

RAD Game Tools Inc., *The RAD Video Tools.* Available from: http://www.radgametools.com/bnkdown.htm [Accessed 10 October 2005].

Rasband, W.S. (1997) *ImageJ,* U. S. National Institutes of Health, Bethesda, Maryland, USA, http://rsb.info.nih.gov/ij/.

Zhang, Z. (1999) *Flexible Camera Calibration By Viewing a Plane From Unknown Orientations,* International Conference on Computer Vision (ICCV'99), Corfu, Greece, pp666-673.

Benchmarking Stiffness of Current Sprint Spikes and Concept Selective Laser Sintered Nylon Outsoles

Daniel Toon[1] Nico Kamperman[2] Uzoma Ajoku[1] Neil Hopkinson[1] and Mike Caine[1]

[1] Wolfson School of Mechanical and Manufacturing Engineering,
 Loughborough University, UK, d.toon@lboro.ac.uk
[2] TNO Science & Industry, De Rondom 1, Box 6235, 5600 HE Eindhoven (NL)

Abstract. Stefanyshyn and Fusco (2004) demonstrated that sprint performance can be improved if the stiffness of standard running spikes is increased. The authors concluded that in order to maximise performance individual tuning of the athlete's shoe stiffness to the athlete's particular characteristics is required. Rapid manufacturing processes, such as selective laser sintering (SLS) of nylon, offer numerous advantages over conventional manufacturing techniques. Key advantages include the ability to produce complex geometries and internal structures in a single process. Of particular relevance to sprint shoe design is the ability to produce customised outsoles that are specific to the requirements of individual athletes. Sprint spike mechanical performance values were measured using a standard test and the data obtained used to produce benchmark values. The results were catalogued into an incremented range of outsole stiffness responses. Additionally, a range of SLS outsoles were subjected to the same benchmarking procedure, their performance is compared to that of commercially available spikes. The future potential to maximise sprint performance by tuning shoe stiffness to the individual characteristics of an athlete is discussed.

1 Introduction

Currently, the manufacture of sprint shoe outsoles is carried out using standard injection moulding techniques. This method restricts the design process due to the inherent constraints imposed by the requirements of tooling. More advanced rapid manufacturing methods, such as selective laser sintering (SLS), enable the realisation of human centred performance enhancements, which are essential to the progression of design in elite sports footwear. SLS is an additive manufacturing process which uses a laser to selectively heat and sinter compacted thermoplastic powder to form a three dimensional object. SLS of nylon demonstrates several key advantages which are particularly relevant to sprint shoe design. Aside from permitting a greater amount of design freedom, the SLS process enables swift and economical production of shoes that are customised to the biometric measurements of an athlete.

Biomechanical studies into the mechanical energy contribution of the joints in the lower extremity (Stefanyshyn and Nigg 1997, 1998) have established that a substantial amount of rotational energy is lost at the metatarsophalangeal joint (MPJ) during running and sprinting. Stefanyshyn and Nigg (2000) used three shoe stiffness values; 0.04, 0.25 and 0.38 Nm·deg^{-1} and discovered that the lost energy at MPJ could be decreased by increasing the bending stiffness of the shoes midsole. The authors directly correlated a reduction in lost energy at the MPJ with improved jump height performance. A more pertinent study by Stefanyshyn et al. (2004) investigated the affects of shoe bending stiffness on sprinting performance. Carbon fibre plates with bending stiffness of 49, 90 and 120N·mm^{-1} were inserted into the athletes' own standard running spikes and times were recorded for 20m sprints. As shoe bending stiffness was increased sprint performance improved. However, the authors asserted that individual tuning of the athlete's shoe stiffness to the athlete's particular characteristics was required to maximise performance.

Whilst the above study highlighted the need to increase the bending stiffness of commercially available sprint spikes, there are currently no benchmark data available for existing sprint spikes. Thus, this paper describes a benchmarking procedure and reports the primary mechanical performance measurements of sprint spikes manufactured by some of the leading brands. The same benchmarking procedure is also used to validate a range of SLS nylon outsoles.

2 Experimental Overview

Eleven pairs of sprint spikes were selected to represent a cross-section of the leading sports brands and to characterise an array of perceived quality and price points. Shoe selection was finalised following discussions with competitive sprinters. The sprint spikes tested are detailed in Table 1.

Shoe	Brand	Model	RRP (£)	size (cm)	Weight(g)	Weight 1 spike (g)	No. Spike Placements	Toe Spring (deg)
A	Adidas	Demolisher	80	27	204.22	0.44	6	30
B	Adidas	Meteor Sprint	50	27	196.88	1.04	6	29
C	Asics	Cyberflash	85	26.5	180.4	1.33	8	21
D	Asics	Hyper Sprint	45	27.5	199.45	1.3	7	25
E	Brooks	F1	55	27.5	234.92	1.35	8	25
F	Brooks	Twitch	40	27.5	168.97	1.36	7	28
G	Mizuno	Tokyo Sprint II	60	28	205.5	1.22	6	21
H	Nike	Zoom Monster Fly	75	27.5	201.26	Fixed Spikes	7	23
I	Nike	Zoom Super Shift	60	26.5	181.3	0.56	10	28
J	Puma	Complete Theseus	60	27	188.24	1.15	10	31
K	Reebok	Foster Millenium	40	27	204.64	1.5	8	30

Table 1. Sprint Spikes Tested

In total 18 samples (11 commercially available complete sprint spikes and 7 SLS nylon outsoles) were subjected to mechanical testing based on a shoe bending stiffness experimental procedure detailed by Oleson, Adler and Goldsmith (2005). The authors compare the stiffness of the forefoot to that of running shoes. The experimental setup for this investigation is shown in Fig. 1.

Each sprint spike was secured to the apparatus using a rigid rearfoot last. The last was fixed to a vertical member and the complete fixture was bolted to a base plate. Each sprint spike was anchored in place using a clamping arrangement, which also assisted in defining the bend region of the outsole. The sprint spike bending stiffness test was performed about a region which corresponds with the MPJ. The flex fulcrum was defined as a line perpendicular to the shoe centre line and at a distance of 70% of the total shoe length from the rearmost part of the heel counter (ASTM F-911

– 85). The authors modified this standard to account for actual MPJ flexion, which was approximated by an angle of 10° from the perpendicular centre line, representing an axis from the first to fifth metatarsal heads. Size specific contoured rigid lasts were used to define this point.

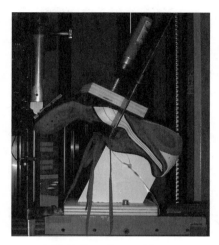

Fig. 1. Experimental Setup

Vertical forces were applied to the sprint spikes using a tensile testing machine (Zwick 1474) and measured using a 10kN load cell. Prior to each test cycle the horizontal distance between the point of flex and the point of application of vertical force and the initial angle between the vertical axis and the outsole of each sprint spike were measured. For each trial the shoe was raised a minimum 25mm vertically against the probe at a constant speed of 2.5mm·s⁻¹ then held for 5s before descending at a constant speed of 2.5mm·s⁻¹ back to the start position. Vertical force was recorded throughout the cycle, five cycles were performed on each shoe, the data presented are a mean of the last two cycles.

3 Results

The test results are listed according to quasi-static bending stiffness in Table 2. Fig. 2 depicts the bending moment plotted against the shoe bending angle for three of the eleven shoes and three of the seven SLS nylon outsoles, describing a range in mechanical performance. The curve for each sample details the loading and unloading phases and the hysteresis loop in the curve represents the loss in energy between phases.

Shoe	Brand	Model	RRP (£)	Vertical Force (N)	Max Angle (deg)	Quasi-static Stiffness (Nm/deg)
A	Adidas	Demolisher	80	81.0	19.4	0.40
F	Brooks	Twitch	40	76.5	23.5	0.28
E	Brooks	F1	55	52.9	21.0	0.24
C	Asics	Cyberflash	85	49.4	21.0	0.24
H	Nike	Zoom Monster Fly	75	56.4	21.1	0.22
D	Asics	Hyper Sprint	45	42.7	20.8	0.20
G	Mizuno	Tokyo Sprint II	60	35.0	20.5	0.18
J	Puma	Complete Theseus	60	29.1	17.0	0.18
I	Nike	Zoom Super Shift	60	31.2	20.4	0.15
B	Adidas	Meteor Sprint	50	27.0	16.8	0.15
K	Reebok	Foster Millenium	40	25.0	19.4	0.12

Table 2. Incremented Range of Sprint Shoe Quasi-static Stiffness

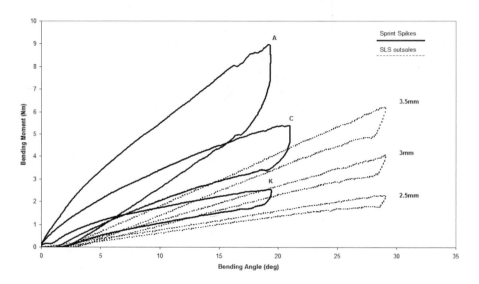

Fig. 2. Comparison of Bending Moments for Sprint Spikes and SLS Nylon outsoles

Fig. 3 illustrates the bending stiffness curves, derived from the bending moment curves, calculated as the ratio of bending moment to bending angle. The quasi-static bending stiffness for each shoe can be approximated as the average between the two asymptotes, as denoted by the dashed lines in Fig. 3. The mean±SD stiffness of all the sprint spikes is 0.21 ± 0.08Nmm·deg^{-1}, the maximum stiffness is 0.4Nmm·deg^{-1} and the minimum stiffness is 0.12Nmm·deg^{-1}.

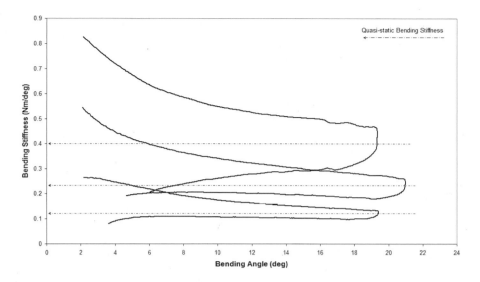

Fig. 3. Quasi-Static Bending Stiffness vs. Bending Angle

4 Discussion

This investigation benchmarked a range of sprint spikes from leading manufacturers according to their bending stiffness about the primary point of flex in the outsole. The measured values for bending stiffness do not necessarily relate directly to quality or performance, but research by Stefanyshyn et al. (2004) indicates that bending stiffness of currently available sprint shoes may be too low.

The hysteresis loops of the bending moment curves show that the loading force required to flex each shoe was always higher than the unloading force. The level of hysteresis is a direct measure of energy loss and the results show that the level of hysteresis is higher in stiffer shoes aimed at high performance sprinting. Stefanyshyn et al. (1997, 1998) determined that the MPJ was inefficient and later established (2000) that lost energy at the MPJ can be reduced by increasing shoe stiffness. Conversely, the results of this investigation imply that the stiffer shoes are less efficient. This could be intentional, as a reduction in responsiveness of the outsole may be required to prevent the sprint shoe from affecting an athlete's kinematics.

The bending stiffness values obtained vary considerably, spanning a range previously investigated by Stefanyshyn et al. (2000) who used carbon fibre inserts to produce a 'control shoe', a 'stiff shoe' and a 'very stiff shoe' with measured stiffness values of 0.04, 0.25 and 0.38Nm·deg^{-1} respectively. The five stiffest sprint spikes in the current investigation are produced by four different brands and it is assumed that these shoes are designed for athletes of a higher standard. Within this range, the mean±SD stiffness is 0.28±0.07Nm·deg^{-1}, emphasising a lack of consistency between brands.

Precise comparisons between the complete sprint spikes and the SLS manufactured outsoles cannot be made with confidence since the SLS outsoles are simplified designs that are not attached to an upper. These differences may explain why the bending moment curves for the SLS nylon outsoles are more linear than the curves for the sprint shoes. The bending moment curves also highlight the distinct difference in hysteresis between the shoes and the SLS nylon outsoles, the requirements of which have yet to be defined from a functional perspective. The presence of noise and deviance from linearity exhibited by the sprint shoes is likely to be due to forces and energy losses associated with the complexity of a complete shoe. However, it is noteworthy that the stiffness levels exhibited by the thicker SLS manufactured outsoles are comparable with the majority of the sprint shoes tested.

5 Conclusions

The benchmarking exercise reported here has enabled the disparity in mechanical properties between currently available sprint spikes to be established. Furthermore, it is apparent that SLS nylon outsoles have the potential to match or exceed the stiffness characteristics exhibited by current market leading sprint shoe brands. It is reasonable to assume that considered structural design on the outsole and the attachment of an upper will increase SLS outsoles stiffness further. The authors are aware that quasi-static measurements are not representative of the actual material behaviour during sprinting. The next investigation phase will include dynamic material testing based on parameters that emulate sprinting conditions. A similar benchmarking procedure of commercially available sprint shoes and SLS nylon outsoles will be undertaken.

Acknowledgments

The authors wish to thank TNO Industries Eindhoven for supporting the project.

References

Oleson, M., Adler,D. and Goldsmith, P. (2005) A comparison of forefoot stiffness in running and running shoe bending stiffness. Journal of Biomechanics 38, 1886-1894.

Stefanyshyn, D. J. and Fusco, C. (2004) Increased Bending Stiffness Increases Sprint Performance. Sports Biomechanics 3, 55-66.

Stefanyshyn, D. J. and Nigg, B. M. (1997) Mechanical energy contribution of the metatarsophalangeal joint to running and sprinting. Journal of Biomechanics. 30, 1081-1085.

Stefanyshyn, D. J. and Nigg, B. M. (1998) Contribution of Lower Extremity Joints to Mechanical Energy in Running Vertical Jumps and Running Long Jumps. Journal of Sports Sciences 16, 177-186.

Stefanyshyn, D. J. and Nigg, B. M. (2000) Influence of Midsole Bending Stiffness on Joint Energy and Jump Height Performance. Medicine and Science in Sports and Exercise 32(2), 471-476.

Relationship Between Shoe Dimensions, Ground Reaction Forces and Perception of Stability During Locomotion

Maxime Roux[1], Céline Puyaubreau[2], Philippe Gorce[3]

[1] Toulon University, Laboratoire d'Ergonomie Sportive et Performance; Centre de Tests et de Recherche de Décathlon, maxime.roux@decathlon.com
[2] Centre de Tests et de Recherche de Décathlon, celine.puyaubreau@free.fr
[3] Toulon University, Laboratoire d'Ergonomie Sportive et Performance, gorce@univ-tln.fr

Abstract. *Objective:* This paper investigates the influence of footwear geometry on ground reaction forces (GRF) and perception of stability during locomotion.

Method: 12 subjects walked on a treadmill ergometer ADAL wearing 5 different shoes models. A barefoot situation was added to the protocol, according to the literature as the best condition of stability.

3 data sets were recorded:

- GRF (Fx, Fy, Fz) were recorded at constant speed (CS) of 2, 4 and 6 km.h^{-1}, and during transition phases (0 to 2, 0 to 4, 0 to 6 km.h^{-1}). Transition phases (TP) were studied because gait transitions are characterized by a loss of stability. Two running conditions (8 km.h^{-1} at CS and 0 to 8 km.h^{-1} at transition) have completed this protocol. At CS, forces were recorded on 10 steps. In TP, maximal force peaks were retained.
- Then, the subjects filled in a questionnaire to evaluate, for each shoe, their sensations of: 1) stability during running and walking, 2) foot support, 3) heightening of the rearfoot, 4) width of the shoe at the heel, 5) general comfort, 6) cushioning.
- Geometrical parameters of the shoe were measured: heel counter dimensions, different widths of the sole, sole slopes, heights of upper on external malleolus…

These 3 data groups were crossed with correlation matrix.

Results: GRF exhibited significant differences between shoes: medial force peaks were more important in barefoot condition and with the shoe which was considered by the subjects as stable in the walking condition. When we added the running condition, there were significant differences between this shoe and all the other conditions (with shoes and barefoot). These results could suggest that a better stability could be associated with more important medial force peaks.

Correlations between GRF data, perception of the subjects and shoe measurements brought to the conclusion that there are different parameters of the shoe which could influence the stability: the sole slope, the height of upper on external malleolus and the inside width of the sole at the heel. These results show that stability sensations and biomechanical measurements are bound. Further investigations will be suggested to correlate these preliminary results with center of pressure path in order to validate that GRF could be useful to evaluate stability of the shoe or not.

1 Introduction

According to A. Forner Cordero (2003), the stability of an object is its capacity to maintain a balance and to resume its position (or sway) after a disturbance. So the stability of the shoe can signify its capacity to limit involuntary imbalances of the foot and of the ankle.

During shod running, the foot biomechanics is modified compared with running in barefoot condition: a greater torsion around Chopart and Lisfranc joints and a less important pronation movement is observed in barefoot condition (Stacoff et al. 1989; Stacoff et al. 1991).

The geometry and the materials of the shoe soles can have an influence on the rearfoot movements. Clarke et al. (1983) has shown that shoes with soft midsoles allowed more maximum pronation and total rearfoot movement than shoe with either medium or hard midsoles. Shoes with 0° heel flare allowed more maximum pronation and total rearfoot movement than shoes with either 15° or 30° heel flares. Stacoff et al. (1996) also measured the movement of the heel inside the shoe during turning using small windows cut in the heel counter. He demonstrated the importance of the shoe sole design and upper design in reducing the risk of ankle injury.

To evaluate the footwear stability, some methodologies can be used: a standard two-dimensional film analysis of the rearfoot could be used (Luthi et al. 1986; Stacoff et al. 1996). Areblad et al. (1990) measures the stability more exactly with a three-dimensional film analysis. The Center of pressure (CoP) path study can be another mean to evaluate the stability. Among all these methodologies to measure the stability, little were interested in its study via forces peaks.

So, the aim of this paper is to investigate the influence of the footwear geometry on ground reaction forces (GRF) and perception of stability during locomotion.

2 Methods and procedures

Twelve subjects, taking sizes 42, 43 and 44, were chosen for participation in the study. These three sizes were distributed in a homogeneous way within the group.

Firstly, the subjects walked on a treadmill ergometer ADAL validated by Belli et al. (2001) wearing five different shoes models. This protocol is not an indicator known to evaluate the stability; so the choice of the footwears was based itself on the considerations of the specialists to take shoes known as being stable (2 shoes models), and others known as being less stable (2 shoes models). One of these shoes (called here Shoe 1) is known for its capacity to stabilize the foot. A barefoot situation was added to the protocol, according to the literature as the best condition of stability (Stacoff et al. 1991; Stacoff et al. 1996).

With each shoe, GRF (Fx, Fy, Fz) were recorded at constant speed (CS) of 2, 4 and 6 km.h^{-1}. Gait transitions are characterized by a loss of stability (Diedrich et al. 1995). So GRF were recorded during transition phases (TP) (0 to 2, 0 to 4 and 0 to 6 km.h^{-1}) too. Two running conditions (8 km.h^{-1} at CS and 0 to 8 km.h^{-1} in TP) have completed this method. At CS, GRF were recorded on ten steps; in TP, maximal force peaks were retained.

Then, after the test of each shoe, the subjects filled in a questionnaire to evaluate their sensations of: 1) stability during running and walking, 2) foot support, 3) heightening of the rearfoot, 4) width of the shoe at the heel, 5) general comfort, 6) cushioning. This sensorial analysis has been organized by Rosa at al. (2003).

Finally, geometrical parameters of the shoe were measured, such as heel counter dimensions (height and width), different widths of the sole, sole slopes, height of upper on external malleolus, etc.... All this parameters were measured according to the own Decathlon's conception rules which are confidential.

To determinate the influence of the shoe on the force peaks evolution in Fx, Fy and Fz, Anova in Repeated Measures were used. Then, a post-hoc test HDS of Tukey was applied. The analysis method of the sensorial questionnaire was done by the fuzzy subset theory. The three data groups (biomechanical data, sensorial data and shoe parameters measurements) were crossed with correlation matrix.

3. Results

3.1 Influence of the shoe on force peaks

At CS and in TP, as shown by Figs. 1-2, medial force peaks are more important in barefoot condition than all the shoes in walking condition, except Shoe 1.

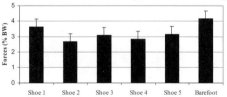

Fig. 1. Medial force peaks at CS

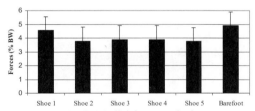

Fig. 2. Medial force peaks in TP

When the running condition is added, medial force peaks are more important with the Shoe 1 than all the other conditions (barefoot or shod conditions) (Figs. 3-4).

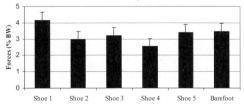

Fig. 3. Medial force peaks at CS with the running condition added

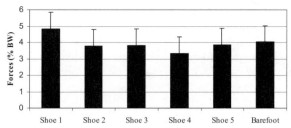

Fig. 4. Medial force peaks in TP with the running condition added

3.2 Sensorial analysis

The questionnaire processing by the fuzzy subset theory gives conclusions that the Shoe 1 provides the least of cushioning and is the least comfortable. But the subjects consider it as among the most stable and like the shoe which has a sole slope the least important.

The Shoe 3 appears like the most comfortable and provides the most of cushioning; but it is among the least stable of the models used for the study.

3.3 Crossing of the data

When biomechanical data and sensorial data are crossed, a correlation can be relieved between stability and comfort sensations during walking, and lateral force peaks at CS (Fig. 5).

When measurements of the shoe and biomechanical data are crossed, a correlation between sole slope and medial force peaks at CS and in TP can be observed (Fig. 6).

A last correlation can be relieved between medial force peaks in TP and the height of upper on external malleolus (Fig. 7).

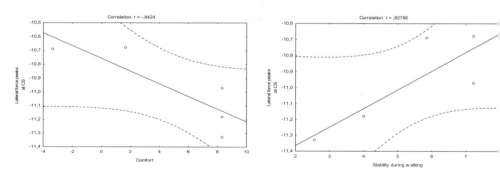

Fig. 5. Correlation between comfort (on the left) and stability (on the right) sensations during walking and lateral force peaks at CS

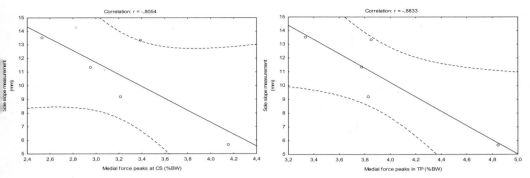

Fig. 6. Correlation between sole slope and medial force peaks at CS (on the left) and in transition phase (on the right)

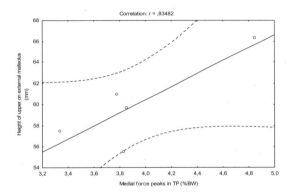

Fig.7. Correlation between the height of the upper on external malleolus and medial force peaks in transition phase

4 Discussion

The results about the influence of the shoe on force peaks showed significant differences in walking condition between barefoot condition and all the shoes, except for Shoe 1 at CS and in TP (Figs. 1-2). In running condition (Figs. 3-4), the difference between Shoe 1 and the barefoot condition could result from the harder EVA of this footwear (the other shoes have a softer EVA).

Studies of Stacoff having shown that the best condition of stability is in barefoot situation (Stacoff et al. 1991; Stacoff et al. 1996), and Shoe 1 being known for its capacity to stabilize the foot, these results suggest that a better stability can be translate by medial force peaks more important.

The crossing between sensorial data and biomechanical data gives the conclusion that the increase of the lateral force peaks gives to the subjects better comfort sensations and worse stability sensations. So what is the relation between comfort and stability? Indeed, the results of this study show that, except for Shoe 4, all these footwears are either stable, or comfortable.

The correlations between biomechanical data and shoe parameters measurements bring to the conclusion that different parameters of the shoe could influence the stability: it seems that a less sole slope, a higher upper on external malleolus and a less important inside width of the sole at the heel increase the stability of the shoe.

5 Conclusion

This study shows that stability sensations and biomechanical measurements are bound and that there are different parameters of the shoe which can influence the stability. It could be interesting to complete this work with other shoes in the aim of having more points in the correlations. In addition, further investigations will be suggested to correlate these preliminary results with Center of Pressure path in order to validate that GRF could be useful to evaluate stability of the shoe or not.

References

Areblad, M., Nigg, B.M., Ekstrand, J., Olsson, K.O. and Ekström, H. (1990). Three-dimensional measurement of rearfoot motion during running. *Journal of Biomechanics; 23 (9): 933-940.*

Belli, A., Bui, P., Berger, A., Geyssant, A. and Lacour, J.R. (2001). A treadmill ergometer for three-dimensional ground reaction forces measurement during walking. *Journal of Biomechanics; 34 (1): 105-112.*

Clarke, T.E., Frederick, E.C. and Hamill, C.L. (1983). The effect of shoe design parameters on rearfoot control in running. *Medicine and Science in Sports and Exercise; 15 (5): 376-381.*

Diedrich, F.J. and Warren, W.H. Jr (1995). Why change gaits? Dynamics of the walk-run transition. *Journal of Experimental Psychology: Human Perception and Performance; 21 (1): 183-202.*

Forner Cordero, A. (2003). Human gait, stumble and... fall? Mechanical limitations of the recovery from a stumble. *ISBN 90-365-1912-8: 5-8.*

Luthi, S., Frederick, E.C., Hawes, M.R. and Nigg, B.M. (1986). Influence of shoe construction on lower extremity kinematics and load during lateral movements in tennis. *International Journal of Sports Biomechanics; 2: 166-174.*

Rosa, V., Guinle, V., Lofi, A., Belluye, N. and Gerville-Reache, L. (2003). Quality Assessment about Sports Tackles: Contribution of the fuzzy Set Theory. *Longevity, Aging and Degradation Models in reliability, Public Health, Medicine and Biology; 1: 249-256.*

Stacoff, A., Kaelin, X., Stuessi, E. and Segesser, B. (1989). The torsion of the foot in running. *International Journal of Sport Biomechanics; 5: 375-389.*

Stacoff, A., Kaelin, X. and Stuessi, E. (1991). The effect of shoes on the torsion and rearfoot motion in running. *Medicine and Science in Sports and Exercise; 23 (4): 482-490.*

Stacoff, A., Steger, J., Stuessi, E. and Reinschmidt, C. (1996). Lateral stability in sideward cutting movements. *Medicine and Science in Sports and Exercise; 28 (3): 350-358.*

Mechanical Properties of Orthopaedic Insoles under Cyclic Loads and Correlation with Daily Use

Nicola Petrone, Emanuela Faggiano and Roberto Meneghello
University of Padova, Italy, nicola.petrone@unipd.it

Abstract. Aim of the work was the definition of an experimental procedure enabling to evaluate the mechanical properties of orthopaedic insoles used in daily and sport activities. Customers and manufacturers can be interested in knowing which is the durability of the insoles for a correct substitution of an exhausted sole together with for product liability. A cyclic test was developed for the measure of stiffness and damping properties of different types of insoles under pulsated compression loads applied by means of a servo hydraulic test machine. Accelerated fatigue tests were performed by applying repeated blocks of compression overloads and measuring periodically the mechanical properties of the insoles to estimate the insole durability. Finally, a group of subjects wearing insoles during daily life was involved in the study and two pairs of identical insoles were prepared for each subject: one pair underwent the fatigue test and the other pair was used by the subject in his daily activity (normal or sport) during a given period of time. Final comparison between the mechanical properties of used and tested insoles enabled to define the number of equivalent damage cycles per month of each type of customer.

1 Introduction

The adoption of orthopaedic insoles is widely prescribed in rehabilitation and sport applications. Such devices enable to modify the pressure distribution between the foot and the shoe and can produce several positive effects if correctly prescribed and applied (Mundermann, 2004). Few works addressed the mechanical behaviour of the insoles usually testing local areas of the insoles (Alcantara, 2001, Dixon, 2001): load cycles were intended to estimate the material recovery in gait (Alcantara, 2001) or material degradation after 40000 running impacts (Dixon, 2001). Recent results showed that, despite a significant increase in mechanical stiffness (depending on the type of material), comfort, kinematicks, kinetics and muscle activities seemed not to change over a three week period of recreational running.

Manufacturers need to know which is the durability of the insoles to prescribe their proper substitution. Furthermore, the safety standards introduce product liability issues because it is up to the manufacturers to define the tests methods and perform the activity to insure the safety of the device. Orthotics like the orthopaedic insoles are supposed to comply shortly to this type of requirements, even if at present there are no tests methods included in the relevant standards (UNI EN 12523 2001).

Aim of the present work was to define an experimental procedure to evaluate the mechanical properties of orthopaedic insoles, based on the functional requirements of the insoles. In analogy with other test methods developed for prosthetic devices (ISO 10328 1996), a cyclic test was defined for measuring the stiffness and damping properties of different types of insoles under pulsated compression loads during an accelerated fatigue tests. Finally, a comparison was made between the properties of the tested insoles and those used by several subjects during daily walking or running activities in order to define the number of cycles per month of each type of activity.

2 Materials and Methods

2.1 Materials and instrumentation

A set of 7 types of insoles (Fig. 1.a) was prepared by different manufacturers from the same plaster cast taken from an healthy adult and shared as in a round-robin project. A resin dummy foot with internal steel frame was prepared and used to apply cyclic compression loads.

Structural testing of insoles was performed on a MTS Minibionics test machine after designing a suitable apparatus enabling to load the insoles between a lower stiff aluminium plate and the upper resin dummy foot hinged at the test head as in Fig. 1.b. No lateral restraint was given to the insoles. The real gait mechanics, with forces moving from the heel to the toe during a step, were not introduced at this stage of the study. The insoles were loaded in compression by means of the dummy foot hinged in the middle of its length and width, in order to apply a contact pressure over the entire insole surface, dependently only from the local thickness and compliance of the different areas of the sole.

A group of 3 subjects was also involved in the study. For each subject, plaster casts of both feet were taken and two couples of insoles were prepared by the same manufacturer, together with a couple of resin dummy foots with internal steel frame. The two pair of insoles were dedicated to different use: one pair (named "Used") was given to the subject for the intended use, the other pair (named "Tested") was kept in the Laboratory for structural testing. One subject (N1) was living normal daily activity, the two other subjects (S1 and S2) were amateur runners and used the insoles in their usual training sessions.

The three subjects involved in the daily use of the insoles were asked to wear the insoles and perform their normal working or running activity during a known period: they were also asked to record each day the type of use and the number of hours of each activity.

TESTER	TYPE	AGE	H [cm]	m [kg]	M/F	USE (months)
N1	Normal	58	180	73	M	Daily (2m)
S1	Sport	24	163	53	F	Track Running (2m)
S2	Sport	23	182	63	M	Cross country (2m)

Table. 1. Information about the three subject involved in daily use testing.

2.2 Test protocol

Mechanical properties were measured in a Cyclic Test within 10 cycles after 40 cycles of settling with load F controlled between 200 and 1000 N at 1Hz and measuring displacement x. The cyclic test was performed not only in the as-manufactured state but was introduced periodically as a measure block in the fatigue test.

To simulate the gait cycle stressing the insoles a Fatigue Test was defined: a fatigue block of 50000 cycles with load ranging from 200 N to 2500 N at 5 Hz was repeated for 6 times for a total number of 300000 cycles. The measure block was performed after each fatigue block to monitor the durability of the insole in terms of variation of insole stiffness, damping and permanent setting.

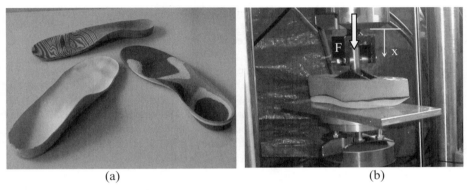

(a)   (b)

Fig. 1. (a) Samples of tested insoles. (b) Experimental setup with indication of load F and x.

2.3 Data Analysis

Mechanical properties were calculated from the F-x cycles during the measure blocks. Insole materials showed typical loops as in Figure 2.a, cycling clockwise. The insoles showed a stiffening behaviour when loaded and some damping when unloaded. From the cyclic response, average stiffness K (N/mm) of the insole was defined as the slope of the linear approximation across the 10 recorded loops, as shown in Fig. 2.a, after having evaluated the dummy foot stiffness and having calculated the insole stiffness as a spring in series with the foot. Damping coefficient C was evaluated after calculation of the loop width at the loop centroid.

During the fatigue tests, F-x loops were analysed after each fatigue block and showed usually a change in stiffness K, damping C and mean values of displacement x_C, as appreciable from Fig. 2.b: the shift of the loop apex was considered as an index of progressing damage to the insole and used to estimate a "permanent set s_N" as the shift of the loop apex after a certain number of cycles N.

A fatigue diagram was obtained for each tested insole as shown in Fig. 3, by plotting the generic insole stiffness at cycle N, $K_T(N)$, normalized with respect to the initial stiffness K_{0T}, versus the number of fatigue cycles. The comparison between the different insole materials can be appreciated in the normalized diagram.

Finally, the insoles worn by the subjects were tested after the intended period of use and submitted to the Cyclic Test as the tested ones. The stiffness value of the used sole, named K_U were compared with the fatigue curves $K_T(N)$ of the tested insoles in order to estimate the number of fatigue cycles corresponding to the same damage (stiffness increase). The number of fatigue cycles in the laboratory needed to apply a fatigue damage equivalent to the damage of one month of typical use in the field was defined N_{eq}: the value of N_{eq} can be used to define the duration of a standard durability test for insoles intended for different uses.

3 Results and Discussion

The results of cyclic tests for the 10 insoles are reported in Table 2: for each sole, the commercial name of the constitutive material is listed and the initial absolute values of the insole stiffness K_{0T}, the damping coefficient C and the final permanent set s_N after 300000 cycles are reported.

As it can be appreciated, great differences appeared in the mechanical properties of the first 7 insoles, ranging from F1 to F7, that were all prepared for the same subject and intended for a normal daily activity. The highest value of stiffness K_{0T} was obtained for carbon fibre (F2) insole as expected, whereas the lowest values was shown by latex insoles (F4): on the contrary, these two materials showed lowest values of permanent set s_N, as well as quite stable stiffness fatigue curves in Fig. 3.

The analysis of the normalized fatigue curves in Fig. 3 shows that all the insoles exhibit a stiffening behaviour, with highest stiffening of more than 4.5 times at the end of the tests. Most curves were monotonically increasing, but some insoles like F2, N1, S1 and S2 showed a tendency to have a central peak value or a plateau. Most stiffening material F6 showed a curve with two distinct slopes having a knee at about 150000 cycles. Material F1 showed the highest value of final permanent set s_{Nf}: high values were shown also by F6, F5 and S1. The permanent set could have be chosen as a parameter alternative to the stiffness K for the definition of the fatigue curves, given the fact that the insoles are intended to apply a postural correction to the foot and that the presence of a permanent set reduces the amount of intended correction. In this work, stiffness K was adopted due to its greater repeatability in testing, as well as to its correlation with dynamic correction, that is the correction applied during the gait: low stiffness insoles like F4, despite the very stable fatigue behaviour, may give a poor correction under gait loads.

Fatigue curves of insoles N1, S1 and S2 were recorded with a higher number of measure blocks in the first cycles, to have a better resolution in the evaluation of N_{eq}. As given in Table 2, the number of fatigue cycles equivalent to one month of sport running resulted to be around 10500 fatigue cycles, with about no difference between running on track and cross country running. On the contrary, the number of equivalent cycles N_{eq} for normal use resulted to be 10 times lower, confirming expectations for a lower intensity use. The fact that S1 and S2 showed similar results and that these were very different from normal use can be seen as a confirmation of the adopted approach: these results can be used in the definition of a standard tests based on the same fatigue loads (2500 N is close to 2.5-3 times the typical adult BW) with different requirements for the number of applied fatigue cycles.

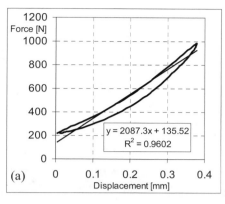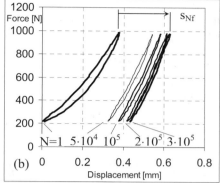

Fig. 2. (a) Typical loop for a cyclic test. (b) Shift of recorded loops after fatigue test.

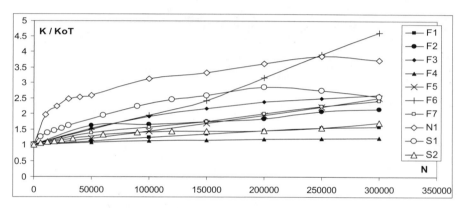

Fig. 3. Normalized stiffness curves versus fatigue cycles for tested insoles

INSOLE	MATERIAL	K_{0T} [N/mm]	C [Ns/rad mm]	s_{Nf} [mm]	Neq
F1	Rubber	1256	41,31	2,06	-
F2	Carbon fibre	5946	69,5	0,28	-
F3	Ludarmed®	984	33,22	0,98	-
F4	Latex Ecolastex®	476	20,51	0,33	-
F5	Lunarflex®	1174	35,62	1,40	
F6	Ecoform 100®	2728	61,22	1,43	-
F7	Ecoform 200®	3024	70,49	0,97	-
N1	Evaform®	3283	102	0,25	1081
S1	AtechKompa.A®	812	69	1,34	11183
S2	Orthofoam®	2857	3	0,84	10126

Table 2. Results of cyclic and fatigue tests on the insoles.

Material properties are undergoing a detailed evaluation in terms of hardness, cyclic stiffness and dumping, temperature and humidity dependence, in order to correlate the test results not only with the commercial names but also with properties.

More studies are ongoing to define the effects on the foot plantar pressure distribution (Fig. 4.a,b) and the "failure" criteria, that is the parameter to be controlled in order to ensure a proper insole functional behaviour. These researches shall investigate the modification of insole geometry in the different areas (Fig.4.c), the variation of pressure distribution on the foot before and after the fatigue testing of insoles and the possibility of defining a two actuators test, like for lower-limb prosthetics (ISO 10328 1996), in order to reproduce more similarly the heel/toe mechanics of the gait.

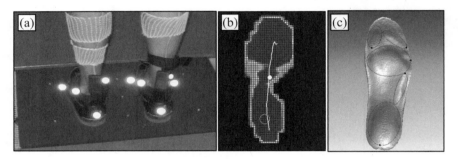

Fig. 4. (a,b) Measure of plantar pressure distribution. (c) 3D insole geometry reconstruction.

3 Conclusions

Mechanical properties of orthopaedic insoles were investigated by introduction of a cyclic test with compression loads between a dummy foot and the insole. Stiffness, damping and permanent set were recorded periodically during the application of fatigue blocks enabling to define fatigue curves. Cyclic tests on insoles worn by normal and sport subjects for a given period allow to define a correlation with daily use.

References

Alcantara E., Gonzalez J.C., Alemany S., Garcia A.C. (2001) Development of a new method to measure the recovery ability of insole materials, by simulating plantar pressures, Proc. 5th Symp. On Footwear Biomechanics, Zurich.
Dixon S.J., House C.M., Waterworth C. (2001) The influence of new and degraded insoles on heel impacts when running in military boots, Proc. 5th Symp. On Footwear Biomechanics, Zurich.
ISO 10328 – (1-8) 1996: Prosthetics - Structural testing of lower-limb prostheses.
Mundermann A., Nigg B.M., Humble N., Stefanyshyn D.J. (2004) Consistent immediate effects of foot orthoses on comfort and lower extremity kinematics, kinetics and muscle activity, J. Appl. Biomech. 20:71-84.
UNI EN 12523:2001. Protesi d'arto esterne e ortesi esterne - Requisiti e metodi di prova.

Author index

434

Subject Index